Lessons from
Nanoelectronics
A New Perspective on Transport

Lessons from Nanoscience: A Lecture Note Series
ISSN: 2301-3354

Series Editors: Mark Lundstrom and Supriyo Datta
(Purdue University, USA)

This series is intended to address a very different need, namely the need to develop conceptual frameworks that can help unify the diverse phenomena being discovered. Such frameworks may not yet be in final form but this series will provide a forum for them to evolve and develop into the textbooks of tomorrow that train our students and guide young researchers as they turn nanoscience into nanotechnology. The focus of the series is on electronics, but volumes in areas of nanoscience and technology broadly related to electronics will be also be considered, as long as they are driven by a quest for unifying principles that embed a diversity of phenomena or techniques.

Published:

Lessons from Nanoscience:
A Lecture Note Series

Vol. 1

Lessons from
Nanoelectronics
A New Perspective on Transport

Supriyo Datta

Purdue University, USA

World Scientific

NEW JERSEY • LONDON • SINGAPORE • BEIJING • SHANGHAI • HONG KONG • TAIPEI • CHENNAI

Published by

World Scientific Publishing Co. Pte. Ltd.

5 Toh Tuck Link, Singapore 596224

USA office: 27 Warren Street, Suite 401-402, Hackensack, NJ 07601

UK office: 57 Shelton Street, Covent Garden, London WC2H 9HE

British Library Cataloguing-in-Publication Data
A catalogue record for this book is available from the British Library.

Lessons from Nanoscience: A Lecture Note Series — Vol. 1
LESSONS FROM NANOELECTRONICS
A New Perspective on Transport

Copyright © 2012 by World Scientific Publishing Co. Pte. Ltd.

ISBN-13 978-981-4335-28-7
ISBN-10 981-4335-28-2
ISBN-13 978-981-4335-29-4 (pbk)
ISBN-10 981-4335-29-0 (pbk)

Printed in Singapore by B & Jo Enterprise Pte Ltd

To
Malika, Manoshi
and
Anuradha

Preface

Everyone is familiar with the amazing performance of a modern smart phone, powered by a billion-plus nanotransistors, each having an active region that is barely a few hundred atoms long.

These lectures, however, are about a less-appreciated by-product of the microelectronics revolution, namely the deeper understanding of current flow, energy exchange and device operation that it has enabled, which forms the basis for what we call the bottom-up approach.

I believe these lessons from nanoelectronics should be of broad relevance to the general problems of non-equilibrium statistical mechanics which pervade many different fields. To make these lectures accessible to anyone in any branch of science or engineering, we assume very little background beyond linear algebra and differential equations. We hope to reach all those who have an interest in basic physics, even if they are not specializing in devices or transport theory.

For dedicated graduate students and the experts, I have written extensively in the past. But they too may enjoy these notes taking a fresh look at a familiar subject, emphasizing the insights from mesoscopic physics and nanoelectronics that are of general interest and relevance.

Finally I should stress that these are *lecture notes* in unfinished form, with typos included to enhance the readers' attention, as my colleague Gerhard likes to put it. With your feedback and suggestions, I hope to have a better version in the future, one that requires less attention!

April 21, 2012   Supriyo Datta
                                                    Purdue University

Acknowledgements

Thanks to World Scientific Publishing Corporation and, in particular, our series editor, Zvi Ruder for joining us in this partnership.

The precursor to this lecture note series, namely the *Electronics from the Bottom Up* initiative on www.nanohub.org was funded by the U.S. National Science Foundation (NSF), the Intel Foundation, and Purdue University.

The nanoHUB-U recently offered its first online course based on these notes and I am thankful for the feedback I received from many online students whom I have never met. We gratefully acknowledge Purdue and NSF support for this program, along with the superb team of professionals who made nanoHUB-U a reality (https://nanohub.org/groups/u).

A special note of thanks to Mark Lundstrom for his leadership that made it all happen and for his encouragement and advice.

I am indebted to Ashraf Alam, Kerem Camsari, Deepanjan Datta, Vinh Diep, Samiran Ganguly, Seokmin Hong, Changwook Jeong, Bhaskaran Muralidharan, Angik Sarkar, Srikant Srinivasan for their valuable feedback and suggestions.

Finally I would like to express my deep gratitude to all who have helped me learn, a list that includes not only those named above, but also many teachers, colleagues and students over the years, starting with the late Richard Feynman whose classic lectures on physics, I am sure, have inspired many like me.

Some Symbols Used

Constants

Electronic charge	$-q$	- 1.6e-19 coul.
Unit of Energy	1 eV	+ 1.6e-19 Joules
Planck's constant	h	6.626e -34 Joule-sec
	$\hbar = h/2\pi$	1.055e -34 Joule-sec
Boltzmann constant	k	1.38e-23 Joule / K
		~ 25 meV / 300K
Free electron mass	m_0	9.11e-31 Kg
Effective mass	m	

Other Symbols

I	Electron Current	amperes (A)
	(See Fig.3.2)	
V	Electron Voltage	volts (V)
U	Electrostatic Potential	eV
μ	Electrochemical Potential	eV
	(also called Fermi level or quasi-Fermi level)	
μ_0	Equilibrium Electrochemical Potential	eV
R	Resistance	Ohms (V/A)
G	Conductance	Siemens (A/V)
$G(E)$	Conductance at 0K with μ_0=E	Siemens (A/V)
$\overline{D}$	Diffusivity	m^2 /sec
$\overline{\mu}$	Mobility	m^2 /V-sec
ρ	Resistivity	Ohm-m (3D), Ohm (2D)
σ	Conductivity	S/m (3D), S (2D)

A	Area	m^2
W	Width	m
L	Length	m
E	Energy	eV
$f(E)$	Fermi Function	Dimensionless
$\left(-\dfrac{\partial f}{\partial E}\right)$	Thermal Broadening Function (TBF)	/ eV
$kT\left(-\dfrac{\partial f}{\partial E}\right)$	Normalized TBF	Dimensionless
$D(E)$	Density of States	/eV
$N(E)$	Number of States with Energy < E Equals Number of Electrons at 0K with μ_0=E	Dimensionless
N	Electron Density (3D or 2D or 1D)	$/m^3$ or $/m^2$ or /m
$M(E)$	Number of Channels *(also called transverse modes)*	Dimensionless
T	Temperature	degrees Kelvin (K)
t	Transfer Time	seconds
ν	Transfer Rate	/second
$\gamma \equiv \hbar\nu$	Energy Broadening	eV
$[X]^+$	Complex conjugate of transpose of matrix	$[X]$
H	(Matrix) Hamiltonian	eV
$G^R(E)$	(Matrix) Retarded Green's function	/eV
$G^A(E)=[G^R(E)]^+$	(Matrix) Advanced Green's function	/eV
$G^n(E)/2\pi$	(Matrix) Electron Density	/eV, per gridpoint
$A(E)/2\pi$	(Matrix) Density of States	/eV, per gridpoint
$\Gamma(E)$	(Matrix) Energy Broadening	eV

Contents

Detailed Contents

I. The New Ohm's Law

Lecture 1

The Bottom-up Approach

"Everyone" has a computer these days, and each computer has more than a billion transistors, making transistors more numerous than anything else we could think of. Even the proverbial ants, I am told, have been vastly outnumbered.

There are many types of transistors, but the most common one in use today is the Field Effect Transistor (FET), which is essentially a resistor consisting of a "channel" with two large contacts called the "source" and the "drain" (Fig. 1.1a).

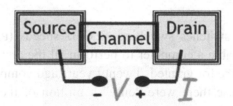

Fig.1.1a.
The Field Effect Transistor (FET) is essentially a resistor consisting of a "channel" with two large contacts called the "source" and the "drain", across which we attach the two terminals of a battery.

The resistance R = Voltage (V) / Current (I) can be switched by several orders of magnitude through the voltage V_G applied to a third terminal called the "gate" (Fig.1.1b) typically from an "OFF" state of ~100 megohms to an "ON" state of ~10 kilohms.

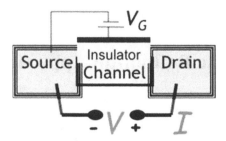

Fig.1.1b.
The resistance $R = V/I$ can be changed by several orders of magnitude through the gate voltage V_G.

Actually, the microelectronics industry uses a complementary pair of transistors such that when one changes from 100M to 10K, the other changes from 10K to 100M. Together they form an inverter whose output is the "inverse" of the input: A low input voltage creates a high output voltage while a high input voltage creates a low output voltage as shown in Fig.1.2.

A billion such switches switching at GHz speeds (that is, once every nanosecond) enable a computer to perform all the amazing feats that we have come to take for granted. Twenty years ago computers were far less powerful, because there were "only" a million of them, switching at a slower rate as well.

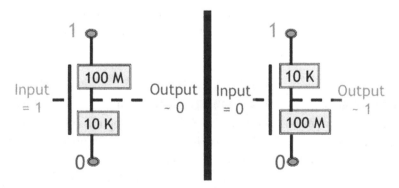

Fig.1.2.
A complementary pair of FET's form an inverter switch.

Both the increasing number and the speed of transistors are consequences of their ever-shrinking size and it is this continuing miniaturization that has driven the industry from the first four-function calculators of the 1970's to the modern laptops. For example, if each transistor takes up a space of say 10 µm x 10 µm, then we could fit 3000 x 3000 = 9 million of them into a chip of size 3cm x 3cm, since

$$3\,cm \,/\, 10\,\mu m \;=\; 3000$$

That is where things stood back in the ancient 1990's. But now that a transistor takes up an area of ~ 1 µm², we can fit 900 million (nearly a billion) of them into the same 3cm x 3cm chip. Where things will go from here remains unclear, since there are major roadblocks to continued miniaturization, the most obvious of which is the difficulty of dissipating the heat that is generated. Any laptop user knows how hot it gets when it is working hard, and it seems difficult to increase the number of switches and/or their speed too much further.

These Lectures, however, are not about the amazing feats of microelectronics or where the field might be headed. They are about a less-appreciated by-product of the microelectronics revolution, namely the deeper understanding of current flow, energy exchange and device operation that it has enabled, based on which we have proposed what we call the bottom-up approach. Let me explain what we mean.

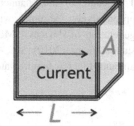

According to Ohm's law, the resistance R is related to the cross-sectional area A and the length L by the relation

$$R \equiv \frac{V}{I} \;=\; \frac{\rho L}{A} \tag{1.1a}$$

ρ being a geometry-independent property of the material that the channel is made of.

The reciprocal of the resistance is the conductance

$$\frac{I}{V} = \frac{\sigma A}{L}$$

$$(1.1b)$$

which is written in terms of the reciprocal of the resistivity called the conductivity.

Our conventional view of electronic motion through a solid is that it is "diffusive," which means that the electron takes a random walk from the source to the drain, traveling in one direction for some length of time before getting scattered into some random direction as sketched in Fig.1.3. The mean free path, λ that an electron travels before getting scattered is typically less than a micrometer (also called a micron = 10^{-3} mm, denoted µm) in common semiconductors, but it varies widely with temperature and from one material to another.

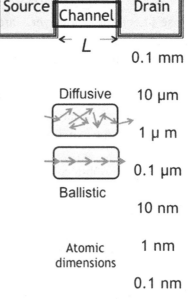

Fig.1.3.
The length of the channel of an FET has progressively shrunk with every new generation of devices ("Moore's Law") and stands today (2010) at ~ 50 nm, which amounts to a few hundred atoms!

Length units:
1 mm = 1000 µm
and 1 µm = 1000 nm

It seems reasonable to ask what would happen if a resistor is shorter than a mean free path so that an electron travels ballistically ("like a bullet") through the channel. Would the resistance still obey Ohm's law? Would it still make sense to talk about its resistance? These questions have intrigued scientists for a long time, but even twenty five years ago one could only speculate about the answers. Today the answers are quite clear and experimentally well established. Even the transistors in commercial laptops now have channel lengths $L \sim 50$ nm, corresponding to a few hundred atoms in length! And in research laboratories people have even measured the resistance of a hydrogen molecule.

It is now clearly established that the resistance of a ballistic conductor can be written in the form

$$R_B = \underbrace{\frac{h}{q^2}}_{\sim 25\,K\Omega} \frac{1}{M} \tag{1.2}$$

where h/q^2 is a fundamental constant and M represents the number of effective channels available for conduction. Note that here we are using the word "channel" not to denote the physical channel in Fig.1.3, but in the sense of parallel paths whose meaning will be clarified in the next few lectures. In future we will refer to M as the number of "modes".

This result is now fairly well-known, but the common belief is that it applies only to short conductors and belongs in a course on special topics like mesoscopic physics or nanoelectronics. What is not as well-known is that the resistance for both long and short conductors can be written in the form (λ : mean free path)

$$R = \underbrace{\frac{h}{q^2 M}}_{R_B} \left(1 + \frac{L}{\lambda}\right) \tag{1.3}$$

Ballistic and diffusive conductors are not two different worlds, but rather a continuum as the length L is increased. For $L << \lambda$, Eq.(1.3) reduces to the ballistic result in Eq.(1.2), while for $L >> \lambda$, it morphs into Ohm's law in Eq.(1.1). Indeed we could rewrite Eq.(1.3) in the form

$$R = \frac{\rho}{A}(L+\lambda) \qquad (1.4)$$

with a new expression

$$\rho = \frac{h}{q^2}\frac{A}{M\lambda} \qquad (1.5)$$

that provides a different view of resistivity in terms of the number of modes per unit area and the mean free path.

This is the result we will try to establish in the first few lectures and it illustrates the essence of our bottom-up approach, viewing short conductors not as an aberration but as the starting point to understanding long conductors. For historical reasons, the subject of conduction is always approached top-down, from large complicated conductors down to hydrogen molecules. As long as there was no experimental evidence for what the resistance of a small conductor might be, it made good sense to start from large conductors where the answers were clear. But now that the answers are clear at both ends, a bottom-up view seems called for, at least to complement the top-down view. After all that is how we learn most things, from the simple to the complex: quantum mechanics, for example, starts with the hydrogen atom, not with bulk solids.

But there is a deeper reason why the bottom-up approach can be particularly useful in transport theory and this is the "new perspective" we are seeking to convey in these lectures. One of the major conceptual issues posed by the ballistic resistance R_B in Eq.(1.2), is the question of "where is the heat". Current flow through any resistance R leads to the generation of an amount of heat $VI = I^2R$, commonly known as Joule heating. A ballistic resistance R_B too must generate a heat of I^2R_B.

But how can a ballistic resistor generate heat? Heat generation requires interactions whereby energetic electrons give up their excess energy to the surrounding atoms. A conductor through which electrons zip through without exchanging energy cannot possibly be generating any heat. It is now generally accepted that in such a resistor, all the Joule heat would be dissipated in the contacts as sketched in Fig.1.4. There is experimental evidence that real nanoscale conductors do approach this ideal and a significant fraction of the Joule heat is generated in the contacts.

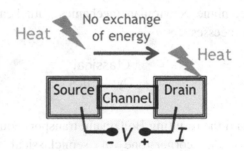

Fig.1.4. **The ideal elastic resistor** with the Joule heat $VI = I^2R$ generated entirely in the contacts as sketched. Many nanoscale conductors are believed to be close to this ideal.

In a sense this seems obvious as my colleague Ashraf often points out. After all a bullet dissipates most of its energy to the object it hits, rather than to the medium it flies through. And yet in the present context, this does seem like a somewhat counter-intuitive result. Clearly the flow of electrons and hence the resistance is determined by the area of the narrow channel that electrons have to squeeze through and not by the large area contacts. But the associated Joule heat occurs in the contacts. And this would be true even if the channel were full of "potholes" that scattered the electrons, as long as the interaction with the electrons is purely *elastic*, that is does not involve any transfer of energy,

The point we wish to make, however, is that flow or transport always involves two fundamentally different types of processes, namely elastic transfer and heat generation, belonging to two distinct branches of physics. The first involves frictionless mechanics of the type described by Newton's laws or the Schrödinger equation. The second involves the generation of heat described by the laws of thermodynamics. The first is driven by forces or potentials and is reversible. The second is driven by entropy and is irreversible. Viewed in reverse, such processes look absurd, like heat flowing spontaneously from a cold to a hot surface or an electron accelerating spontaneously by absorbing heat from its surroundings.

Normally the two processes are intertwined and a proper description of current flow in electronic devices requires the advanced methods of non-equilibrium statistical mechanics that integrate mechanics with thermodynamics. Over a century ago Boltzmann taught us how to combine Newtonian mechanics with heat generating or entropy-driven processes

$$\text{Classical Dynamics} \; + \; \text{⚡} \; = \; \text{BTE}$$

and the resulting Boltzmann transport equation (BTE) is widely accepted as the cornerstone of semiclassical transport theory. The word semiclassical is used because some quantum effects have also been incorporated approximately into the same framework.

A full treatment of quantum transport requires a formal integration of quantum dynamics described by the Schrödinger equation with heat generating processes. This is exactly what is achieved in the non-equilibrium Green function (NEGF) method

$$\text{Quantum Dynamics} \; + \; \text{⚡} \; = \; \text{NEGF}$$

originating in the 1960's from the seminal works of Martin and Schwinger (1959), Kadanoff and Baym (1962), Keldysh (1965) and others (see Lecture 19).

The BTE takes many semesters to master and the full NEGF formalism, even longer. Much of this complexity, however, comes from the difficulty of combining mechanics with distributed heat-generating processes which are a key part of the physics of resistance in long conductors.

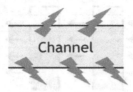

Fig.1.5. Resistance in long conductors primarily arise from distributed heat generating processes along the channel. Prior to 1990, papers dealing with basic transport theory seldom considered the actual physical contacts.

The modern developments in mesoscopic physics and nanoelectronics give us a different perspective, with the ***elastic resistor*** in Fig.1.4 as the starting point. The operation of the elastic resistor can be understood in far more elementary terms because of the clean spatial separation between the mechanical and the heat-generating processes. The former is confined to the channel and the latter to the contacts. As we will see in the next few lectures, the latter is easily taken care of, indeed so easily that it is easy to miss the profound nature of what is being accomplished.

Even quantum transport can be discussed in relatively elementary terms using this viewpoint. My own research has largely been focused in this area developing the NEGF method, but we will get to it only in Part III after we have "set the stage" in Parts I and II using a semiclassical picture.

But does this viewpoint help us understand long conductors? Short conductors may be elastic and conceptually simple, but don't we finally have to deal with distributed heat generation if we want to understand long conductors?

We argue that many properties of long conductors, especially at low bias can be understood in simple terms by viewing them as a series of elastic resistors as sketched in Fig.1.6.

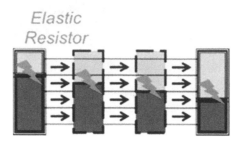

Fig.1.6. Long resistors can be approximately viewed as a series of elastic resistors, as discussed in Section 3.3.

Many well-known results like the conductivity and the thermoelectric coefficients for large conductors, that are commonly obtained from the BTE, can be obtained in a more transparent manner by using this viewpoint, as we will show in the first two parts of these lectures. We will then use this viewpoint in Part III to look at a variety of quantum transport phenomena like resonant tunneling, conductance quantization, the integer quantum Hall effect and spin precession.

In short, the lesson of nanoelectronics we are trying to convey is the utility of the concept of an elastic resistor with its clean separation of mechanics from thermodynamics. The concept was introduced by Rolf Landauer in 1957 and has been widely used in mesoscopic physics ever since the seminal work in the 1980's helped establish its relevance to understanding experiments in short conductors.

What we hope to convey in these lectures is that the concept of an elastic resistor is not just useful for short conductors but provides a fresh new perspective for long conductors as well, that makes a wide variety of devices and phenomena transparent and accessible.

I do not think any of the end results will come as a surprise to the experts. I believe they all follow directly from the BTE or the NEGF and one might well ask whether anything is gained from approximate physical pictures based on elastic resistors. This is a subjective matter that is not easy to argue. Perhaps Feynman (1963) expressed it best in his Lectures on Physics when he said

".. people .. say .. there is nothing which is not contained in the equations .. if I understand them mathematically inside out, I will understand the physics inside out. Only it doesn't work that way. .. A physical understanding is a completely unmathematical, imprecise and inexact thing, but absolutely necessary for a physicist."

I believe the elastic resistor contributes to our physical understanding of the BTE and the NEGF method, without being too "imprecise" or "inexact", and I hope it will facilitate the insights needed to take us to the next level of understanding, discovery and innovation.

Lecture 2

Why Electrons Flow

It is a well-known and well-established fact, namely that when the two terminals of a battery are connected across a conductor, it gives rise to a current due to the flow of electrons across the channel from the source to the drain.

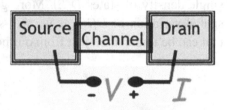

If you ask anyone, novice or expert, what causes electrons to flow, by far the most common answer you will receive is that it is the electric field. However, this answer is incomplete at best. After all even before we connect a battery, there are enormous electric fields around every atom due to the positive nucleus whose effects on the atomic spectra are well-documented. Why is it that these electric fields do not cause electrons to flow, and yet a far smaller field from an external battery does?

The standard answer is that microscopic fields do not cause current to flow, a macroscopic field is needed. This too is not satisfactory, for two reasons. Firstly, there are well-known inhomogeneous conductors like p-n junctions which have large macroscopic fields extending over many micrometers that do not cause any flow of electrons till an external battery is connected.

Secondly, experimentalists are now measuring current flow through conductors that are only a few atoms long with no clear distinction between the microscopic and the macroscopic. This is a result of our progress in nanoelectronics, and it forces us to search for a better answer to the question, "why electrons flow."

2.1 Two Key Concepts

To answer this question, we need two key concepts. First is the ***density of states per unit energy D(E) available for electrons to occupy*** inside the channel (Fig.2.1). For the benefit of experts, I should note that we are adopting what we will call a "point channel model" represented by a single density of states *D(E)*. More generally one needs to consider the spatial variation of *D(E)*, as we will see in Lecture 8, but there is much that can be understood just from our point channel model.

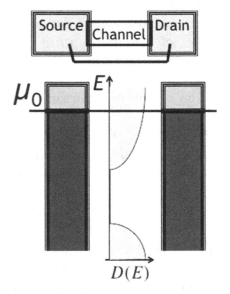

Fig.2.1.
The first step in understanding the operation of any electronic device is to draw the available density of states *D(E)* as a function of energy E, inside the channel and to locate the equilibrium electrochemical potential μ_0 separating the filled from the empty states.

The second key input is the ***location of the electrochemical potential***, μ_0 which at equilibrium is the same everywhere, in the source, the drain and the channel. Roughly speaking (we will make this statement more precise shortly) it is the energy that demarcates the filled states from the empty ones. All states with energy $E < \mu_0$ are filled while all states with $E > \mu_0$ are empty. For convenience I might occasionally refer to the electrochemical potential as just the *"potential"*.

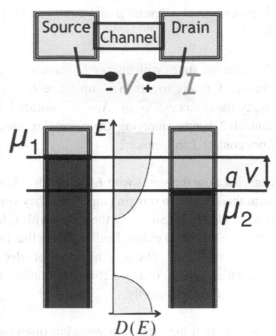

Fig.2.2.
When a voltage is applied across the contacts, it lowers all energy levels at the positive contact (drain in the picture). As a result the electrochemical potentials in the two contacts separate: $\mu_1 - \mu_2 = qV$.

When a battery is connected across the two contacts creating a potential difference V between them, it lowers all energies at the positive terminal (drain) by an amount qV, - q being the charge of an electron ($q = 1.6 \times 10^{-19}$ coulombs) making the two electrochemical potentials separate by qV as shown in Fig.2.2:

$$\mu_1 - \mu_2 = qV \qquad (2.1)$$

Just as a temperature difference causes heat to flow and a difference in water levels makes water flow, a difference in electrochemical potentials causes electrons to flow. Interestingly, only the states in and around an energy window around μ_1 and μ_2 contribute to the current flow, all the states far above and well below that window playing no part at all. Let us explain why.

2.1.1 Energy Window for Current Flow

Each contact seeks to bring the channel into equilibrium with itself, which roughly means filling up all the states with energies E less than its electrochemical potential μ and emptying all states with energies greater than μ.

Consider the states with energy E that are less than μ_1 but greater than μ_2. Contact 1 wants to fill them up since $E < \mu_1$, but contact 2 wants to empty them since $E > \mu_2$. And so contact 1 keeps filling them up and contact 2 keeps emptying them causing electrons to flow continually from contact 1 to contact 2.

Consider now the states with E greater than both μ_1 and μ_2. Both contacts want these states to remain empty and they simply remain empty with no flow of electrons. Similarly the states with E less than both μ_1 and μ_2 do not cause any flow either. Both contacts like to keep them filled and they just remain filled. There is no flow of electrons outside the window between μ_1 and μ_2, or more correctly outside $\pm$ a few kT of this window, as we will discuss shortly.

This last point may seem obvious, but often causes much debate because of the common belief we alluded to earlier, namely that electron flow is caused by the electric field in the channel. If that were true, all the electrons should flow and not just the ones in any specific window determined by the contacts.

2.2 Fermi Function

Let us now make the above statements more precise. We stated that roughly speaking, at equilibrium, all states with energies E below the electrochemical potential μ_0 are filled while all states with $E > \mu_0$ are empty. This is precisely true only at absolute zero temperature. More generally, the transition from completely full to completely empty occurs over an energy range $\sim \pm 2\,kT$ around $E = \mu_0$ where k is the Boltzmann constant (~ 80 μeV/K) and *T* is the absolute temperature. Mathematically this transition is described by the Fermi function :

$$f(E) \;=\; \cfrac{1}{\exp\!\left(\cfrac{E-\mu}{kT}\right)+1}$$

(2.2)

This function is plotted in Fig.2.3 (left panel), though in an unconventional form with the energy axis vertical rather than horizontal. This will allow us to place it alongside the density of states, when trying to understand current flow (see Fig.2.4).

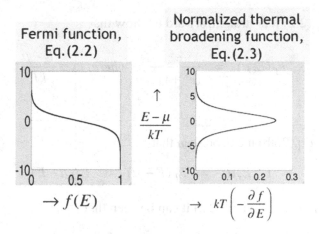

Fig,2.3. Fermi function and the normalized (dimensionless) thermal broadening function.

For readers unfamiliar with the Fermi function, let me note that an extended discussion is needed to do justice to this deep but standard result, and we will discuss it a little further in Lecture 16 when we talk about the key principles of equilibrium statistical mechanics. At this stage it may help to note that what this function (Fig.2.3) basically tells us is that states with low energies are always occupied (*f=1*), while states with high energies are are always empty (*f=0*), something that seems reasonable since we have heard often enough that (1) everything goes to its lowest energy, and (2) electrons obey an exclusion principle that stops them from all getting into the same state. The additional fact that the Fermi function tells us is that the transition from $f=1$ to $f=0$ occurs over an energy range of $\sim \pm 2kT$ around μ_0.

2.2.1. Thermal Broadening Function

Also shown in Fig.2.3 is the derivative of the Fermi function, multiplied by kT to make it dimensionless:

$$F_T(E,\mu) = kT\left(-\frac{\partial f}{\partial E}\right) \tag{2.3a}$$

Using Eq.(2.2) it is straightforward to show that

$$F_T(E,\mu) = \frac{e^x}{(e^x+1)^2}, \quad where \quad x \equiv \frac{E-\mu}{kT} \tag{2.3b}$$

Note:

(1) From Eq.(2.3b) it can be seen that

$$F_T(E,\mu) = F_T(E-\mu) = F_T(\mu-E) \tag{2.4a}$$

(2) From Eqs.(2.3b) and (2.2) it can be seen that

$$F_T = f(1-f) \tag{2.4b}$$

(3) If we integrate F_T over all energy the total area equals kT:

$$\int_{-\infty}^{+\infty} dE\, F_T(E,\mu) = kT \int_{-\infty}^{+\infty} dE\left(-\frac{\partial f}{\partial E}\right)$$

$$= kT\left[-f\right]_{-\infty}^{+\infty} = kT\,(1-0) = kT \tag{2.4c}$$

so that we can approximately visualize F_T as a rectangular "pulse" centered around $E=\mu$ with a peak value of $1/4$ and a width of $\sim 4kT$.

2.3 Non-equilibrium: Two Fermi Functions

When a system is in equilibrium the electrons are distributed among the available states according to the Fermi function. But when a system is driven out-of-equilibrium there is no simple rule for determining the distribution of electrons. It depends on the specific problem at hand making non-equilibrium statistical mechanics far richer and less understood than its equilibrium counterpart.

For our specific non-equilibrium problem, we argue that the two contacts are such large systems that they cannot be driven out-of-equilibrium. And so each remains locally in equilibrium with its own electrochemical potential giving rise to two different Fermi functions (Fig.2.4):

$$f_1(E) = \frac{1}{\exp\left(\dfrac{E - \mu_1}{kT}\right) + 1}$$

(2.5a)

$$f_2(E) = \frac{1}{\exp\left(\dfrac{E - \mu_2}{kT}\right) + 1}$$

(2.5b)

The "little" channel in between does not quite know which Fermi function to follow and as we discussed earlier, the source keeps filling it up while the drain keeps emptying it, resulting in a continuous flow of current.

In summary, what makes electrons flow is the difference in the "agenda" of the two contacts as reflected in their respective Fermi functions, $f_1(E)$ and $f_2(E)$. This is qualitatively true for all conductors, short or long. But for short conductors, the current at any given energy E is quantitatively proportional to

$$I(E) \sim f_1(E) - f_2(E)$$

representing the difference in the probabilities in the two contacts. This quantity goes to zero when E lies way above μ_1, μ_2 since f_1 and f_2 are both zero. It also goes to zero when E lies way below μ_1, μ_2 since f_1 *and* f_2 are both one. Current flow occurs only in the intermediate energy window, as we had argued earlier.

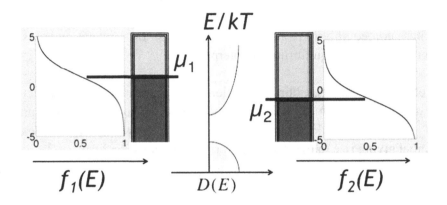

Fig.2.4.
Electrons in the contacts occupy the available states with a probability described by a Fermi function $f(E)$ with the appropriate electrochemical potential μ.

2.4 Linear Response

Current-voltage relations are typically not linear, but there is a common approximation that we will frequently use throughout these lectures to extract the "linear response" which refers to the low bias conductance, dI/dV, as $V \rightarrow 0$.

The basic idea can be appreciated by plotting the difference between two Fermi functions, normalized to the applied voltage

$$F(E) \;=\; \frac{f_1(E) - f_2(E)}{qV/kT} \qquad (2.6)$$

where

$$\mu_1 \;=\; \mu_0 + (qV/2)$$

$$\mu_2 \;=\; \mu_0 - (qV/2)$$

Fig.2.5 shows that the difference function F gets narrower as the voltage is reduced relative to kT. The interesting point is that as qV is reduced below kT, the function F approaches the thermal broadening function F_T we defined (see Eq.(2.3a)) in Section 2.2:

$$F(E) \rightarrow F_T(E), \quad as \quad qV/kT \rightarrow 0$$

so that from Eq.(2.6)

$$f_1(E) - f_2(E) \approx \frac{qV}{kT} F_T(E, \mu_0) = \left(-\frac{\partial f_0}{\partial E}\right) qV \qquad (2.7)$$

if the applied voltage $\mu_1 - \mu_2 = qV$ *is much less than kT.*

Fig.2.5. $F(E)$ from Eq.(2.6) versus $(E-\mu_0)/kT$ for different values of $y=qV/kT$.

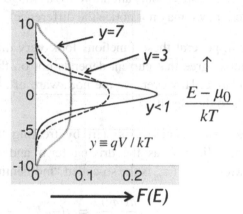

The validity of Eq.(2.7) for $qV << kT$ can be checked numerically if you have access to MATLAB or equivalent. For those who like to see a mathematical derivation, Eq. (2.7) can be obtained using the Taylor series expansion described in Appendix A to write

$$f(E) - f_0(E) \approx \left(-\frac{\partial f_0}{\partial E}\right)(\mu - \mu_0) \qquad (2.8)$$

Eq. (2.8) and Eq. (2.7) which follows from it, will be used frequently in these lectures.

2.5. Difference in "Agenda" Drives the Flow

Before moving on, let me quickly reiterate the key point we are trying to make, namely that current is determined by

$$-\frac{\partial f_0(E)}{\partial E} \quad and\ not\ by \quad f_0(E)$$

The two functions look similar over a limited range of energies

$$-\frac{\partial f_0(E)}{\partial E} \approx \frac{f_0(E)}{kT} \quad if\ E - \mu_0 \gg kT$$

So if we are dealing with a so-called "non-degenerate conductor" where we can restrict our attention to a range of energies satisfying this criterion, we may not notice the difference.

But in general these functions look very different (see Fig.2.3) and the experts agree that current depends not on the Fermi function, but on its derivative. However, we are not aware of any elementary treatment that leads to this result.

Freshman physics texts start by treating the force due to an electric electric field F as the driving term and adding a frictional term to Newton's law (τ_m is the so-called "momentum relaxation time")

$$\underbrace{\frac{d(mv)}{dt}}_{Newton's\ Law} = (-qF) - \underbrace{\frac{mv}{\tau_m}}_{Friction}$$

At steady-state ($d/dt = 0$) this gives a non-zero drift velocity, from which one calculates the current. This elementary approach leads to the Drude formula (discussed in Lecture 5) which played a major historical role in our understanding of current flow. But since it treats electric fields as the driving term, it also suggests that the current depends on the total number of electrons. This is commonly explained away by saying that there are mysterious quantum mechanical forces that prevent electrons in full bands from moving and what matters is the number of "free electrons".

But this begs the question of which electrons are free and which are not, a question that becomes more confusing for atomic scale conductors.

It is well-known that the conductivity varies widely, changing by a factor of $\sim 10^{20}$ going from copper to glass, to mention two materials that are near two ends of the spectrum. But this is not because one has more electrons than the other. The total number of electrons is of the same order of magnitude for all materials from copper to glass.

Whether a conductor is good or bad is determined by the availability of states in an energy window $\sim kT$ around the electrochemical potential μ_0, which can vary widely from one material to another. This is well-known to experts and comes mathematically from the dependence of the conductivity

$$on \quad -\frac{\partial f_0(E)}{\partial E} \quad rather\ than \quad f_0(E)$$

a result that typically requires advanced treatments based on the Boltzmann (Lecture 7) or the Kubo formalism (Lecture 15).

Our bottom-up approach, however, leads us to this result in an elementary way as we have just seen. Current is driven by the difference in the "agenda" of the two contacts which for low bias is proportional to the derivative of the equilibrium Fermi function:

$$f_1(E) - f_2(E) \approx \left(-\frac{\partial f_0}{\partial E} \right) qV$$

There is no need to invoke mysterious forces that stop some electrons from moving, though one could perhaps call $f_1 - f_2$ a mysterious force, since the Fermi function (Eq.(2.2)) reflects the exclusion principle. In Lecture 11 we will see how this approach is readily extended to describe the flow of phonons which is proportional to $n_1 - n_2$, n being the Bose (not Fermi) function which is appropriate for particles that do not have an exclusion principle.

The Elastic Resistor

3.1. How an Elastic Resistor Dissipates Heat
3.2. Conductance of an Elastic Resistor
3.3. Why an Elastic Resistor is Relevant

We saw in the last Lecture that the flow of electrons is driven by the difference in the "agenda" of the two contacts as reflected in their respective Fermi functions, $f_1(E)$ and $f_2(E)$. The negative contact with its larger $f(E)$ would like to see more electrons in the channel than the positive contact. And so the positive contact keeps withdrawing electrons from the channel while the negative contact keeps pushing them in.

This is true of all conductors, big and small. But it is generally difficult to express the current as a simple function of $f_1(E)$ and $f_2(E)$, because electrons jump around from one energy to another and the current flow at different energies is all mixed up.

Fig. 3.1.
An elastic resistor:
Electrons travel along
fixed energy channels.

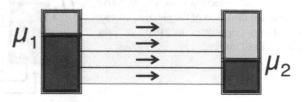

But for the ideal elastic resistor shown in Fig.1.4, the current in an energy range from E to $E+dE$ is decoupled from that in any other energy range, allowing us to write it in the form (Fig.3.1)

$$dI = \frac{1}{q} dE\, G(E)\,(f_1(E) - f_2(E))$$

and integrating it to obtain the total current I. Making use of Eq.(2.7), this leads to an expression for the low bias conductance

$$\frac{I}{V} = \int\limits_{-\infty}^{+\infty} dE \left(-\frac{\partial f_0}{\partial E} \right) G(E) \qquad (3.1)$$

where $(-\partial f_0 / \partial E)$ can be visualized as a rectangular pulse of area equal to one, with a width of $\sim \pm 2kT$ (see Fig.2.3, right panel).

Let me briefly comment on a general point that often causes confusion regarding the direction of the current. As I noted in Lecture 2, because the electronic charge is negative (an unfortunate choice, but something we cannot do anything about) the side with the higher voltage has a lower electrochemical potential. Inside the channel, electrons flow from the higher to the lower electrochemical potential, so that the electron current flows from the source to the drain. The conventional current on the other hand flows from the higher to the lower voltage.

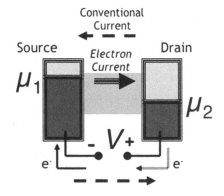

Fig.3.2.
Because an electron carries negative charge, the direction of the electron current is always opposite to that of the conventional current.

Since our discussions will usually involve electron energy levels and the electrochemical potentials describing their occupation, it is also convenient for us to use the electron current instead of the conventional current. For example, in Fig.3.2 it seems natural to say that the current flows from the source to the drain and not the other way around. And that is what I will try to do consistently throughout these Lectures. In short, we will use the current, *I*, to mean ***electron current.***

Getting back to Eq.(3.1), we note that it tells us that for an elastic resistor, we can define a conductance function *G(E)* whose average over an energy range ~ ± *2kT* around the electrochemical potential μ_0 gives the experimentally measured conductance. At low temperatures, we can simply use the value of *G(E)* at $E = \mu_0$.

This energy-resolved view of conductance represents an enormous simplification that is made possible by the concept of an ***elastic resistor*** which is a very useful idealization that describes short devices very well and provides insights into the operation of long devices as well.

Note that by elastic we do not just mean "ballistic" which implies that the electron goes straight from source to drain, "like a bullet." We also include the possibility that an electron takes a more traditional diffusive path ***as long as it changes only its momentum and not its energy along the way:***

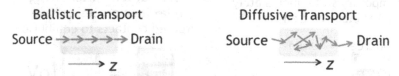

In ***Section 3.2*** we will obtain an expression for the conductance function *G(E)* for an elastic resistor in terms of the density of states *D(E)*.

The concept of an elastic resistor is not only useful in understanding nanoscale devices, but it also helps understand transport properties like the conductivity of large resistors by viewing them as multiple elastic resistors in series, as explained in ***Section 3.3***. This is what makes the bottom-up approach so powerful in clarifying transport problems in general.

But before we talk further about the conductance of an elastic resistor, let us address an important conceptual issue. Since current flow (*I*) through a resistor (R) dissipates a Joule heat of I^2R per second, it seems like a contradiction to talk of an elastic resistor where electrons do not lose energy? The point to note is that while the electron does not lose any energy in the channel of an elastic resistor, it does lose energy both in the source and the drain and that is where the Joule heat gets dissipated. This is a very non-intuitive result that seems to be at least approximately true of nanoscale conductors: *An elastic resistor has a resistance R determined by the channel, but the corresponding heat I^2R is entirely dissipated outside the channel.*

3.1. How an Elastic Resistor Dissipates Heat

How could this happen? Consider a one level elastic resistor having one sharp level with energy ε. Every time an electron crosses over through the channel, it appears as a "hot electron" on the drain side with an energy ε in excess of the local electrochemical potential μ_2 as shown below:

(a) Temporary state immediately after electron transfer

(b) Final state after energy relaxation processes have returned contacts to equilibrium

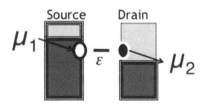 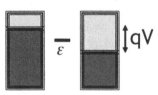

Energy dissipating processes in the contact quickly make the electron get rid of the excess energy $(\varepsilon - \mu_2)$. Similarly at the source end an empty spot (a "hole") is left behind with an energy ε that is much less than the local electrochemical potential μ_1, which gets quickly filled up by electrons dissipating the excess energy $(\mu_1 - \varepsilon)$.

In effect, every time an electron crosses over from the source to the drain, *an energy $(\mu_1 - \varepsilon)$ is dissipated in the source*

an energy $(\varepsilon - \mu_2)$ is dissipated in the drain

The total energy dissipated is

$$\mu_1 - \mu_2 \;=\; qV$$

which is supplied by the external battery that maintains the potential difference μ_1 - μ_2. The overall flow of electrons and heat is summarized in Fig.3.3 below.

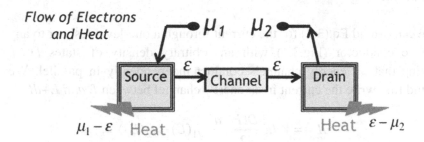

Fig.3.3. Flow of electrons and heat in a one-level elastic resistor having one level with $E = \varepsilon$.

If N electrons cross over in a time t

$$Dissipated\ power \;=\; qV * N/t \;=\; V * I$$

since
$$Current \;=\; q * N/t$$

Note that $V*I$ is the same as I^2R and V^2G.

The heat dissipated by an "elastic resistor" thus occurs in the contacts. As we will see next, the detailed mechanism underlying the complicated process of heat transfer in the contacts can be completely bypassed simply by legislating that the contacts are always maintained in equilibrium with a fixed electrochemical potential.

3.2. Conductance of an Elastic Resistor

Consider first the simplest elastic resistor having just one level with energy $\mathcal{E}$ in the energy range of interest through which electrons can squeeze through from the source to the drain. We can write the resulting current as

$$I_{one\ level} = \frac{q}{t}\left(f_1(\mathcal{E}) - f_2(\mathcal{E})\right) \tag{3.2}$$

where t is the time it takes for an electron to transfer from the source to the drain.

We can extend Eq.(3.2) for the current through a one-level resistor to any elastic conductor (Fig.3.1) with an arbitrary density of states $D(E)$, noting that all energy channels conduct independently in parallel. We could first write the current in an energy channel between E and $E+dE$

$$dI = dE \frac{D(E)}{2} \frac{q}{t}(f_1(E) - f_2(E))$$

since an energy channel between E and $E+dE$ contains $D(E)dE$ states, half of which contribute to carrying current from source to drain.

Integrating we obtain an expression for the current through an elastic resistor:

$$I = \frac{1}{q} \int_{-\infty}^{+\infty} dE\, G(E)(f_1(E) - f_2(E)) \tag{3.3}$$

where
$$G(E) = \frac{q^2 D(E)}{2t(E)} \tag{3.4}$$

If the applied voltage $\mu_1 - \mu_2 = qV$ is much less than kT, we can use Eq.(2.7) to write

$$I = V \int\limits_{-\infty}^{+\infty} dE \left(-\frac{\partial f_0}{\partial E} \right) G(E)$$

which yields the expression for conductance stated earlier in Eq.(3.1).

3.2.1. Degenerate and Non-Degenerate Conductors

Eq. (3.1) is valid in general, but depending on the nature of the conductance function $G(E)$ and the thermal broadening function $-\partial f_0 / \partial E$, two distinct physical pictures are possible. The first is case A where the conductance function $G(E)$ is nearly constant over the width of the broadening function.

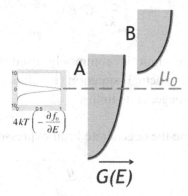

We could then pull $G(E)$ out of the integral in Eq.(3.1) to write

$$\frac{I}{V} \approx G(E = \mu_0) \int\limits_{-\infty}^{+\infty} dE \left(-\frac{\partial f_0}{\partial E} \right) = G(E = \mu_0) \qquad (3.5)$$

This relation suggests an operational definition for the conductance function $G(E)$: *It is the conductance measured at low temperatures for a channel with its electrochemical potential μ_0 located at E.*

Case A is a good example of the so-called degenerate conductors. The other extreme is the non-degenerate conductor shown in case B where the electrochemical potential is located at an energy many kT's below the energy range where the conductance function is non-zero. As a result over the energy range of interest where $G(E)$ is non-zero, we have

$$x \equiv \frac{E - \mu_0}{kT} \gg 1$$

and it is common to approximate the Fermi function with the Boltzmann function

$$\frac{1}{1 + e^x} \approx e^{-x}$$

so that
$$\frac{I}{V} \approx \int\limits_{-\infty}^{+\infty} \frac{dE}{kT} G(E) \, e^{-(E-\mu_0)/kT}$$

This non-degenerate limit is commonly used in the semiconductor literature though the actual situation is often intermediate between degenerate and non-degenerate limits.

We will generally use the degenerate limit expressed by Eq.(3.5) writing

$$G = \frac{q^2 D}{2t}$$

with the understanding that the quantities D and t are evaluated at $E = \mu_0$ and depending on the nature of $G(E)$ may need to be averaged over a range of energies using $-\partial f_0 / \partial E$ as a "weighting function" as prescribed by Eq.(3.1).

Eq.(3.4) seems quite intuitive: it says that the conductance is proportional to the product of two factors, namely ***the availability of states (D) and the ease with which electrons can transport through them (1/t)***. This is the key result that we will use in subsequent Lectures.

3.3. Why an Elastic Resistor is Relevant

The elastic resistor model is clearly of great value in understanding nanoscale conductors, but the reader may well wonder how an elastic resistor can capture the physics of real conductors which are surely far from elastic? In long conductors inelastic processes are distributed continuously through the channel, inextricably mixed up with all the elastic processes (Fig.3.4). Doesn't that affect the conductance and other properties we are discussing?

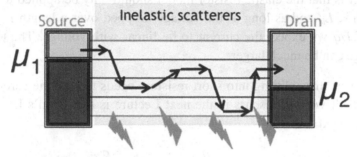

Fig.3.4
Real conductors have inelastic scatterers distributed throughout the channel.

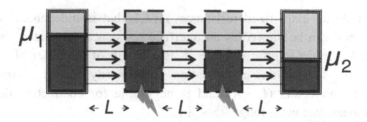

Fig.3.5
A hypothetical series of elastic resistors as an approximation to a real resistor with distributed inelastic scattering as shown in Fig.3.4.

One way to apply the elastic resistor model to a large conductor with distributed inelastic processes is to break up the latter conceptually into a sequence of elastic resistors (Fig.3.5), each much shorter than the physical length L, having a voltage that is only a fraction of the total voltage V. We could then argue that the total resistance is the sum of the individual resistances.

This splitting of a long resistor into little sections of length shorter than L_{in} (L_{in}: length an electron travels on the average before getting inelastically scattered) also helps answer another question one may raise about the elastic resistor model. We obtained the linear conductance by resorting to a Taylor's series expansion (see Eq.(2.6)). But keeping the first term in the Taylor's series can be justified only for voltages $V <$ kT/q, which at room temperature equals 25 mV. But everyday resistors are linear for voltages that are much larger. How do we explain that? The answer is that the elastic resistor model should only be applied to a short length $< L_{in}$ and as long as the voltage dropped over a length L_{in} is less than kT/q we expect the current to be linear with voltage. The terminal voltage can be much larger.

However, this splitting into short resistors needs to be done carefully. A key result we will discuss in the next Lecture is that Ohm's law should be modified

$$ from \quad \underbrace{R = \frac{\rho}{A}L}_{Eq.(1.1)} \quad to \quad \underbrace{R = \frac{\rho}{A}(L+\lambda)}_{Eq.(1.4)} $$

to include an extra fixed resistance $\rho\lambda/A$ that is independent of the length and can be viewed as an interface resistance associated with the channel- contact interfaces. Here λ is a length of the order of a mean free path, so that this modification is primarily important for near ballistic conductors ($L \sim \lambda$) and is negligible for conductors that are many mean free paths long ($L \gg \lambda$).

Conceptually, however, this additional resistance is very important if we wish to use the hypothetical structure in Fig.3.5 to understand the real structure in Fig.3.4. The structure in Fig.3.5 has too many interfaces that are not present in the real structure of Fig.3.4 and we have to remember to exclude the resistance coming from these conceptual interfaces.

For example, if each section in Fig.3.5 is of length L having a resistance of

$$R = \frac{\rho(L+\lambda)}{A}$$

then the correct resistance of the real structure in Fig.3.4 of length $3L$ is given by

$$R = \frac{\rho(3L+\lambda)}{A} \quad \text{and NOT by} \quad R = \frac{\rho(3L+3\lambda)}{A}$$

Clearly we have to be careful to separate the interface resistance from the length dependent part. This is what we will do next.

Lecture 4

Ballistic and Diffusive Transport

4.1. Ballistic and Diffusive Transfer Times
4.2. Channels for Conduction

We saw in the last Lecture that the resistance of an elastic resistor can be written as

$$G = \frac{q^2 D}{2t}$$

(see Eq.(3.4))

In this Lecture I will first argue that the transfer time t across a resistor of length L for diffusive transport with a mean free path λ can be related to the time t_B for ballistic transport by the relation (*Section 4.1*)

$$t = t_B \left(1 + \frac{L}{\lambda} \right)$$

(4.1)

Combining with Eq.(3.4) we obtain

$$G = \frac{G_B \lambda}{L + \lambda}$$

(4.2)

$$G_B \equiv \frac{q^2 D}{2 t_B}$$

where

(4.3)

We could invert Eq.(4.2) to write the new Ohm's law

$$G = \frac{\sigma A}{L + \lambda}$$

(4.4a)

39

where $$\sigma A = G_B \lambda$$

So far we have only talked about three dimensional resistors with a large cross-sectional area A. Many experiments involve two-dimensional resistors whose cross-section is effectively one-dimensional with a width W, so that the appropriate equations have the form

$$G = \frac{\sigma W}{L + \lambda} \tag{4.4b}$$

where $$\sigma W = G_B \lambda$$

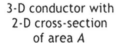

3-D conductor with 2-D cross-section of area A 2-D conductor with 1-D cross-section of area W 1-D conductor

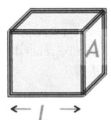 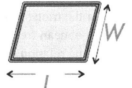 

Fig.4.1. 3-D, 2-D and 1-D conductors

Finally we have one-dimensional conductors for which

$$G = \frac{\sigma}{L + \lambda} \tag{4.4c}$$

where $$\sigma = G_B \lambda$$

We could collect all these results and write them compactly in the form

$$G = \frac{\sigma}{L + \lambda}\{1, \quad W, \quad A\} \tag{4.5}$$

with
$$\sigma = G_B\lambda\left\{1, \quad \frac{1}{W}, \quad \frac{1}{A}\right\} \tag{4.6}$$

where the three items in parenthesis correspond to 1-D, 2-D and 3-D conductors. Note that the conductivity σ has different dimensions in 1-D, 2-D and 3-D, while both G_B and λ have the same dimensions, namely Siemens (S) and meters (m) respectively.

The standard Ohm's law predicts that the resistance will approach zero as the length L is reduced to zero. Of course no one expects it to become zero, but the common belief is that it will approach a value determined by the interface resistance which can be made arbitrarily small with improved contacting technology.

What is now well established experimentally is that even with the best possible contacts, there is a minimum interface resistance determined by the properties of the channel, independent of the contact. The modified Ohm's law in Eq.(4.5) reflects this fact: Even a channel of zero length with perfect contacts has a resistance equal to that of a hypothetical channel of length λ.

But what does it mean to talk about the mean free path λ of a channel of zero length? The answer is that neither ρ nor λ mean anything for a short conductor, but their product $\rho\lambda$ does. The ballistic resistance has a simple meaning that has become clear in the light of modern experiments as we will see in *Section 4.2*. It is inversely proportional to the number of channels, $M(E)$ available for conduction, which is proportional to, but not the same as, the density of states, $D(E)$.

The concept of density of states has been with us since the earliest days of solid state physics. By contrast, the number of channels (or transverse modes) $M(E)$ is a more recent concept whose significance was appreciated only after the seminal experiments in the 1980's on ballistic conductors showing conductance quantization.

4.1 Ballistic and Diffusive Transport

Consider how the two quantities in

$$G = \frac{q^2 D}{2t}$$

namely the density of states, D and the transfer time t scale with channel dimensions for large conductors. The first of these is relatively easy to see since we expect the number of states to be additive. A channel twice as big should have twice as many states, so that the density of states $D(E)$ for large conductors should be proportional to the volume $(A*L)$.

Regarding the transfer time, t, broadly speaking there are two transport regimes:

Ballistic regime: Transfer time $t \sim L$

Diffusive regime: Transfer time $t \sim L^2$

Consequently the ballistic conductance is proportional to the area (note that $D \sim A*L$ as discussed above), but **independent of the length**. This "non-ohmic" behavior has indeed been observed in short conductors. It is only diffusive conductors that show the "ohmic" behavior $G \sim A/L$.

These two regimes can be understood as follows. In the ballistic regime electrons travel straight from the source to the drain "like a bullet," taking a time

$$t_B = \frac{L}{\bar{u}} \qquad (4.7)$$

Ballistic Transport

Source $\rightarrow\rightarrow\rightarrow\rightarrow\rightarrow$ Drain

where $\bar{u} = \langle |v_z| \rangle$

$\longrightarrow z$

is the average velocity of the electrons in the z-direction.

But conductors are typically not short enough for electrons to travel "like bullets." Instead they stumble along, getting scattered randomly by various defects along the way taking much longer than the ballistic time in Eq.(4.7). We could write

$$t = \frac{L}{\bar{u}} + \frac{L^2}{2\bar{D}}$$

Diffusive Transport

Source ➙⟋⟋⟍⟍⟍⟍➙ Drain

➙ z

viewing it as a sort of "polynomial expansion" of the transfer time t in powers of L. We could then argue that the lowest term in this expansion must equal the ballistic limit $L/\bar{u}$, while the highest term should equal the diffusive limit well-known from the theory of random walks. This theory (see for example, Berg, 1983) identifies the coefficient $\bar{D}$ as the diffusion constant

$$\bar{D} = \left\langle v_z^2 \tau \right\rangle$$

τ being the mean free time.

We could use Eq.(4.7) to rewrite the expression for the transit time in Eq.(4.8) in the form

$$t = t_B \left(1 + \frac{L\bar{u}}{2\bar{D}} \right)$$

which agrees with Eq.(4.1) if the mean free path is given by

$$\lambda = \frac{2\bar{D}}{\bar{u}}$$

In defining the two constants $\bar{D}, \bar{u}$ we have used the symbol $\langle \cdots \rangle$ to denote an average over the angular distribution of velocities which yields a different numerical factor depending on the dimensionality of the conductor (see Appendix B).

For *d = {1, 2, 3}* dimensions

$$\bar{u} \;=\; \left\langle |v_z| \right\rangle \;=\; v(E)\left\{1,\; \frac{2}{\pi},\; \frac{1}{2}\right\}$$

(4.8a)

and

$$\bar{D} \;=\; \left\langle v_z^2\tau \right\rangle \;=\; v^2\tau(E)\left\{1,\; \frac{1}{2},\; \frac{1}{3}\right\}$$

(4.8b)

so that

$$\lambda \;=\; \frac{2\bar{D}}{\bar{u}} \;=\; v\tau\left\{2,\; \frac{\pi}{2},\; \frac{4}{3}\right\}$$

(4.9)

Note that our definition of the **mean free path** includes a dimension-dependent numerical factor over and above the standard value of $v\tau$. Couldn't we simply use the standard definition? We could, but then the new Ohm's law would not simply involve replacing L with L plus λ. Instead it would involve L plus a dimension-dependent factor times λ. Instead we have chosen to absorb this factor into the definition of λ.

Interestingly, even in one dimensional conductors the factor is not one, but two. This is because τ is the mean free time after which an electron gets scattered. Assuming the scattering to be isotropic, only half the scattering events will result in an electron traveling towards the drain to head towards the source. The mean free time for backscattering is thus 2τ, making the mean free path $2v\tau$ rather than $v\tau$.

Next we obtain an expression for the **ballistic conductance** by combining Eq.(4.3) with Eq.(4.7) to write

$$G_B \;\equiv\; \frac{q^2 D\bar{u}}{2L}$$

and then make use of Eq.(4.8a) to write

$$G_B \;\equiv\; \frac{q^2 D v}{2L}\left\{1,\; \frac{2}{\pi},\; \frac{1}{2}\right\}$$

(4.10)

Finally we can use Eqs.(4.9) and (4.10) in Eq.(4.6) and make use of Eq.(4.8b) to obtain an expression for the **conductivity**:

$$\sigma = q^2 \bar{D} \frac{D}{L} \left\{ 1, \frac{1}{W}, \frac{1}{A} \right\} \qquad (4.11)$$

We have thus obtained expressions for the conductance in the ballistic regime as well as the conductivity in the diffusive regime, starting from our expression

$$G = \frac{q^2 D}{2t}$$

based on the expression for the ballistic and diffusive transfer times

$$t = \frac{L}{\bar{u}} + \frac{L^2}{2\bar{D}}$$

which some readers may not find completely satisfactory. But this approach has the advantage of getting us to the new Ohm's law (Eq.(4.5)) very quickly using simple algebra. In Lecture 6 we will re-derive Eq.(4.5) more directly by solving a differential equation, without invoking the transfer time.

4.2 Channels for Conduction

Eq.(4.10) tells us that the ballistic conductance depends on D/L, the density of states per unit length. Since D is proportional to the volume, the ballistic conductance is expected to be proportional to the cross-sectional area A in 3-D conductors (or the width W in 2-D conductors).

Numerous experiments since the 1980's have shown that for small conductors, the ballistic conductance does not go down linearly with the area A. Rather it goes down in integer multiples of the **conductance quantum**

$$G_B \equiv \underbrace{\frac{q^2}{h}}_{38\,\mu S} \underbrace{M}_{integer}$$

$$(4.12)$$

How can we understand this relation and what does the integer M represent? This result cannot come out of our elementary treatment of electrons in classical particle-like terms, since it involves Planck's constant h. Some input from quantum mechanics is clearly essential and this will come in Lecture 5 when we evaluate $D(E)$.

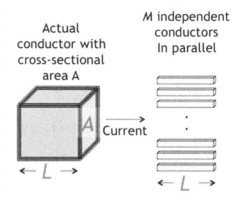

Actual conductor with cross-sectional area A

M independent conductors In parallel

For the moment we note that heuristically Eq.(4.8) suggests that we visualize the real conductor as M independent channels in parallel whose conductances add up to give Eq.(4.12) for the ballistic conductance.

This suggests that we use Eqs.(4.10) and (4.12) to define a quantity $M(E)$

$$M \equiv \frac{hDv}{2L}\left\{1, \ \frac{2}{\pi}, \ \frac{1}{2}\right\} \tag{4.13}$$

which should provide us a measure of the number of conducting channels. From Eqs.(4.6) and (4.12) we can write the conductivity in terms of M and the mean free path λ :

$$\sigma = \frac{q^2}{h}M\lambda\left\{1, \ \frac{1}{W}, \ \frac{1}{A}\right\} \tag{4.14}$$

In the next Lecture we will use a simple model that incorporates the wave nature of electrons to show that for a one-dimensional channel the quantity M indeed equals one showing that it has only one channel, while for two- and three-dimensional conductors the quantity M represents the number of de Broglie wavelengths that fit into the cross-section, like the modes of a waveguide.

Lecture 5

Conductivity

5.1. *E(p) or E(k) Relations*
5.2. *Counting States*
5.3. *Drude Formula*
5.4. *Is Conductivity proportional to Electron Density?*
5.5. *Quantized Conductance*

A common expression for conductivity is the Drude formula relating the conductivity to the electron density n, the effective mass m and the mean free time

$$\sigma \equiv \frac{1}{\rho} = \frac{q^2\, n\, \tau}{m}$$

(5.1a)

This expression is very well-known since even freshman physics texts start by deriving it. It also leads to the widely used concept of mobility

$$\bar{\mu} = \frac{q\tau}{m}$$

(5.1b)

such that

$$\sigma = qn\bar{\mu}$$

(5.1c)

On the other hand, in Lecture 4 we obtained two equivalent expressions for the conductivity, one as a product of the density of states D and the diffusion coefficient $\bar{D}$ (see Eq.(4.11))

$$\sigma(E) = q^2 \bar{D} \frac{D}{L} \left\{ 1, \frac{1}{W}, \frac{1}{A} \right\}$$

(5.2a)

and the other as a product of the number of modes M and the mean free path λ:

$$\sigma(E) = \frac{q^2}{h} M \lambda \left\{ 1, \ \frac{1}{W}, \ \frac{1}{A} \right\}$$

$$(5.2b)$$

Note that, like the conductance (see Eq.(3.1)), these expressions for the energy-dependent conductivity also have to be averaged over an energy range of a few kT's around $E=\mu_0$, using the thermal broadening function,

$$\bar{\sigma} = \int_{-\infty}^{+\infty} dE \left(-\frac{\partial f_0}{\partial E} \right) \sigma(E)$$

$$(5.3a)$$

It is this averaged conductivity $\bar{\sigma}$ that should be compared to the Drude conductivity in Eq.(5.1). But for degenerate conductors (see Section 3.2.1) the averaged conductivity $\bar{\sigma}$ is approximately equal to the conductivity at an energy $E = \mu_0$:

$$\bar{\sigma} \approx \sigma(E = \mu_0)$$

$$(5.3b)$$

and so we can compare $\sigma(E = \mu_0)$ from Eq.(5.2) to Eq.(5.1).

Although Eq.(5.2b) is not very well-known, the equivalent version in Eq.(5.2a) is a standard result that is derived in many textbooks. However, the usual derivation of Eq.(5.2a) requires advanced concepts like the Boltzmann or the Kubo formalism and so appears much later than Eq.(5.1) in any solid-state physics text. Not surprisingly, most people remember Eq.(5.1) and not Eq.(5.2).

But the point we wish to stress is that while Eq.(5.1) is often very useful, it is a result of limited validity that can be obtained from Eq.(5.2) by making suitable approximations based on a specific model. But when these approximations are not appropriate, we can still use Eq.(5.2) which is *far more generally applicable*. For example, Eq.(5.2) gives sensible answers even for materials like graphene whose non-parabolic bands make the meaning of mass somewhat unclear, causing considerable confusion when using Eq.(5.1). In general we should really use Eq.(5.2), and not Eq.(5.1), to shape our thinking about conductivity.

There is a fundamental difference between Eq.(5.2) and (5.1). The averaging implied in Eq.(5.3) makes the conductivity a "Fermi surface property", that is one that depends only on the energy levels close to $E=\mu_0$. By contrast, Eq.(5.1) depends on the total electron density n integrated over all energy. But this dependence on the total number is true only in a limited sense.

Experts know that n only represents the density of "free" electrons and have an instinctive feeling for what it means to be free. They know that there are p-type semiconductors which conduct better when they have fewer electrons, but in that case they know that n should be interpreted to mean the number of "holes". For beginners, all this appears confusing and much of this confusion can be avoided by using Eq.(5.2) instead of (5.1).

Interestingly, Eq.(5.2a) was used in a seminal paper to obtain Eq.(3.4)

$$G = \frac{q^2 D}{2t} \qquad \text{(same as Eq.(3.4))}$$

(see Eq.(1) of Thouless (1977)). Instead we have used the concept of an elastic resistor to first obtain Eq.(3.4) from elementary arguments, and then used it to obtain Eq.(5.2a).

Eq.(5.2) stresses that the essential factor determining the conductivity is the density of states around $E=\mu_0$. Materials are known to have conductivities ranging over many orders of magnitude from glass to copper. And the basic fact remains that they all have approximately the same number of electrons. Glass is not an insulator because it is lacking in electrons. It is an insulator because it has a very low density of states or number of modes around $E=\mu_0$.

So when does Eq.(5.2) reduce to (5.1)? Answer: If the electrons are described by a "single band effective mass model" as I will try to show in this Lecture. So far we have kept our discussion general in terms of the density of states, $D(E)$ and the velocity, $v(E)$ without adopting any specific models. These concepts are generally applicable even to amorphous materials and molecular conductors. A vast amount of literature both in condensed matter physics and in solid state devices, however, is devoted to crystalline solids with a periodic arrangement of atoms because of the major role they have played from both basic and applied points of view.

For such materials, energy levels over a limited range of energies are described by a *E(p)* relation and we will show in this Lecture that irrespective of the specific *E(p)* relation, at any energy E the density of states *D(E)*, velocity *v(E)* and momentum *p(E)* are related to the total number of states *N(E)* with energy less than E by the relation (*d*: number of dimensions)

$$D(E)v(E)p(E) = N(E).d \qquad (5.4)$$

We can combine this relation with Eq.(5.2a) and make use of Eq.(4.8b), $\bar{D} = v^2 \tau / d$, to write

$$\sigma(E) = \frac{q^2 \tau(E)}{m(E)} \left\{ \frac{N(E)}{L}, \ \frac{N(E)}{WL}, \ \frac{N(E)}{AL} \right\} \qquad (5.5)$$

where we have defined mass as

$$m(E) = \frac{p(E)}{v(E)} \qquad (5.6)$$

For parabolic *E(p)* relations, the mass is independent of energy, but in general it could be energy-dependent.

Eq.(5.5) indeed looks like Drude expression (Eq.(5.1a)) if we identify the quantity in parenthesis {N/L, N/WL, N/AL} as the electron density, n per unit length, area and volume in 1D, 2D and 3D respectively. At low temperatures, this is easy to justify since the energy averaging in Eq.(5.3) amounts to looking at the value at $E = \mu_0$ and $N(E)$ at $E = \mu_0$ represents the total number of electrons (Fig.5.1).

At non-zero temperatures one needs a longer discussion which we will get into later in the Lecture. Indeed as will see, some subtleties are involved even at zero temperature when dealing with differently shaped density of states.

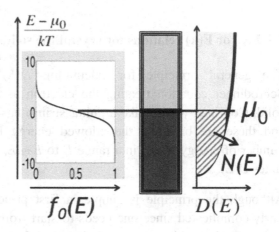

Fig.5.1.
Equilibrium Fermi function $f_0(E)$, Density of states $D(E)$ and integrated density of states $N(E)$.

Note, however, that the key to reducing our conductivity expression (Eq.(5.2)) to the Drude-like expression (Eq.(5.5)) is Eq.(5.4) which is an interesting relation for it relates $D(E)$, $v(E)$ and $p(E)$ at a given energy E, to the total number of states $N(E)$ obtained by integrating $D(E)$

$$N(E) = \int_{-\infty}^{E} dE\, D(E)$$

How can the integrated value of $D(E)$ be uniquely related to the value of quantities like $D(E)$, $v(E)$ and $p(E)$ at a single energy ? The answer is that this relation holds only as long as the energy levels are given by a single $E(p)$ relation. It may not hold in an energy range with multiple bands of energies or in an amorphous solid not described by an $E(p)$ relation. Eq.(5.2) is then not equivalent to Eq.(5.5), and *it is Eq.(5.2) that can be trusted.*

With that long introduction let us now look at how single bands described by an $E(p)$ relation leads to Eq.(5.4) and helps us connect our conductivity expression (Eq.(5.2)) to the Drude formula (Eq.(5.1)). This will also lead to a different interpretation of the quantity $M(E)$ introduced in the last Lecture, that will help understand why it is an integer representing the number of channels.

5.1 E(p) or E(k) relations for crystalline solids

The general principle for calculating $D(E)$ is to start from the Schrodinger equation treating the electron as a wave confined to the solid. Confined waves (like a guitar string) have resonant "frequencies" and these are basically the allowed energy levels. By counting the number of energy levels in a range E to $E+dE$, we obtain the density of states $D(E)$.

Although the principle is simple, a first principles implementation is fairly complicated since one needs to start from a Schrödinger equation including the nuclear potential that the electrons feel inside the solid. One of the seminal concepts in solid state physics is the realization that in crystalline solids electrons behave as if they are in vacuum, but with an effective mass different from their natural mass, so that the energy-momentum relation can be written as

Parabolic Dispersion

$$E(p) = E_c + \frac{p^2}{2m} \qquad (5.7a)$$

where E_c is a constant. The momentum p is equated to $\hbar k$, providing the link between the energy-momentum relation $E(p)$ associated with the particle viewpoint and the dispersion relation $E(k)$ associated with the wave viewpoint. Here we will write everything in terms of p, but they are easily translated in terms of $k = p/\hbar$.

Eq.(5.7a) is generally referred to as a parabolic dispersion relation and is commonly used in a wide variety of materials from metals like copper to semiconductors like silicon, because it often approximates the actual

E(p) relation fairly well in the energy range of interest. But it is by no means the only possibility. Graphene, a material of great current interest, is described by a linear relation:

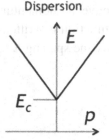

Linear Dispersion

$$E = E_c + v_0\, p \qquad\qquad (5.7b)$$

where v_0 is a constant. Note that p denotes the magnitude of the momentum and we will assume that the $E(p)$ relation is isotropic, which means that it is the same regardless of which direction the momentum vector points.

For any given isotropic $E(p)$ relation, the velocity points in the same direction as the momentum, while its magnitude is given by

$$v \equiv \frac{dE}{dp} \qquad\qquad (5.8)$$

This is a general relation applicable to arbitrary energy-momentum relations for classical particles. On the other hand, in wave mechanics it is justified as the group velocity for a given dispersion relation $E(k)$.

5.2 Counting states

One great advantage of this principle is that it reduces the complicated problem of electron waves in a solid to that of waves in vacuum, where the allowed energy levels can be determined the same way we find the resonant frequencies of a guitar string: simply by requiring that an integer number of wavelengths fit into the solid. Noting that the de Broglie principle relates the electron wavelength to the Planck's constant divided by its momentum, h/p, we can write

$$\frac{L}{h/p} \;=\; Integer \;\Rightarrow\; p \;=\; Integer * \left(\frac{h}{L}\right)$$

where L is the length of the box. This means that the allowed states are uniformly distributed in p with each state occupying a "space" of

$$\Delta p \; = \; \frac{h}{L} \tag{5.9}$$

Let us define a function $N(p)$ that tells us the total number of states that have a momentum less than a given value p. In *one dimension* this function is written down by dividing the total range of $2p$ (from $-p$ to $+p$) by the spacing h/L:

$$N(p) \; = \; \frac{2p}{h/L} \; = \; 2L\left(\frac{p}{h}\right) \qquad \text{1D}$$

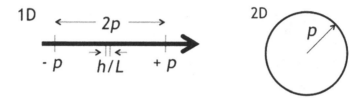

In two dimensions we divide the area of a circle of radius p by the spacing $h/L * h/W$, L and W being the dimensions of the two dimensional box.

$$N(p) \; = \; \frac{\pi p^2}{(h/L)(h/W)} \; = \; \pi W L \left(\frac{p}{h}\right)^2 \qquad \text{2D}$$

In three dimensions we divide the volume of a sphere of radius p by the spacing $h/L * h/W_1 * h/W_2$, L, W_1 and W_2 being the dimensions of the three dimensional box. Writing $A = W_1 * W_2$ we have

$$N(p) \; = \; \frac{(4\pi/3)p^3}{(h/L)(h^2/A)} \; = \; \frac{4\pi}{3} AL\left(\frac{p}{h}\right)^3 \qquad \text{3D}$$

We can combine all three results into a single expression for $d = \{1, 2, 3\}$ *dimensions:*

$$N(p) \; = \; \left\{ 2\frac{L}{h/p}, \; \pi\frac{LW}{(h/p)^2}, \; \frac{4\pi}{3}\frac{LA}{(h/p)^3} \right\} \tag{5.10}$$

We could use a given $E(p)$ relation to turn this function $N(p)$ into a function of energy $N(E)$ that tells us the total number of states with energy less than E.

5.2.1. Density of states, D(E)

This function $N(E)$ that we have just obtained must equal the density of states $D(E)$ **integrated** up to an energy E, so that $D(E)$ can be obtained from the derivative of $N(E)$:

$$N(E) = \int_{-\infty}^{E} dE' \, D(E') \quad \rightarrow \quad D(E) = \frac{dN}{dE}$$

Hence from Eq.(5.10),

$$D(E) = \frac{dN}{dp} \frac{dp}{dE} = K_N \frac{dp}{dE} \frac{p^{d-1} d}{h^d}$$

Making use of Eqs.(5.8) and (5.10), we obtain the relation stated earlier

$$D(E)v(E)p(E) = N(E) . d \quad \text{(same as Eq.(5.4))}$$

which is completely general *independent of the actual E(p) relation.*

5.3 Drude formula

As noted earlier, using this relation we can rewrite Eq.(5.2) in the form

$$\sigma(E) = \frac{q^2 \tau(E)}{m(E)} \left\{ \frac{N(E)}{L}, \frac{N(E)}{WL}, \frac{N(E)}{AL} \right\} \quad \text{(same as Eq.(5.5))}$$

with an energy-dependent mass $m(E)$ defined in Eq.(5.6)). As we have seen, it is straightforward to connect this relation to the Drude formula (Eq.(5.1)) at low temperatures where the energy averaging in Eq.(5.3) amounts to looking at the value at a single energy $E=\mu_0$. What about non-zero temperatures?

5.3.1. n-type Conductors

Using Eq.(5.5) and assuming m and τ to be energy-independent we have

$$\bar{\sigma} = \frac{q^2\tau}{m} \int\limits_{-\infty}^{+\infty} dE\left(-\frac{\partial f_0}{\partial E}\right) N(E)\left\{\frac{1}{L}, \ \frac{1}{WL}, \ \frac{1}{AL}\right\} \qquad (5.11)$$

The integral can be carried out "by parts" to yield

$$\int\limits_{-\infty}^{+\infty} dE\left(-\frac{\partial f_0}{\partial E}\right) N(E) = \left[-N(E)f_0(E)\right]_{-\infty}^{+\infty} + \int\limits_{-\infty}^{+\infty} dE\frac{dN}{dE} f_0(E)$$

$$= [0-0] + \int\limits_{-\infty}^{+\infty} dE\, D(E)\, f_0(E)$$

$$= \quad Total\ Number\ of\ Electrons$$

since $dE\, D(E)\, f_0(E)$ tells us the number of electrons in the energy range from E to $E+dE$. When integrated it gives us the total number of electrons.

Eq.(5.11) then reduces to

$$\bar{\sigma} = \frac{q^2\tau}{m} \ (Number\ of\ electrons)\left\{\frac{1}{Length}, \ \frac{1}{Area}, \ \frac{1}{Volume}\right\}$$

which is the Drude formula stated in Eq.(5.1a).

5.3.2. p-type conductors

An interesting subtlety is involved when we consider a p-type conductor for which the $E(p)$ relation extends downwards, say something like

$$E(p) = E_v - \frac{p^2}{2m}$$

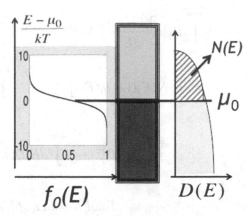

Fig.5.2: Equilibrium Fermi function $f_0(E)$, Density of states $D(E)$ and integrated density of states $N(E)$: p-type conductor.

Instead of

$$N(E) = \int_{-\infty}^{E} dE'\, D(E')$$

we now have (see Fig.5.2)

$$N(E) = \int_{E}^{+\infty} dE'\, D(E') \quad \rightarrow \quad D(E) = -\frac{dN}{dE}$$

This is because we defined the function $N(E)$ from $N(p)$ which represents the total number of states with momenta less than p, which means energies greater than E for a p-type dispersion relation.

Now if we carry out the integration by parts as before

$$\int_{-\infty}^{+\infty} dE \left(-\frac{\partial f_0}{\partial E}\right) N(E) = \left[-N(E)f_0(E)\right]_{-\infty}^{+\infty} + \int_{-\infty}^{+\infty} dE\, \frac{dN}{dE}\, f_0(E)$$

we run into a problem because the first term does not vanish at the lower limit where both $N(E)$ and $f_0(E)$ are both non-zero.

We can get around this problem by writing the derivative in terms of $1\text{-}f_0$ instead of f_0:

$$\int\limits_{-\infty}^{+\infty} dE \left(\frac{\partial(1-f_0)}{\partial E} \right) N(E)$$

$$= \left[-N(E)(1-f_0(E)) \right]_{-\infty}^{+\infty} + \int\limits_{-\infty}^{+\infty} dE \frac{dN}{dE} (1-f_0(E))$$

$$= \begin{bmatrix} 0-0 \end{bmatrix} + \int\limits_{-\infty}^{+\infty} dE \, D(E)\,(1-f_0(E))$$

$$= \quad Total \ Number \ of \ "holes", P$$

What this means is that with p-type conductors we can use the Drude formula

$$\sigma = q^2 n \tau / m$$

but the *n* now represents the density of empty states or holes. A larger *n* really means fewer electrons.

5.3.3. *"Double-ended" density of states*

How would we count *n* for a density of states $D(E)$ that extends in both directions as shown in Fig.5.3 (left panel). This is representative of graphene, a material of great interest (recognized by the 2010 Nobel prize in physics), whose $E(p)$ relation is commonly approximated by

$$E = \pm v_0 \, p.$$

People usually come up with clever ways to handle such "double-ended" density of states so that the Drude formula can be used. For example they divide the total density of states into an n-type and a p-type component

$$D(E) = D_n(E) + D_p(E)$$

as shown in Fig.5.3 and the two components are then handled separately, using a prescription that is less than obvious: The conductivity due to the upper half D_n depends on the number of occupied states (electrons), while that due to the lower half depends on the number of unoccupied states (holes).

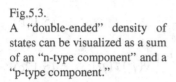

Fig.5.3.
A "double-ended" density of states can be visualized as a sum of an "n-type component" and a "p-type component."

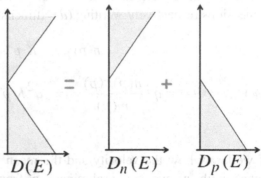

But the point we would like to stress is that there is really no particular reason to insist on using a Drude formula and keep inventing clever ways to make it work. One might just as well use Eq.(5.2) which reflects the correct physics of conduction, ***namely that it takes place in a narrow band of energies around μ_0.***

5.4. Is conductivity proportional to electron density?

Experimental conductivity measurements are often performed as a function of the electron density and the common expectation based on the Drude formula (Eq.(5.1)) is that conductivity should be proportional to the electron density and any non-linearity must be a consequence of the energy-dependence of the mean free time. What is not often recognized is that for non-parabolic dispersion relations, the mass itself defined as p/v can be energy-dependent and this will affect the conductivity- electron density relation.

First we note that from Eq.(5.10)

$$n(p) \;=\; \left\{ 2\frac{p}{h}, \;\; \pi\frac{p^2}{h^2}, \;\; \frac{4\pi}{3}\frac{p^3}{h^3} \right\} \tag{5.12a}$$

where we have defined n as N/L or N/WL or N/AL in 1, 2 and 3 dimensions respectively. Writing (d = dimensions, K = constant)

$$n(p) \;=\; K\,p^d \tag{5.12b}$$

we have $\qquad \sigma = q^2\,\dfrac{n(p)\tau(p)}{m(p)} \;=\; q^2 K\,p^{d-1}\,v(p)\tau(p) \tag{5.12c}$

If we know how the velocity and the mean free time vary with E (and hence with p) we could eliminate p from the expressions for the conductivity, σ and the electron density, n to obtain a direct relation between them (for degenerate conductors, as explained in the introduction)

For example, with graphene, $E = \pm v_0 p$, so that the velocity dE/dp is a constant (v_0), independent of p. If we assume an energy independent mean free time τ, we obtain

$$\sigma \;=\; \frac{q^2}{h}\,\lambda\,\sqrt{\frac{4n}{\pi}} \tag{5.13}$$

after a little algebra, noting that graphene is two-dimensional and making use of Eq.(4.9) for the mean free path.

To compare with experiments, we need to modify Eq.(5.13) a little to account for the degeneracy factor g which denotes the number of equivalent states. For example all non-magnetic materials have two spin states with identical energies, which would make $g=2$. Certain materials also have equivalent "valleys" having identical energy momenta relations so that the N we calculate for one valley has to be multiplied by g when relating to the experimentally measured electron densities. For graphene, $g = 2*2 = 4$.

Eq.(5.13) applies to a single spin and valley for which the conductivity and the electron density are each $1/g$ times the actual, so that

$$\frac{\sigma}{g} = \frac{q^2}{h} \lambda \sqrt{\frac{4n/g}{\pi}} \quad \rightarrow \quad \sigma = \frac{q^2}{h} \lambda \sqrt{\frac{4gn}{\pi}} \tag{5.14}$$

The calculated results from Eq.(5.14) with $\lambda = 2$ μm and with $\lambda = 300$ nm compares well with the experimental data on graphene reported in Bolotin et al. (2008). Note that the values of the mean free path indicated in the paper are half the values we have used. This is because our definition of mean free path

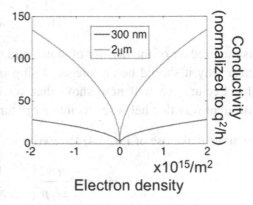

differs from the standard one by a dimension-dependent factor (see Eq.(4.9)).

I should mention, however, that long graphene samples often show a conductivity $\sim n$ and not $\sim \sqrt{n}$. This is believed to be because the mean free time and hence the mean free path λ due to charged impurity scattering is $\sim E \sim \sqrt{n}$. Eq.(5.14) then predicts a conductivity $\sim n$. It is only for an energy-independent mean free path that Eq.(5.14) predicts a $\sim \sqrt{n}$ dependence of the conductivity and this is only seen in short near ballistic samples for which the mean free path plays no role.

5.5. Quantized Conductance

I noted in the last Lecture 4 that the ballistic conductance is given by

$$G_B \equiv \underbrace{\frac{q^2}{\underbrace{h}_{38\ \mu S}}}\ \underbrace{M}_{integer} \qquad \text{(same as Eq.(4.12))}$$

and that experimentally M is found to be an integer in low dimensional conductors at low temperatures. However, in the last lecture we defined M (see Eq.(4.13))

$$M \equiv \frac{hD\,v}{2L}\left\{1,\ \frac{2}{\pi},\ \frac{1}{2}\right\} \qquad (5.15)$$

as the product of the density of states and the velocity and it is not at all clear why it should be an integer. Using the $E(p)$ relations discussed in this Lecture we will now show that we can interpret $M(p)$ in a very different way that helps see its integer nature.

First we make use of Eq.(5.4) to rewrite Eq.(5.15) in the form

$$M = \frac{hN}{2Lp}\left\{1,\ \frac{4}{\pi},\ \frac{3}{2}\right\} \qquad (5.16)$$

where $N(p)$ is the total number of states with a momentum that is less than p and we have seen that it is equal to the number of wavelengths that fit into the solid. Making use of Eq.(5.10) for $N(p)$, we obtain from Eq.(5.16)

$$M(p) = \left\{1,\ 2\frac{W}{h/p},\ \pi\frac{A}{(h/p)^2}\right\} \qquad (5.17)$$

Just as $N(p)$ tells us the number of wavelengths that fit into the volume, *M(p) tells us the number that fit into the cross-section* and this result is independent of the actual E(p) relation, since we have not made use of any specific relationship.

Now we are ready to look at the origin of conductance quantization. If we evaluate our expressions for $N(p)$ and $M(p)$ for a given sample we will in general get a fractional number. However, since these quantities represent the number of states, we would expect them to be integers and if we obtain say *201.59*, we should take the lower integer *201*.

This point is commonly ignored in large conductors at high temperatures, where experiments do not show this quantization because of the energy averaging over $\mu_0 \pm 2kT$ associated with experimental measurements. For example, if over this energy range, *M(E)* varies from say *201.59* to *311.67*, then it seems acceptable to ignore the fact that it really varies from *201* to *311*.

But in small structures where one or more dimensions is small enough to fit only a few wavelengths the integer nature of M is observable and shows up in the quantization of the ballistic conductance. We should then rewrite Eq.(5.17) as

$$M(p) = Int\left\{1, \; 2\frac{W}{h/p}, \; \pi\frac{A}{(h/p)^2}\right\} \tag{5.18}$$

where *Int (x)* represents the largest integer less than or equal to *x*.

For one dimensional conductors the number of modes is equal to g, which is the number of spins times the number of valleys. Ballistic conductors have a resistance of

$$\frac{h}{q^2 M} \approx \frac{25 \; K\Omega}{M}$$

so that the resistance of a 1D ballistic conductor is approximately equal to *25 KΩ* divided by g. This has indeed been observed experimentally. Most metals and semiconductors like GaAs have *g=2*, and the 1D ballistic resistance ~ *12.5 KΩ*. But carbon nanotubes have two valleys as well making g=4 and exhibit a ballistic resistance ~ *6.25 KΩ*.

For two- and three-dimensional conductors, Eq.(5.17) is not quite right, because it is based on the heuristic idea of counting modes by counting the number of wavelengths that fit into the solid (see Eq.(5.5)). Mathematically it can be justified only if we assume periodic boundary conditions, that is if we assume that the cross-section is in the form of a ring rather than a flat sheet for a 2D conductor. For a 3D conductor it is hard to visualize what periodic boundary conditions might look like though it is easy to impose it mathematically as we have been doing.

Ring-shaped conductor

Most real conductors do not come in the form of rings, yet periodic boundary conditions are widely used because it is mathematically convenient and people believe that the actual boundary conditions do not really matter. But this is true only if the cross-section is large. For small area conductors the actual boundary conditions do matter and we cannot use Eq.(5.10).

Flat Conductor

Interestingly a conductor of great current interest has actually been studied in both forms: a ring-shaped form called a carbon nanotube and a flat form called graphene. If the circumference or width is tens of nanometers they have much the same properties, but if it is a few nanometers their properties are observably different including their ballistic resistances.

Lecture 6

Diffusion Equation for Ballistic Transport

6.1. Electrochemical Potentials Out of Equilibrium
6.2. Currents in Terms of Non-Equilibrium Potentials

The title of this Lecture may sound contradictory, like the elastic resistor. Doesn't the diffusion equation describe diffusive transport? How can one use it for ballistic transport? An important idea we are trying to get across with our bottom-up approach is the essential unity of these two regimes of transport and hopefully this lecture will help.

The diffusion equation relates the current to the slope of the electrochemical potential $\mu(z)$

$$\frac{I}{A} = -\frac{\sigma}{q}\frac{d\mu}{dz}$$

(6.1a)

where σ is the conductivity (Eq.(5.2)) from the last Lecture.

We can obtain this equation by viewing a long conductor as a series of elastic resistors as discussed in Section 3.3:

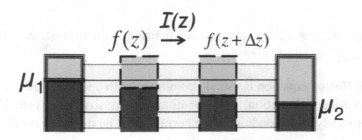

Using Eq.(3.3) we can write the current $I(z)$ in a section of the conductor as

$$I(z) \; = \; \frac{1}{q} \int_{-\infty}^{+\infty} dE \, G(E) \, (f(z,E) - f(z + \Delta z, E))$$

From Eq.(4.5) we could write

$$\frac{1}{G(E)} = \rho \frac{\Delta z + \lambda}{A}$$

but the point to note is that part of this resistance represents the interface resistance, which should not be included since there are no actual interfaces except at the very ends. Omitting the interface resistance we can write (Note: $\sigma = 1/\rho$, Eq.(1.1))

$$G(E) = \frac{\sigma A}{\Delta z}$$

Combining this with our usual linear expansion for small potential differences from Eq.(2.7)

$$f(z,E) - f(z + \Delta z, E) \approx \left(-\frac{\partial f_0}{\partial E} \right) \left(\mu(z) - \mu(z + \Delta z) \right)$$

and defining the conductivity σ as the thermal average of $\sigma(E)$ (Eq.(5.3)), we obtain

$$I(z) \; = \; \frac{1}{q} \frac{\sigma A}{\Delta z} (\mu(z) - \mu(z + \Delta z))$$

letting $\Delta z \rightarrow 0$, we obtain the diffusion equation stated above in Eq.(6.1a).

The diffusion equation is usually combined with a second equation called the continuity equation. For one-dimensional structures (see Fig.6.1), under steady-state conditions, the current must be the same at all z:

$$\frac{dI}{dz} = 0$$

(6.1b)

The reason is easy to see. If we have a current of *25* electrons per second entering a section of the conductor and only *10* electrons per second leaving it, then the number of electrons will be building up in this section at the rate of *25-10=15* per second. That is a transient condition, not a steady-state one. Under steady-state conditions the current has to be the same at all points along the z-axis as required by Eq.(6.1b).

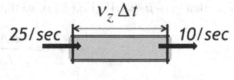

The standard approach is to solve Eqs.(6.1a,b) with the boundary conditions

$$\mu(z=0) = \mu_1$$

(6.2a)

$$\mu(z=L) = \mu_2$$

(6.2b)

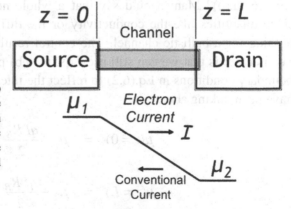

Fig.6.1. Solution to Eqs.(6.1a,b) with the boundary conditions in Eq.(6.2). Note that we are using I to represent the electron current as explained earlier (see Fig.3.2).

It is easy to see that the linear solution sketched in Fig.6.1 meets the boundary conditions in Eq.(6.2) and at the same time satisfies both Eqs.(6.1a,b) since a linear $\mu(z)$ has a constant $d\mu/dz$

$$\frac{d\mu}{dz} = -\frac{\mu_1 - \mu_2}{L}$$

so that from Eq.(6.1a) we have a constant current with $dI/dz = 0$:

$$I = \frac{\sigma A}{q} \frac{\mu_1 - \mu_2}{L}$$

Note that $\mu_1 - \mu_2 = qV$ (Eq.(2.1)), so that

$$I = \frac{\sigma A}{L} V \qquad\qquad (6.3a)$$

which is the standard Ohm's law and not the generalized one we have been discussing

$$I = \frac{\sigma A}{L + \lambda} V \qquad\qquad (6.3b)$$

that includes ballistic channels as well.

Can we obtain this result (Eq.(6.3b)) from the diffusion equation (Eqs.(6.1a,b))? Many would say that a whole new approach is needed since quantities like the conductivity or the diffusion coefficient mean nothing for a ballistic channel. The central result I wish to establish in this Lecture is that we can still use Eqs.(6.1a,b) provided we modify the boundary conditions in Eq.(6.2) to reflect the interface resistance that we have been talking about:

$$\mu(z = 0) = \mu_1 - \frac{qI\,R_B}{2} \qquad\qquad (6.4a)$$

$$\mu(z = L) = \mu_2 + \frac{qI\,R_B}{2} \qquad\qquad (6.4b)$$

R_B being the inverse of the ballistic conductance G_B discussed earlier (see Eqs.(4.6), (4.12)):

$$R_B = \frac{\lambda}{\sigma A} = \frac{h}{q^2 M} \qquad (6.5)$$

The new boundary conditions in Eqs.(6.4a,b) can be visualized in terms of lumped resistors $R_B/2$ at the interfaces as shown in Fig.6.2. leading to additional potential drops as shown.

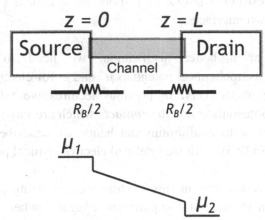

Fig.6.2. Eqs.(6.1a,b) can be used to model both ballistic and diffusive transport provided we modify the boundary conditions in Eq.(6.2) to reflect the two interface resistances, each equal to $R_B/2$.

It is straightforward to see that this **new boundary condition** applied to a uniform resistor leads to the new Ohm's law in Eq.(6.3b). Since $\mu(z)$ varies linearly from $z=0$ to $z=L$, the current is obtained from Eq.(6.1a)

$$I = \frac{\sigma A}{q} \frac{\mu(0) - \mu(L)}{L}$$

Using Eqs.(6.4a,b)

$$I = \frac{\sigma A}{q} \left(\frac{\mu_1 - \mu_2}{L} - \frac{qI R_B}{L} \right)$$

Since $\sigma A R_B = \lambda$, $\qquad I\left(1+\dfrac{\lambda}{L}\right) = \dfrac{\sigma A}{q}\left(\dfrac{\mu_1 - \mu_2}{L}\right)$

Noting that $\mu_1 - \mu_2 = qV$ (Eq.(2.1)), this yields Eq.(6.3b).

But how do we justify this new boundary condition (Eqs.(6.4a,b))? It follows from the new Ohm's law (Eq.(6.3b)) if we assume that the extra resistance $\sigma A / \lambda$ corresponding to $L=0$ is equally divided between the two interfaces.

For a better justification, we need to introduce two different electrochemical potentials μ^+ and μ^- for electrons moving along $+z$ and $-z$ respectively. In previous lectures we talked about electrochemical potentials inside the **contacts** which are large regions that always remain close to equilibrium and hence are described by Fermi functions (see Eq.(2.5)) with well-defined electrochemical potentials.

By contrast in this Lecture we are using $\mu(z)$ to represent quantities inside the out-of-equilibrium **channel**, where it is at best an approximate concept since the electron distribution among the available states need not follow a Fermi function. Even if it does, electronic states carrying current along $+z$ must be occupied differently from those carrying current along $-z$, or else there would be no net current.

This difference in occupation is reflected in different electrochemical potentials μ^+ and μ^- and we will show that the current is proportional to the difference (**Section 6.2**)

$$I = \frac{q}{h} M \left(\mu^+(z) - \mu^-(z) \right)$$

(6.6a)

which can also be rewritten in the form

$$I = \frac{1}{qR_B} \left(\mu^+(z) - \mu^-(z) \right)$$

(6.6b)

$$= \frac{\sigma A}{q\lambda}\left(\mu^+(z)-\mu^-(z)\right) \tag{6.6c}$$

using Eq.(4.12). The correct boundary conditions for μ^+ and μ^- are

$$\mu^+(z=0) = \mu_1,$$
$$\mu^-(z=L) = \mu_2 \tag{6.7}$$

which can be understood by noting that at $z=0$ the electrons moving along $+z$ have just emerged from the left contact and hence have the same distribution and electrochemical potential, μ_1. Similarly at $z=L$ the electrons moving along $-z$ have just emerged from the right contact and thus have the same potential μ_2 (Fig.6.3).

Fig.6.3.
Spatial profile of electrochemical potentials μ^+, μ^- across a diffusive channel.

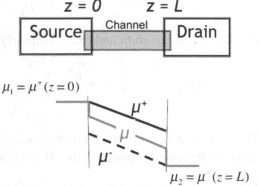

In the next Lecture I will show that the current is related to the potentials μ^+ and μ^- by an equation

$$I = -\frac{\sigma A}{q}\frac{d\mu^+}{dz} = -\frac{\sigma A}{q}\frac{d\mu^-}{dz} \tag{6.8}$$

that looks just like the diffusion equation (Eq.(6.1a)) which applies to the average potential:

$$\mu(z) \;=\; \frac{\mu^+(z)+\mu^-(z)}{2} \tag{6.9}$$

Eq.(6.8) can be solved with the boundary conditions in Eq.(6.7) to obtain the plot shown in Fig.6.3 for μ^+, μ^- and their average indeed looks like Fig.6.2 for μ with its discontinuities at the ends.

However, it is not necessary to abandon the traditional diffusion equation (Eq.(6.1a)) in favor of the new diffusion equation (Eq.(6.8)). We can obtain the same results simply by modifying the boundary conditions for $\mu(z)$ as follows:

$$\mu(z=0) \;=\; \left(\frac{\mu^+ + \mu^-}{2}\right)_{z=0} \;=\; \left(\mu^+ - \frac{\mu^+ - \mu^-}{2}\right)_{z=0}$$

$$=\; \mu_1 \;-\; (qIR_B/2)$$

making use of Eqs.(6.6) and (6.7). Similarly

$$\mu(z=L) \;=\; \left(\mu^- + \frac{\mu^+ - \mu^-}{2}\right)_{z=L} \;=\; \mu_2 \;+\; \frac{qIR_B}{2}$$

These are exactly the new boundary conditions for the standard diffusion equation that we mentioned earlier (Eqs.(6.4a,b)).

Let me finish up this Lecture by establishing the key result we stated without proof in the above discussion, namely, Eq.(6.6) (Section 6.2). But first let me say a few words about how the non-equilibrium potentials μ^+ and μ^- are defined. (Section 6.1).

6.1. Electrochemical Potentials Out of Equilibrium

As I mentioned earlier, it is conceptually straightforward to talk about electrochemical potentials inside the **contacts** which are large regions that always remain close to equilibrium and hence are described by

Fermi functions (see Eq.(2.5)) with well-defined electrochemical potentials. But in an out-of-equilibrium *channel*, the electron distribution among the available states need not follow a Fermi function.

In general one has to solve a full-fledged transport equation like the semiclassical Boltzmann equation to be introduced in the next Lecture which allows us to calculate the full occupation factors $f(z;E)$. More generally for quantum transport one can use the non-equilibrium Green's function (NEGF) formalism to be introduced in Part three to solve for the quantum version of $f(z;E)$. Can we really represent these distribution functions using electrochemical potentials $\mu^+(z)$ and $\mu^-(z)$?

Interestingly for a perfectly ballistic channel with good contacts, such a representation in terms of $\mu^+(z)$ and $\mu^-(z)$ is exact and not just an approximation. All drainbound electrons (traveling along +z, see Fig.6.4) are distributed according to the source contact with $\mu^+ = \mu_1$

$$f^+(z;E) = f_1(E) \equiv \frac{1}{1+\exp\left(\dfrac{E-\mu_1}{kT}\right)}$$

(6.10a)

while all sourcebound electrons (traveling along −z) are distributed according to the drain contact with $\mu^- = \mu_2$:

$$f^-(z;E) = f_2(E) \equiv \frac{1}{1+\exp\left(\dfrac{E-\mu_2}{kT}\right)}$$

(6.10b)

This is justified by noting that the drainbound channels from the source are filled only with electrons originating in the source and so these channels remain in equilibrium with the source with a distribution function $f_1(E)$. Similarly the sourcebound channels from the drain are in equilibrium with the drain with a distribution function $f_2(E)$.

Suppose at some energy $f_1(E) = 1$, and $f_2(E) = 0$, so that there are lots of electrons waiting to get out of the source, but none in the drain. We would then expect the drainbound lanes of the electronic highway to be completely full ("bumper-to-bumper traffic"), while the sourcebound lanes would all be empty as shown below in Fig.6.4.

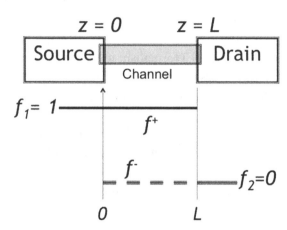

Fig.6.4.
Spatial profile of the occupation factors f⁺, f⁻ across a ballistic channel.

Of course this assumes that electrons do not turn around either along the way or at the ends. This means ballistic channels with good contacts where there are so many channels available that electrons can exit smoothly with a very low probability of turning around. If we either have bad contacts or diffusive channels, the solution in Eq.(6.10a,b) wouldn't work. In Lecture 14 on spin valves we will see some consequences of bad contacts, but for the moment let us talk about diffusive channels with good contacts.

Eqs.(6.10a,b) suggest a plausible guess for what we might expect the distributions to look like in a diffusive channel. We assume the same Fermi-like function but with spatially varying electrochemical potentials reflecting the fact that electrons from the drainbound channels continually transfer over to the sourcebound lanes:

$$f^+(z;E) \; = \; \cfrac{1}{1+\exp\left(\cfrac{E-\mu^+(z)}{kT}\right)}$$

$$(6.11a)$$

$$f^{-}(z;E) \;=\; \cfrac{1}{1+\exp\left(\cfrac{E-\mu^{-}(z)}{kT}\right)}$$

(6.11b)

Note that the potentials are in general energy-dependent and could be written as $\mu^{\pm}(z;E)$. In an elastic resistor, every energy is independent and in general each one could exhibit a different spatial variation in the potential if the mean free path is energy-dependent. But for simplicity, we will ignore this point assuming some average energy-independent mean free path.

But if we accept these forms for the occupation factors, then it is straightforward to translate a plot of occupation factors f (like the one in Fig.6.4) into a corresponding plot for the electrochemical potentials by noting that at low bias, the deviation of f from a reference value f_0 is proportional to the deviation of the corresponding μ from the corresponding reference value of μ_0 :

$$f(E)-f_0(E) \;\approx\; \left(-\frac{\partial f_0}{\partial E}\right)(\mu-\mu_0)\,(\text{same as Eq. (2.8)})$$

This relation, for example, can be used to translate Fig.6.4 into Fig.6.5.

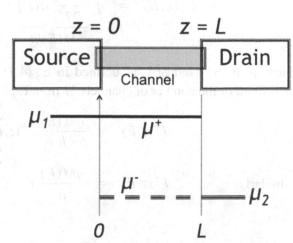

Fig.6.5.
Spatial profile of the electrochemical potentials μ^{+}, μ^{-} across a ballistic channel, obtained from Fig.6.4 by translating f's into μ's using Eq.(2.8).

6.2. Current in Terms of Non-Equilibrium Potentials

Usually we talk about the net current I which can be expressed as the difference between the drainbound flux I^+ and the sourcebound flux I^-:

$$I(z) = I^+(z) - I^-(z) \qquad (6.12)$$

The current I^+ equals the amount of charge exiting from the right per unit time. In a time Δt, all the charge in a length $v_z \Delta t$ exits, so that

$$I^+ = q * Electrons\, per\, unit\, length * v_z$$

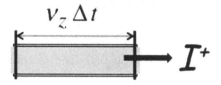

The number of electrons per unit length is equal to half the density of states (since only half the states carry current to the right) per unit length, $D(E)/2L$, times the fraction f^+ of occupied states, so that

$$I^+(z;E) = q \underbrace{\frac{D(E)}{2L} \bar{u}(E)}_{M(E)/h} f^+(z;E)$$

Here $\bar{u}$ is the average v_z as defined in Eq.(4.7) and making use of the definition of the number of channels M from Eq.(4.13) we have

$$I^+(z;E) = \frac{qM(E)}{h} f^+(z;E) \qquad (6.13a)$$

Similarly

$$I^-(z;E) = \frac{qM(E)}{h} f^-(z;E) \qquad (6.13b)$$

This allows us to write the current from Eq.(6.12)

$$I(z) = \int_{-\infty}^{+\infty} dE \left(I^+(z;E) - I^-(z;E) \right)$$

$$= \frac{q}{h} \int_{-\infty}^{+\infty} dE \left(f^+(z;E) - f^-(z;E) \right) M(E)$$

$$(6.14)$$

Once again, to get from distribution functions $f^\pm$ to electrochemical potentials $\mu^\pm$, we make use of the low bias result (Eq.(2.8)) to write

$$f^+(z;E) - f^-(z;E) = \left(-\frac{\partial f_0}{\partial E} \right) \left(\mu^+(z) - \mu^-(z) \right) \qquad (6.15)$$

so that from Eq.(6.14) we obtain Eq.(6.6a)

$$I(z) = \frac{q}{h} \left(\mu^+(z) - \mu^-(z) \right) \underbrace{\int_{-\infty}^{+\infty} dE \left(-\frac{\partial f_0}{\partial E} \right) M(E)}_{\equiv M} \qquad (6.16)$$

provided we identify M with the thermally averaged $M(E)$ as indicated in Eq.(6.16).

What about Drift?

Interestingly in our Lectures so far we have hardly ever mentioned the electric field, in contrast to most treatments of electronic transport which start by considering the electric field induced force as the driving term. It may seem paradoxical that we could obtain the conductivity without ever mentioning the electric field!

Electric fields are typically visualized as the gradient of an electrostatic potential U/q. By contrast, we have been using the electrochemical potential μ as the basis for our discussions. It is important to recognize the difference between the two "potentials":

$$\underbrace{\mu}_{Electrochemical} = \underbrace{(\mu - U)}_{Chemical} + \underbrace{U}_{Electrostatic} \qquad (7.1)$$

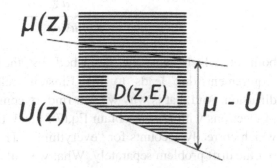

Fig.7.1.
The two potentials: Electrostatic U/q and electrochemical μ/q. $D(z;E)$ denotes the spatially varying density of states.

79

μ is a measure of the energy upto which the states are filled, while U determines the energy shift of the available states, so that μ - U is a measure of the degree to which the states are filled and hence the number of electrons.

In the last chapter we obtained the equation

$$\frac{I}{A} = -\sigma\frac{d(\mu/q)}{dz}$$

(7.2)

But what we really showed was that

$$\frac{I}{A} = -\sigma\frac{d(\mu-U)/q}{dz}$$

(7.3)

assuming zero electric field, $dU/dz = 0$. So how do we know what the correct equation is, when we include U?

It would seem that we needed to solve a whole new problem including the effect of the field $(= d(U/q)/dz)$ on electrons. However, this is unnecessary because the basic principles of equilibrium statistical mechanics require the current to be zero for a constant μ, just as there can be no heat current if the temperature is constant. Hence the current expression must have the form given in Eq.(7.2) which can be written as the sum of a drift term and a diffusion term

$$\frac{I}{A} = \underbrace{-\sigma\frac{d(\mu-U)/q}{dz}}_{Diffusion} \underbrace{-\sigma\frac{dU/q}{dz}}_{Drift}$$

(7.4)

both of which must be described by the same coefficient σ, a requirement that leads to the Einstein relation between drift and diffusion. And that is why we can find σ considering only the diffusion of electrons with $U = 0$, obtain Eq.(7.3) and just replace it with Eq.(7.2) which correctly accounts for "everything." There is really no need work out the drift problem separately. What we called the diffusion equation is really the ***drift-diffusion equation*** even though we did not consider drift explicitly.

Couldn't we instead have neglected diffusion completely and just gone with the drift term? That way we could stick to the view that current is driven by electric fields and not have to bother with electrochemical potentials. The problem is that if we take this view then one has to invoke mysterious quantum mechanical forces to explain why all electrons are not affected by the field. In our discussion the energy window for transport (F_T, see Fig.2.3) arises naturally from the difference in the "agenda" of the two contacts (see Eqs.(2.7), (2.8))

$$f_1(E) - f_2(E) \;=\; \left(-\frac{\partial f_0}{\partial E}\right)(\mu_1 - \mu_2)$$

as discussed in Lectures 2, 3. The point is that regardless of which potential we choose to work with, it finally affects transport through the occupation factor, f.

In this Lecture we will justify our neglect of drift more explicitly by introducing the Boltzmann Transport Equation (BTE) which is the standard starting point for all discussions of the transport of particles. We too could have used it as the starting point for but we did not do so because it is harder to digest with its multiple independent variables, compared to the ordinary differential equation in Lecture 6, which follows from relatively elementary arguments.

Even in this Lecture we will not really do justice to the BTE. We will introduce it briefly and use it to show that for low bias, the current indeed depends only on $d\mu/dz$ and not on dU/dz, thus putting our discussion of steady-state, low bias transport without electric fields on a firmer footing and identifying possible issues with it.

Note the two qualifying phrases, namely "steady-state" and "low bias." We will show later in this lecture that for time varying transport, the neglect of electric fields can lead to errors, but we will not discuss it further in these lectures. However, even under steady-state conditions, electric fields can play an important role in determining the full current-voltage characteristics, once we go beyond low bias, as we will discuss in the next Lecture.

7.1 Boltzmann Transport Equation, BTE

In Lecture 6 we introduced electron distribution functions $f^{\pm}$ and electrochemical potentials describing the drainbound and sourcebound currents $I^{\pm}$. Both the drainbound and sourcebound current, however, is composed of electrons traveling at different angles having different z-momemtum p_z, even though they all have the same energy (we are still talking about an elastic resistor) and hence the same total momentum. To include the effect of the electric field we need "momentum-resolved" distribution functions $f^{\pm}(z, p_z, t)$.

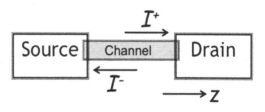

The BTE describes the evolution of such "momentum-resolved" distribution functions $f(z,p_z,t)$ that tell us the occupation of states with a given momentum p_z and velocity v_z at a location z at time t :

$$\frac{\partial f}{\partial t} + v_z \frac{\partial f}{\partial z} + F_z \frac{\partial f}{\partial p_z} \;=\; S_{op} f$$

$$(7.5)$$

where F_z is the force on the electrons, and $S_{op} f$ symbolically represents the complex scattering processes that continually redistribute electrons among the available velocity states.

The BTE with the right hand side set to zero (that is without scattering processes)

$$\frac{\partial f}{\partial t} + v_z \frac{\partial f}{\partial z} + F_z \frac{\partial f}{\partial p_z} = 0 \qquad (7.6)$$

is completely equivalent to describing a set of particles each with position $z(t)$ and momenta $p_z(t)$ that evolve according to the semiclassical laws of motion:

$$v_z \equiv \frac{dz}{dt} = \frac{\partial E}{\partial p_z} \qquad (7.7a)$$

$$F_z \equiv \frac{dp_z}{dt} = -\frac{\partial E}{\partial z} \qquad (7.7b)$$

where $E(z,p_z,t)$ is the total energy.

Eqs.(7.7a,b) describe semiclassical dynamics in single particle terms where the position $z(t)$ and momenta $p_z(t)$ for each of the electrons is a dependent variable evolving in time. By contrast, the BTE provides a collective description with all three independent variables z, p_z,t on an equal footing.

To get from Eqs.(7.7) to (7.6) we start by noting that in the absence of scattering, we can write

$$f(z, p_z, t) = f(z - v_z \Delta t, p_z - F_z \Delta t, t - \Delta t)$$

reflecting the fact that any electron with a momentum

$$p_z \text{ at } z \text{ at time } t,$$

must have had a momentum of

$$p_z - F_z \Delta t \text{ at } z - v_z \Delta t \text{ a little earlier at time } t - \Delta t.$$

Next we expand the right hand side to the first term in a Taylor series to write

$$f(z, p_z, t) \;=\; f(z, p_z, t) - \frac{\partial f}{\partial z} v_z \Delta t - \frac{\partial f}{\partial p_z} F_z \Delta t - \frac{\partial f}{\partial t} \Delta t$$

Eq.(7.6) follows readily on canceling out the common terms.

The left hand side of the BTE thus represents an alternative way of expressing the laws of motion. What makes it different from mere mechanics, however, is the stochastic scattering term on the right which makes the distribution function f approach the equilibrium Fermi function when external driving terms are absent. This last point of course is not meant to be obvious. It requires an extended discussion of the scattering operator S_{op} that we talk a little more about in Lecture 16 when we discuss the second law.

For our purpose it suffices to note that a common approximation for the scattering term is the relaxation time approximation (RTA)

$$S_{op} f \;\cong\; -\frac{f - f_0}{\tau} \tag{7.8}$$

which assumes that the effect of the scattering processes is proportional to the degree to which a given distribution f differs from the equilibrium distribution f_0.

One comment about why we call this approach **semiclassical.** The BTE is **classical** in the sense that it is based on a particle view of electrons. But it is not **fully** classical, since it typically includes quantum input both in the scattering operator S_{op} and in the form of the energy-momentum relation. For example, graphene is often described by a linear energy-momentum relation

$$\vec{E} = v_0 \vec{p}$$

a result that is usually justified in terms of the bandstructure of the graphene lattice requiring quantum mechanics that Boltzmann did not live to see. But once we accept that, many transport properties of

graphene can be understood in classical particulate terms using the BTE
that Boltzmann taught us to use.

7.2 Diffusion equation from BTE

We start by combining the RTA (Eq.(7.8)) with the full BTE (Eq.(7.5))
to obtain for steady-state ($\partial / \partial t = 0$),

$$v_z \frac{\partial f}{\partial z} + F_z \frac{\partial f}{\partial p_z} = -\frac{f - f_0}{\tau}$$

(7.9)

In the presence of an electric field we can write the total energy as

$$E(z, p_z) = \varepsilon(p_z) + U(z)$$

(7.10)

where $\varepsilon(p_z)$ denotes the energy-momentum
relation with $U=0$ and this gets shifted
locally by $U(z)$ as sketched in Fig.7.2.

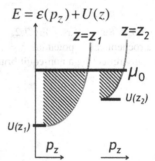

Fig.7.2. The energy momentum relation with $U=0$ is
shifted locally by $U(z)$. At equilibrium the
electrochemical potential μ_0 is spatially constant.

The first point to note is that the equilibrium distribution with a constant
electrochemical potential μ_0

$$f_0(z, p_z) = \frac{1}{\exp\left(\dfrac{E(z, p_z) - \mu_0}{kT}\right) + 1}$$

(7.11)

satisfies the BTE in Eq.(7.9). The right hand side of Eq.(7.9) is obviously
zero, but it takes a little differential calculus to see that the left hand side
is zero too.

Defining

$$X_0 \;\equiv\; E(z, p_z) - \mu_0 \;=\; \varepsilon(p_z) + U(z) - \mu_0 \qquad (7.12)$$

we have

$$v_z \frac{\partial f_0}{\partial z} + F_z \frac{\partial f_0}{\partial p_z} \;=\; \left(\frac{\partial f_0}{\partial X_0}\right)\left(v_z \frac{\partial X_0}{\partial z} + F_z \frac{\partial X_0}{\partial p_z}\right)$$

$$=\; \left(\frac{\partial f_0}{\partial X_0}\right)\left(v_z \frac{\partial E}{\partial z} + F_z \frac{\partial E}{\partial p_z}\right) \;=\; 0$$

making use of Eqa.(7.7a,b).

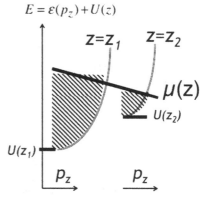

$$E = \varepsilon(p_z) + U(z)$$

Fig.7.3. Same as Fig.7.2, but the electrochemical potential $\mu(z)$ varies spatially reflecting a non-equilibrium state.

Out of equilibrium, we assume the distribution function $f(z,p_z)$ to have the same form as Eq.(7.11) but with a spatially varying electrochemical potential $\mu(z)$:

$$f(z, p_z) \;=\; \frac{1}{\exp\left(\dfrac{E(z, p_z) - \mu(z)}{kT}\right) + 1} \qquad (7.13)$$

Using Eq.(7.13), the left hand side of BTE (see Eq.(7.9)) reduces to

$$\left(\frac{\partial f}{\partial X}\right)\left(v_z \frac{\partial X}{\partial z} + F_z \frac{\partial X}{\partial p_z}\right) \;=\; \left(\frac{\partial f}{\partial X}\right)\left(-v_z \frac{d\mu}{dz}\right)$$

$$X \equiv E(z, p_z) - \mu(z)$$

where
$$= X_0(z, p_z) + \mu_0 - \mu(z) \qquad (7.14)$$

We now assume small deviations in $\mu(z)$ from the equilibrium value so that we can write the left hand side as

$$\left(\frac{\partial f}{\partial X}\right)_{X=X_0} \left(-v_z \frac{d\mu}{dz}\right)$$

and use our standard Taylor series expansion (see Eq.(2.8)) to write the right hand side of BTE as

$$-\frac{f - f_0}{\tau} \cong \left(\frac{\partial f}{\partial X}\right)_{X=X_0} \frac{\mu(z) - \mu_0}{\tau}$$

Combining the two sides

$$v_z \frac{d\mu}{dz} = -\frac{\mu(z) - \mu_0}{\tau} \qquad (7.15)$$

We now introduce two separate electrochemical potentials μ^+ and μ^- for the right-moving ($v_z > 0$) and left-moving ($v_z < 0$) electrons to write

$$\frac{d\mu^+}{dz} = -\frac{\mu^+ - \mu_0}{v_z \tau} \quad , \quad \frac{d\mu^-}{dz} = -\frac{\mu^- - \mu_0}{v_z \tau}$$

Assuming $\mu_0 = (\mu^+ + \mu^-)/2$, we obtain

$$\frac{d\mu^+}{dz} = -\frac{\mu^+ - \mu^-}{\lambda} = \frac{d\mu^-}{dz} \qquad (7.16)$$

with $\lambda = 2v_z \tau$. Combining with Eq.(6.6a) for the current, we obtain the result (Eq.(6.8)) stated without proof in the last Lecture. Note that we have included electric fields explicitly and shown that their effect cancels out.

7.3. Equilibrium Fields Do Matter

However, we believe there is an important subtlety worth pointing out. Although the externally applied electric field does not affect the low bias conductance, any inbuilt fields that exist within the conductor under equilibrium conditions can affect its low bias conductance. Let me explain.

Note that in our treatment above we assumed that under non-equilibrium conditions, the electrochemical potential is a function of z (Eq.(7.13)) and the resulting linearized equation (Eq.(7.15)) does not involve the field $F_z = dU/dz$. However, the field term would not have dropped out so nicely if we were to assume that the electrochemical potential is not just a function of z, but of both z and p_z. Instead of Eq.(7.15) we would then obtain

$$v_z \frac{\partial \mu}{\partial z} + F_z \frac{\partial \mu}{\partial p_z} = - \frac{\mu(z, p_z) - \mu_0}{\tau}$$

$$(7.17)$$

However, the additional term involving the field F_z does not play a role in determining linear conductivity because it is $\sim V^2$, V being the applied voltage. At equilibrium with $V=0$, $\mu = \mu_0$, so that both derivatives appearing on the left are zero. Under bias, in principle, both could be non-zero and to first order $\sim V$. But the point is that while v_z is a constant, the applied field F_z is also $\sim V$. So while the first term on the left is $\sim V$, the second term is $\sim V^2$.

But this argument would not hold if F_z were not the applied field, but internal inbuilt fields independent of V that are present even at equilibrium. Equilibrium requires a constant μ and NOT a constant U.

The equilibrium condition depicted in Fig.7.2 (also shown here for ease of reference) is quite common in real conductors, with varying $U(z)$ corresponding to non-zero fields F_z. Indeed this picture could also represent an interface between dissimilar materials (called "heterostructures") where the discontinuity in band edges is often modeled with effective fields.

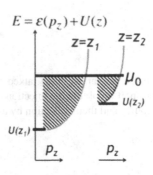

The point is that such equilibrium fields can and do affect the low bias conductance. For an ideal homogeneous conductor we do not have such fields. But even then we need to make two contacts in order to measure the resistance. Each such contact represents a heterostructure qualitatively similar to that shown in Fig.7.2 with inbuilt effective (if not real) fields. I think these fields give rise to the interface resistance distinguishing the new Ohm's law from the standard one, but I have not checked.

7.4. The Two Potentials

In these Lectures we will generally focus on steady-state transport involving the injection of electrons from a source and their collection by a drain (Fig.7.4). We have seen that the low bias conductance can be understood in terms of the electrochemical potential μ, without worrying about the electrostatic potential U.

However, we would like to briefly consider ac transport through a nanowire far from any contacts where we have a local voltage $V(z,t)$ and current $I(z,t)$ (Fig.7.5), because this provides a contrasting example where it is important to pay attention to the difference between the two potentials even for low bias, in order to obtain the correct inductance and capacitance.

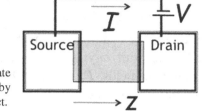

Fig.7.4. So far we have talked of steady-state transport involving the injection of electrons by a source and their collection by a drain contact.

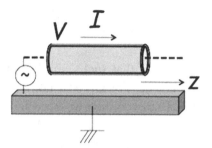

Fig.7.5.
Ac or time varying transport along a nanowire can be described in terms of a voltage $V(z,t)$ and a current $I(z,t)$.

For this problem too we start from the BTE with the RTA approximation as in the last section, but we do not set $\partial/\partial t = 0$,

$$\frac{\partial f}{\partial t} + v_z \frac{\partial f}{\partial z} + F_z \frac{\partial f}{\partial p_z} = -\frac{f - f_0}{\tau}$$

and linearize it assuming a distribution of the form (compare Eq.(7.13))

$$f(z, p_z, t) = \frac{1}{\exp\left(\dfrac{E(z, p_z, t) - \mu(z, t)}{kT}\right) + 1} \qquad (7.18)$$

Compared to the steady-state problem (Eq.(7.15)) we now have two extra terms involving the time derivatives of E and μ:

$$\frac{\partial \mu}{\partial t} + v_z \frac{\partial \mu}{\partial z} - \frac{\partial E}{\partial t} = -\frac{\mu(z, t) - \mu_0}{\tau} \qquad (7.19)$$

As we did in the last Section with Eq.(7.15), we can separate Eq.(7.19) into two equations for μ^+ and μ^-, whose sum and difference are identified with voltage and current to obtain a set of equations

$$\frac{\partial(\mu/q)}{\partial z} = -(L_K + L_M)\frac{\partial I}{\partial t} - \frac{I}{\sigma A}$$

(7.20a)

$$\frac{\partial(\mu/q)}{\partial t} = -\left(\frac{1}{C_Q} + \frac{1}{C_E}\right)\frac{\partial I}{\partial z}$$

(7.20b)

that look just like the transmission line equations with a distributed series inductance and resistance and a shunt capacitance.

The algebra getting from Eq.(7.19) to Eqs.(7.20a,b) is a little long-winded and since time-varying transport is only incidental to our main message we have relegated the details to Appendix D. Those who are really interested can look at the original paper on which this discussion is based (Salahuddin et al., 2005).

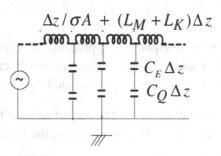

But note the two inductors and the two capacitors in series. The **kinetic inductance** L_K and the **quantum capacitance** C_Q per unit length, arise from transport-related effects

$$L_K = \frac{h}{q^2}\frac{1}{\langle 2Mv_z \rangle}$$

(7.21a)

$$C_Q = \frac{q^2}{h}\left\langle \frac{2M}{v_z}\right\rangle$$

(7.21b)

while the L_M and the C_E are just the normal **magnetic inductance** and the **electrostatic capacitance** from the equations of magnetostatics and electrostatics.

The point I wish to make is that the fields enter the expression for the energy $E(z,p_z,t)$ and if we ignore the fields we would miss the $\partial E / \partial t$ term in Eq.(7.19) to obtain

$$\frac{\partial \mu}{\partial t} + v_z \frac{\partial \mu}{\partial z} = - \frac{\mu(z,t) - \mu_0}{\tau}$$

and after working through the algebra obtain instead of Eqs.(7.20a,b)

$$\frac{\partial (\mu / q)}{\partial z} = - L_K \frac{\partial I}{\partial t} - \frac{I}{\sigma A} \tag{7.22a}$$

$$\frac{\partial (\mu / q)}{\partial t} = - \left(\frac{1}{C_Q} \right) \frac{\partial I}{\partial z} \tag{7.22b}$$

Do these equations approximately capture the physics? Not unless we are considering wires with very small cross-sections so that M is a small number making $L_K \gg L_M$ and $C_Q \ll C_E$.

We could recover the correct answer from Eqs.(7.22a,b) by replacing the μ in with $\mu - U$ and then using the laws of electromagnetics to replace

$$\frac{\partial U}{\partial t} \quad \text{with} \quad \frac{1}{C_E} \frac{\partial I}{\partial z} \quad \textbf{and} \quad \frac{\partial U}{\partial z} \quad \text{with} \quad L_M \frac{\partial I}{\partial t}$$

But these replacements may not be obvious and it is more straightforward to go from Eq. (7.19) to (7.20) as spelt out in Appendix D.

Note that if we specialize to steady-state ($\partial / \partial t = 0$), both Eqs.(7.20) and (7.22) give us back our old diffusion equation (Eq.(6.1)). As we argued earlier, for low bias steady-state transport, the applied electric field can be treated as incidental.

Lecture 8

Electrostatics is Important

8.1. The Nanotransistor
8.2. Why the Current Saturates
8.3. Role of Charging
8.4. Rectifier Based on Electrostatics
8.5. Extended Channel Model

In the last Lecture we tried to justify our "field-less" approach to conductivity which comes as a surprise to many since it is commonly believed that currents are driven by electric fields. However, we hasten to add that the field can and does play an important role once we go beyond low bias and our purpose in this lecture is to discuss the role of the electrostatic potential and the corresponding electric field on the current-voltage characteristics beyond low bias.

To illustrate these issues, I will use the nanotransistor, an important device that is at the heart of microelectronics. As we noted at the outset the nanotransistor is essentially a voltage-controlled resistor whose length has shrunk over the years and is now down to a few hundred atoms. But as any expert will tell you, it is not just the low bias resistance, but the entire shape of the current-voltage characteristics of a nanotransistor that determines its utility. And this shape is controlled largely by its electrostatics, making it a perfect example for our purpose.

I should add, however, that this Lecture does not do justice to the nanotransistor as a device. This will be discussed in a separate volume in this series written by Lundstrom, whose model is widely used in the field and forms the basis of our discussion here. We will simply use the nanotransistor to illustrate the role of electrostatics in determining current flow.

93

We have seen that the elastic transport model characterized by the current formula

$$I = \frac{1}{q} \int_{-\infty}^{+\infty} dE\, G(E)\left(f_1(E) - f_2(E) \right)$$

(see Eq.(3.3))

In this Lecture I will use the nanotransistor to illustrate some of the issues that need to be considered at high bias, some of which can be modeled with a simple extension of Eq.(3.3)

$$I = \frac{1}{q} \int_{-\infty}^{+\infty} dE\, G(E-U)\left(f_1(E) - f_2(E) \right) \qquad (8.1)$$

to include an appropriate choice of the potential U in the channel which is treated as a single point. We call this the *point channel* model to distinguish it from the standard and more elaborate extended channel model which we will introduce at the end of the Lecture.

8.1 The nanotransistor

The nanotransistor is a three-terminal device (Fig.8.1), though ideally no current should flow at the gate terminal whose role is just to control the current. In other words, the current-drain voltage, I- V_D, characteristics are controlled by the gate voltage, V_G (see Fig.8.2). The low bias current and conductance can be understood based on the principles we have already discussed. But currents at high V_D involve important new principles.

The basic principle underlying an FET is straightforward (see Fig.8.3). A positive gate voltage V_G changes the potential in the channel, lowering all the states down in energy, which can be included by replacing Eq.(8.1) with Eq.(8.2) and setting $U = qV_G$.

Fig.8.1.
Sketch of a field effect transistor (FET): Channel length, L; Transverse width, W (Perpendicular to page).

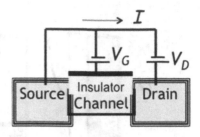

Fig.8.2.
Typical current-voltage, I-V_D characteristic and its variation with V_G for an FET.

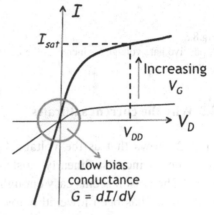

For an n-type conductor this increases the number of available states in the energy window of interest around μ_1 and μ_2 as shown. Of course for a p-type conductor (see Fig.7.2) the reverse would be true leading to a complementary FET (see Fig.0.2) whose conductance variation is just the opposite of what we are discussing. But we will focus here on n-type FET's.

We will not discuss the low bias conductance since these involve no new principles. Instead we will focus on the current at high bias, specifically on why the current-voltage, I- V_D characteristic is (1) non-linear, and (2) "rectifying," that is different for positive and negative V_D.

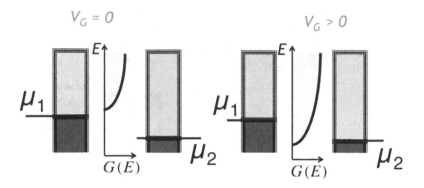

Fig.8.3.
A positive gate voltage V_G increases the current in an FET by moving the states down in energy.

8.2 Why the current saturates

Fig.8.2 shows that as the voltage V_D is increased the current does not continue to increase linearly. Instead it levels off tending to saturate. Why? The reason seems easy enough. Once the electrochemical potential in the drain has been lowered below the band edge the current does not increase any more (Fig.8.4).

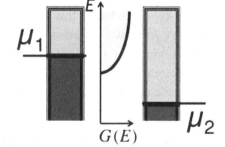

Fig.8.4.
The current saturates once μ_2 drops below the band-edge.

The saturation current can be written from Eq.(8.1)

$$I_{sat} = \frac{1}{q} \int_{-\infty}^{+\infty} dE\, G(E-U)\, f_1(E) \qquad (8.2)$$

by dropping the second term $f_2(E)$ assuming μ_2 is low enough that $f_2(E)$ is zero for all energies where the conductance function is non-zero. In the simplest approximation

$$U^{(1)} = -qV_G$$

The superscript *1* is included to denote that this expression is a little too simple, representing a first step that we will try to improve.

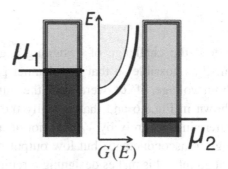

Fig.8.5.
The current does not saturate completely because the states in the channel are also lowered by the drain voltage.

If this were the full story the current would have saturated completely as soon as μ_2 dropped a few *kT* below the band edge. In practice the current continues to increase with drain voltage as sketched in Fig.8.6. The reason is that when we increase the drain voltage we do not just lower μ_2, but also lower the energy levels inside the channel (Fig.8.5) similar to the way a gate voltage would. The result is that the current keeps increasing as the conductance function $G(E)$ slides down in energy by a fraction α (< *1*) of the drain voltage V_D, which we could include in our model by choosing

$$U^{(2)} = \alpha(-qV_D) + \beta(-qV_G) \equiv U_L \qquad (8.3)$$

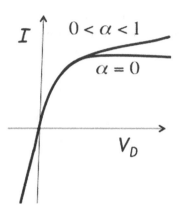

Fig.8.6.
Current in an FET would saturate perfectly if the channel potential were unaffected by the drain voltage.

Indeed the challenge of designing a good transistor is to make α as small as possible so that the channel potential is hardly affected by the drain voltage. If α were zero the current would saturate perfectly as shown in Fig.8.6 and that is really the ideal: a device whose current is determined entirely by V_G and not at all by V_D or in technical terms, a high transconductance but low output conductance. For reasons we will not go into, this makes designing circuits much easier.

To ensure that V_G has far greater control over the channel than V_D it is necessary to make the insulator thickness a small fraction of the channel length. This means that for a channel length of a few hundred atoms we need an insulator that is only a few atoms thick in order to ensure a small α. This thickness has to be precisely controlled since thinner insulators would leak unacceptably. We mentioned earlier that today's laptops have a billion transistors. What is even more amazing is that each has an insulator whose thickness is precisely controlled down to a few atoms!

8.3 Role of charging

There is a second effect that leads to an increase in the saturation current over what we get using Eq.(8.3) in (8.1). Under bias, the occupation of the channel states is less than what it is at equilibrium. This is because at equilibrium both contacts are trying to fill up the channel states, while under bias only the source is trying to fill up the states while the drain is trying to empty it. Since there are fewer electrons in the channel, it tends to become positively charged and this will lower the states in the channel as shown in Fig.8.5, even for perfect electrostatics ($\alpha = 0$) resulting in an increase in the current.

This effect can be captured within the point channel model (Eq.(8.1)) by writing the channel potential as

$$U = U_L + U_0(N - N_0) \tag{8.4}$$

where U_L is given by our previous expression in Eq.(8.3). The extra term represents the change in the channel potential due to the change in the number of electrons in the channel, N under non-equilibrium conditions relative to the equilibrium number N_0, U_0 being the change in the channel potential energy per electron.

To use Eq.(8.4), we need expressions for N_0, N. N_0 is the equilibrium number of channel electrons, which can be calculated simply by filling up the density of states, $D(E)$ according to the equilibrium Fermi function $f_0(E)$.

$$N_0 = \int_{-\infty}^{+\infty} dE \, D(E - U) \, f_0(E) \tag{8.5a}$$

while the number of electrons, N in the channel under non-equilibrium conditions is given by

$$N = \int_{-\infty}^{+\infty} dE \, D(E - U) \frac{f_1(E) + f_2(E)}{2} \tag{8.5b}$$

assuming that the channel is "equally" connected to both contacts. Note that the calculation is now a little more intricate than what it would be if U_0 were zero. We now have to obtain a solution for U and N that satisfy both Eqs.(8.4) and (8.5) simultaneously through an iterative procedure as shown schematically in Fig.8.7.

Once a self-consistent U has been obtained, the current is calculated from Eq.(8.1), or an equivalent version that is sometimes more convenient numerically and conceptually.

$$I \;=\; \frac{1}{q} \int\limits_{-\infty}^{+\infty} dE\, G(E)\,(f_1(E+U)-f_2(E+U)) \qquad (8.6)$$

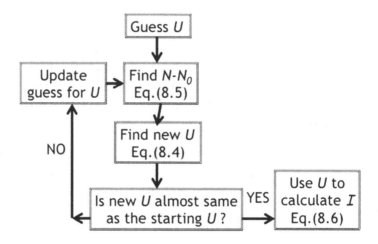

Fig.8.7. Self-consistent procedure for calculating the channel potential U in point channel model.

This simple point channel model often provides good agreement with far more sophisticated models as discussed in Rahman et al. (2003).

Fig.8.8 shows the current versus voltage characteristic calculated numerically (MATLAB code included in Appendix) assuming a 2-D channel with a parabolic dispersion relation for which the density of states is given by (L: Length, W: Width)

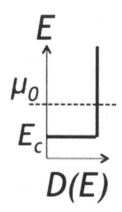

$$D(E) = g\frac{mLW}{2\pi\hbar^2}\,\vartheta(E - E_c)$$

where ϑ represents the unit step function. The numerical results are obtained using $g=2$, $m = 0.2*9.1e\text{-}31\ Kg$, $\beta=1$, $\alpha=0$ and $U_0 = 0\ or\ \infty$ as indicated with $L=1\ \mu m$, $W=1\ \mu m$ assuming ballistic transport, so that

$$G(E) = \frac{q^2}{h}M(E),$$

$M(E)$ being the number of modes given by

$$M(E) = g\frac{2W}{h}\sqrt{2m(E - E_c)}\,\vartheta(E - E_c)$$

The current-voltage characteristics in Fig.8.8 has two distinct parts, the initial linear increase followed by a saturation of the current. Although these results were obtained numerically, both the slope and the saturation current can be calculated analytically, especially if we make the low temperature approximation that the Fermi functions change abruptly from 1 to 0 as the energy E crosses the electrochemical potential μ. Indeed we used a kT of $5\ meV$ instead of the usual 25 meV, so that the numerical results would compare better with simple low temperature estimates.

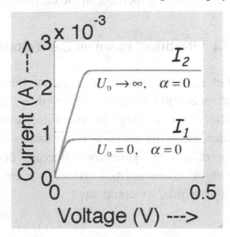

Fig.8.8. Current-voltage characteristics calculated numerically using the self-consistent point channel model shown in Fig.8.7.

There are two key points we wanted to illustrate with this example. Firstly, the initial slope of the current-voltage characteristics is unaffected by the charging energy. This slope defines the low bias conductance that we have been discussing till we came to this Lecture. The fact that it remains unaffected is reassuring and justifies our not bringing up the role of electrostatics earlier.

Secondly, the saturation current is strongly affected by the electrostatics and changes by a factor of ~ 2.8 from a model with zero charging energy to one with a very large charging energy. This is because of the reason mentioned at the beginning of this section. With U_0 = 0, the channel states remain fixed and the number of electrons N is equal to $N_0/2$, since $f_1=1$ and $f_2=0$ in the energy range of interest. With very large U_0, to avoid $U_0(N-N_0)$ becoming

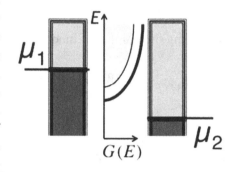

excessive, N needs to be almost equal to N_0 even though the states are only half-filled. This requires the states to move down as sketched with a corresponding increase in the current.

8.4 "Rectifier" Based on Electrostatics

Let us now look at an example that can be handled using the point channel model just discussed though it does not illustrate any issues affecting the design of nanotransistors. I have chosen this example to illustrate a fundamental point that is often not appreciated, namely that an otherwise symmetric structure could exhibit asymmetric current-voltage characteristics (which we are loosely calling a "rectifier"). In other words, we could have

$$I(+V_D) \neq I(-V_D)$$

for a symmetric structure, simply because of *electrostatic asymmetry*.

Consider a nanotransistor having perfect electrostatics represented by α = 0 (Eq.(8.3)), connected (a) in the standard configuration (Fig.8.9a) and (b) with the gate left floating (Fig.8.9b).

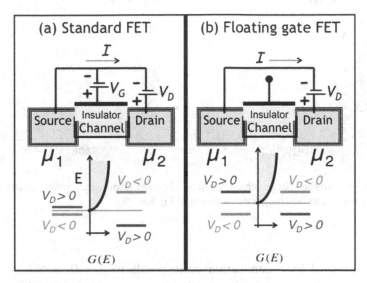

Fig.8.9. (a) Standard FET assuming perfect electrostatics. (b) Floating gate FET

The basic device is assumed physically symmetric, so that one could not tell the difference between the source and drain contacts. This may not be true of real transistors, but that is not important, since we are only trying to make a conceptual point. The configuration in (a) has electrostatic asymmetry, since the gate is held at a fixed potential with respect to the source, but not with respect to the drain. But configuration (b) is symmetric in this respect too, since the gate floats to a potential halfway between the source and the drain.

Fig.8.10 shows the current-voltage characteristics calculated using the model summarized in Fig.8.7 (MATLAB code in Appendix F), for each of the structures shown in (a) and (b). The parameters are the same as those used for the example shown in Fig.8.8, except that the equilibrium electrochemical potential is located exactly at the bottom of the band as shown in Fig.8.9: $\mu_0 = E_c$

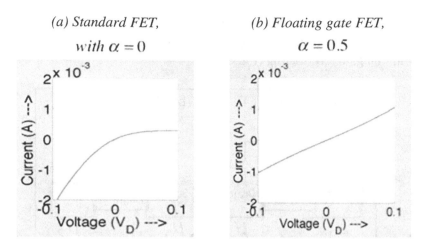

Fig.8.10 Current-voltage characteristics obtained from the point channel model corresponding to the confgurations shown in Fig.8.9.

The standard FET connection corresponds to $\alpha = 0$ assuming perfect electrostatics, while the same physical structure in the floating gate connection corresponds to $\alpha = 0.5$. The former gives a rectifying characteristic, while the latter gives a linear characteristic, often called "Ohmic". The point is that it is not necessary to design an asymmetric channel to get asymmetric *I-V* characteristics. Even the simplest symmetric channel can exhibit non-symmetric $I(V_D)$ characteristic if the electrostatics is asymmetric.

Note also that the linear conductance given by the slope *dI/dV* around V=0 is unaffected by our choice of α and can be predicted without any reference to the electrostatics, even though the overall shape obviously cannot.

8.5 Extended Channel Model

The point channel elastic model that we have described (Eqs.(8.1), (8.2)) integrates our elastic resistor with a simple electrostatic model for the channel potential U/q, allowing it to capture some of the high bias physics that the pure elastic resistor misses. Let me end this Lecture by noting some of the things it misses.

The point channel model ignores *the electric field in the channel* and assumes that the density of states $D(E)$ stays the same from source to drain. In the real structure, however, the electric field lowers the states at the drain end relative to the source as sketched here. Doesn't this change the current?

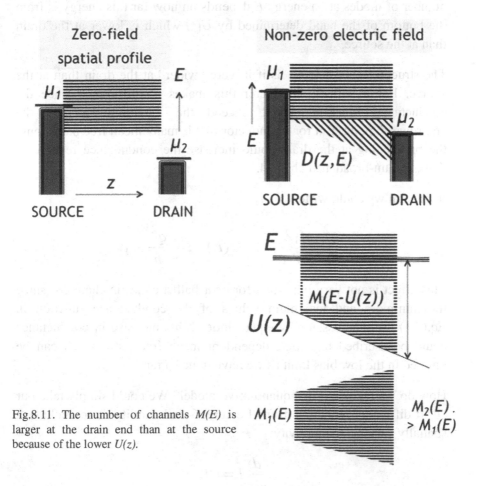

Fig.8.11. The number of channels $M(E)$ is larger at the drain end than at the source because of the lower $U(z)$.

For an elastic resistor one could argue that the additional states with the slanted (rather than horizontal) shading are not really available for conduction since (in an elastic resistor) every energy represents an independent energy channel and can only conduct if it connects all the way from the source to the drain.

But even for an elastic resistor there should be an increase in current because at a given energy E, the number of modes at the drain end is larger than the number of modes at the source end. This is because the number of modes at an energy E depends on how far this energy is from the bottom of the band determined by $U(z)$ which is lower at the drain than at the source.

The structure almost looks as if it were "wider" at the drain than at the source. For a ballistic conductor this makes no difference since the conductance function cannot exceed the maximum set by the "narrowest" point. But for a conductor that is many mean free paths long, the broadening at the drain could increase the conductance relative to that of an un-broadened channel.

In general we could write

$$\frac{q^2}{h} \frac{M_1 \lambda}{L + \lambda} \leq G(E) \leq \frac{q^2}{h} M_1 \qquad (8.7)$$

This effect is not very important for near ballistic elastic channels, since the minimum and maximum values of the conductance function in Eq.(8.6) are then essentially equal. Indeed this increase in conductance could be ascribed to a field-dependent mean free path which can be ignored in the low bias limit as we have done so far.

How do we include it in a quantitative model? We could simply take our "drift-diffusion" equation from Lecture 6 and modify it to include a spatially varying conductivity:

$$\frac{d}{dz} I = 0$$

$$\frac{I}{A} = -\frac{\sigma(z)}{q}\frac{d\mu}{dz} \tag{8.8}$$

What do we use for the conductivity, $\sigma(z)$? Our old expression

$$\sigma = \int_{-\infty}^{+\infty} dE\, \sigma(E)\left(-\frac{\partial f}{\partial E}\right)_{E=\mu_0} \qquad \text{(same as Eq.(5.3))}$$

involved an energy average of $\sigma(E)$ over an energy window of a few kT around $E = \mu_0$.

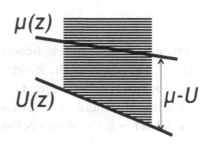

The spatially varying $U(z)$ shifts the available energy states in energy, so that one now has to look at the energy window around $E = \mu(z) - U(z)$ suggesting that we replace Eq.(5.3) with

$$\sigma = \int_{-\infty}^{+\infty} dE\, \sigma(E)\left(-\frac{\partial f}{\partial E}\right)_{E=\mu(z)-U(z)} \tag{8.9}$$

For low bias, we could replace $\mu(z)$ with μ_0 to obtain our earlier result in Eq,(7.12) from obtained by directly linearizing the BTE.

Note that to use Eqs.(8.7), (8.8) we have to determine $\mu(z) - U(z)$ from a self-consistent solution the Poisson equation (ε: Permittivity, n_0, n: electron density per unit volume at equilibrium and out of equilibrium)

$$\frac{d}{dz}\left(\varepsilon\frac{dU}{dz}\right) = q^2(n-n_0) \tag{8.10}$$

The electron density per unit length entering the Poisson equation is calculated by filling up the density of states (per unit length) shifted by the local potential $U(z)$, according to the local electrochemical potential, so that we can write

$$n(z) \equiv \int_{-\infty}^{+\infty} dE \frac{D(E-U(z))}{L} \frac{1}{1+\exp\dfrac{E-\mu(z)}{kT}} \qquad (8.11a)$$

$$n_0 = \int_{-\infty}^{+\infty} dE \frac{D(E)}{L} \frac{1}{1+\exp\dfrac{E-\mu_0}{kT}} \qquad (8.11b)$$

Solving Eq.(8.11)) self-consistently with the Poisson equation (Eq.(8.10)) is indeed the standard approach to obtaining the correct $\mu(z)$, $U(z)$, which can then be used to find the current from Eq.(8.8). We could view this procedure as the extended channel version of the point channel model in Fig.8.7 as shown below.

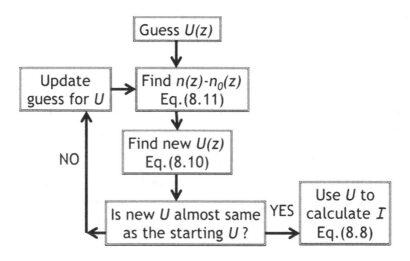

Fig.8.12. Extended channel version of the point channel model in Fig.8.7.

Note that this whole approach is based on the assumption of local electrochemical potentials $\mu^{\pm}(z)$ describing right and left-moving electrons whose average is the $\mu(z)$ appearing in Eq.(9.1). In general, electron distributions can deviate so badly from Fermi functions that an electrochemical potential may not be adequate and one needs the full semiclassical formalism based on the Boltzmann Transport Equation (BTE) and much progress has been made in this direction. However, full-fledged BTE-based simulation is time-consuming and the drift-diffusion equation based on the concept of a local potential $\mu(z)$ continues to be the "bread and butter" of device modeling.

What our bottom-up approach adds is that Eq.(8.8) can be used even to model ballistic channels if the boundary conditions are modified appropriately (Eq.(6.4)) to include the interface resistance, a result that was obtained by carefully accounting for the distinction between $\mu^+(z)$ and $\mu^-(z)$ (Lecture 6).

Lecture 9

Smart Contacts

9.1. Why p-n Junctions are Different
9.2. Contacts are Fundamental

We are now ready to finish up with part one of these lectures, which I entitled "the new Ohm's law" referring to

$$R = \frac{\rho(L+\lambda)}{A}$$

(same as Eq.(4.2))

which includes an extra contact resistance $\rho\lambda / A$ that depends solely on the properties of the channel and cannot be eliminated by better contacting procedures.

As we saw in Lecture 6, the key concept in identifying this interface resistance was the recognition that when a current flows, the electrochemical potentials μ^+ and μ^- for the drainbound and sourcebound states are different (Fig.6.3, also reproduced here for convenience).

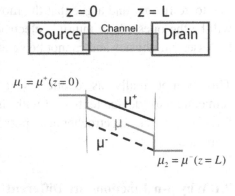

From Eqs.(6.3b) and (6.6c) we could write (Note: $\mu_1 - \mu_2 = qV$)

$$\delta\mu \ \equiv \ \mu^+ - \mu^- \ = \ \frac{\mu_1 - \mu_2}{1 + L/\lambda}$$

(9.1)

111

The contacts held at different potentials μ_1 and μ_2 drive the two groups of states (drainbound and sourcebound) out of equilibrium, while backscattering processes described by the mean free path λ try to restore equilibrium. Eq.(9.1) describes the result of these competing forces.

Normally we do not like to deal with multiple electrochemical potentials. The diffusion equation for example (see Eq.(6.1)),

$$\frac{I}{A} = -\frac{\sigma(z)}{q}\frac{d\mu}{dz}$$

$$(9.2)$$

works in terms of a single potential $\mu(z)$ and what we saw in Lecture 6 was how we could sweep the two potentials $\mu^+(z)$ and $\mu^-(z)$ under the proverbial rug, by defining $\mu(z)$ as the average of the two and including interface resistances into the boundary conditions by replacing Eq.(6.2) with Eq.(6.4).

The point I wish to make in this Lecture is that this separation of the electrochemical potentials for different groups of states is really far more ubiquitous and cannot always be swept under the rug. Indeed I would like to go further and argue that the most interesting devices of the future will be the ones where multiple electrochemical potentials will represent the essential physics and cannot be swept under the rug.

This is not really as exotic as it may sound. For example, all semiconductor device texts start with the p-n junction for which the need for two separate electrochemical potentials is well-recognized. Let me elaborate.

9.1. Why p-n Junctions are Different

Fig.9.1 shows a grayscale plot of the density of states $D(z,E)$. The white band indicates the bandgap with a non-zero DOS both above and below it on each side which are shifted in energy with respect to each other. A positive voltage is applied to the right with respect to the left, so that μ_2 is lower than μ_1 as shown.

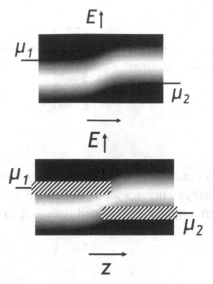

Fig.9.1. Simplified grayscale plot of the spatially varying density of states $D(z,E)$ across a p-n junction.

If we look at a narrow range of energies around μ_1 (see shaded area on the left) it communicates primarily with contact 1. If we look at a narrow range of energies around μ_2 (see shaded area on the right) it communicates primarily with contact 2.

We could draw an *idealized* diagram with each of these two groups communicating just with one contact and cut off from the other as shown in Fig.9.2. In reality of course neither group is completely cut off from either contact, and people who design real devices often go to great lengths to achieve better isolation, but let us not worry about such details.

Would the idealized device in Fig.9.2 allow any current to flow? None at all, if we it were an elastic resistor. There is no energy channel that will let an electron get all the way from left to right. The ones connected to the left are disconnected from the right and those connected to the right are disconnected from the left.

Fig.9.2.   An   idealized version of the p-n junction in Fig.9.1.

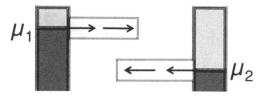

But current can and does flow because of inelastic processes that allow electrons to change energies along the channel. Electrons can then come in from the left, change energy and then exit to the right as sketched in Fig.9.3.

Fig.9.3. Current flow in the idealized device of Fig.9.2 is facilitated by distributed   inelastic processes.

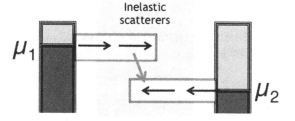

Indeed this is exactly how currents flow in p-n junctions, by transferring from the upper group of states down to the lower group by inelastic processes, which are generally referred to as recombination-generation (R-G) processes, since people like to think in terms of electrons in the upper group recombining with a "hole" in the lower group. But as we mentioned in Lecture 5, this is really an unnecessary complication and one could simply think purely in terms of electrons transferring inelastically from one group of states to another.

The point to note is that this class of devices cannot be described with one electrochemical potential and to capture the correct physics, it is essential to treat the two groups of states separately, introducing *two different electrochemical potentials*, labeled with the index n

$$I_n = -\frac{\sigma_n}{q}\frac{d\mu_n}{dz} \qquad (9.3)$$

These currents are all coupled together by inelastic processes generally called "RG processes" in the context of p-n junctions

$$\frac{dI_n}{dz} = \sum_m [RG]_{m\to n} - [RG]_{n\to m} \qquad (9.4)$$

that take electrons from one group of states m to the other n. This is indeed the way p-n junctions are modeled.

It is well-known that the current in a p-n junction is given by an expression of the form

$$I = I_0 (e^{qV/vkT} - 1) \qquad (9.5)$$

where the number v as well as the coefficient I_0 are determined by the nature of the inelastic or RG processes. The conductivities σ_n of either of the two groups of states plays hardly any role in determining this current.

The physical reason for this is clear. The rate-determining step in current flow is the inelastic process transferring electrons from one group of states to the other. Transport within any of these groups only adds an unimportant resistance in series with the basic device.

Everything we have talked about in these lectures has been about the conductivities σ_n of the homogeneous p-type or n-type materials. And this is exactly the physics that is relevant to the operation of the most popular electronic device today, namely the Field Effect Transistor (FET) whose conductivity is controlled by a gate electrode through the electrostatic potential U.

Lessons from Nanoelectronics

But the p-n junction is a totally different device from the FET both in terms of its current-voltage characteristics and the physics that underlies it. It is the basic device structure used to construct solar cells and the principle it embodies is key to a broad class of energy conversion devices. So let me take a short detour to elaborate on this principle.

9.1.1. Current-Voltage Characteristics

Consider for example the device in Fig.9.4 assuming that the upper group of states (labeled A) is clustered around an energy ε_A while the lower group (labeled B) is clustered around ε_B .

Fig.9.4. Same as Fig.9.3 with the two groups of states labeled A and B. Electronic transitions between A and B are facilitated by inelastic interactions.

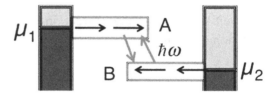

The essential physics of such p-n junction like devices is contained not in Eq.(9.3), but in Eq.(9.4) which for two levels A and B can be written as

$$I \sim \overbrace{D_{B \leftarrow A}\, f_A(\varepsilon_A)\,(1 - f_B(\varepsilon_B))}^{A \rightarrow B}$$

$$- \overbrace{D_{A \leftarrow B}\, f_B(\varepsilon_B)\,(1 - f_A(\varepsilon_A))}^{B \rightarrow A} \quad (9.6)$$

where the coefficients D_{BA} and D_{AB} denote the strength of the inelastic processes inducing the transitions from A to B and from B to A respectively (note that the transition occurs from the second subscript to the first).

Interestingly these two rates D_{AB} and D_{BA} are generally NOT equal. D_{AB} involves absorbing an amount of energy

$$\hbar\omega = \varepsilon_A - \varepsilon_B$$

from the surroundings, while D_{BA} involves giving up the same amount of energy. A fundamental principle of equilibrium statistical mechanics (see Lecture 16) is that if the entity causing the inelastic scattering is at equilibrium with a temperature T_0, then it is always harder to absorb energy from it than it is give up energy to it and the ratio of the two processes is given by

$$\frac{D_{AB}}{D_{BA}} = \exp\left(-\frac{\hbar\omega}{kT_0}\right) \tag{9.7}$$

We can write the current from Eq.(9.4) in the form

$$I \sim D_{AB}\, f_B(\varepsilon_B)(1 - f_A(\varepsilon_A))\,(X - 1) \tag{9.8}$$

where

$$X \equiv \frac{D_{BA}}{D_{AB}} \frac{f_A(\varepsilon_A)}{1 - f_A(\varepsilon_A)} \frac{1 - f_B(\varepsilon_B)}{f_B(\varepsilon_B)} \tag{9.9}$$

Making use of Eq.(9.8), Eq.(9.9) and the following property of Fermi functions (Eq.(2.2))

$$\frac{1 - f_0(\varepsilon)}{f_0(\varepsilon)} = \exp\left(\frac{\varepsilon - \mu_0}{kT}\right) \tag{9.10}$$

we can rewrite Eq.(9.9) as

$$X = \exp\left(\frac{\hbar\omega}{kT_0} - \frac{\hbar\omega}{kT}\right)\exp\left(\frac{\mu_A - \mu_B}{kT}\right) \tag{9.11}$$

Since Level A is connected to contact 1 and Level B to contact 2, if the inelastic processes taking electrons from A to B are not too strong, level A is almost in equilibrium with contact 1 and level B with contact 2 , so that

$$\mu_A - \mu_B \;\cong\; \mu_1 - \mu_2 \;=\; qV$$

If $T_0 = T$, we can write the current from Eq.(9.8) as

$$I \;\sim\; (X-1) \;\sim\; e^{qV/kT} - 1$$

which is the standard *I-V* relation for p-n junctions stated earlier (see Eq.(9.5)) with $\nu=1$. Other values of ν would be obtained if we consider more elaborate RG processes rather than the direct "band-to-band" processes considered here.

But the more important point I want to stress is that this device can be used for **energy conversion**. If the scatterers are at a temperature different from that of the device ($T_0 \neq T$) then one can have a current flowing even without any applied voltage. This short circuit current is given by

$$I_{sc} \;\equiv\; I(V=0) \;\sim\; \exp\frac{\hbar\omega}{k}\left(\frac{1}{T_0}-\frac{1}{T}\right) - 1 \tag{9.12}$$

One could in principle use a device like this to convert a temperature difference ($T_0 \neq T$) into an electrical current. The short circuit current has the opposite sign for $T_0 > T$ and for $T > T_0$. Readers familiar with Feynman's ratchet and pawl lecture (Feynman 1963, cited in Lecture 17 of these notes) may notice the similarity. The ratchet reverses direction depending on whether its temperature is lower or higher than the ambient.

One could view more practical devices like solar cells as embodiments of the same principle, the light from the sun having a temperature $T_0 \sim$

6000^0C characteristic of the surface of the sun, much larger than the ambient temperature.

From Eq.(9.8) it is easy to see that under open circuit conditions $(I=0)$, we must have $X=1$, so that from Eq.(9.11) we have

$$\frac{qV_{oc}}{\hbar\omega} = 1 - \frac{T}{T_0}$$

The left hand side represents the energy extracted per photon under very low current (near open circuit) conditions, so that this could be called the Carnot efficiency of a solar cell viewed as a "heat engine". However, since $T_0 >> T$, this Carnot efficiency is very close to 100% and my colleague Ashraf often points out that other factors related to the small angular spectrum of solar energy are important in lowering the ideal efficiency to much lower values.

9.2. Contacts Are Fundamental

The point I want to make is how important the discriminating contacts are in the design of this class of devices which we could generally refer to as "solar cells" (Fig.9.5a). The external source raises electrons from the B states to the A states from where they exit through the left contact, while the empty state left behind in B is filled up by an electron that comes in through the right contact. Every electron raised from B to A thus causes an electron to flow in the external circuit.

But if the contacts are connected "normally" injecting and extracting equally from either group (Figure 9.5b) then we cannot expect any current to flow in the external circuit, from the sheer symmetry of the arrangement. After all, why should electrons flow from left to right any more that they would flow from right to left?

It is this asymmetric contacting that makes p-n junctions fundamentally different from the Field Effect Transistor (FET) that we started our lectures with, both in terms of the current-voltage characteristics and the physics underlying it. It is of course well recognized that the physics of p-n junctions demands two different electrochemical potentials. What is not as well recognized is the generic nature of this phenomenon. Let me explain.

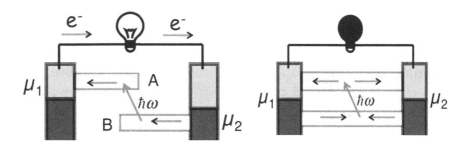

Fig.9.5. (a) Asymmetric contacts are central to the operation of the "solar cell". (b) If contacted symmetrically no electrical output is obtained.

For most of these lectures we have discussed how the contacts in an ordinary device drive drainbound and sourcebound states out of equilibrium faster than backscattering processes can restore equilibrium. In p-n junctions we just saw how the contacts drive the two bands out of equilibrium, faster than R-G processes can restore equilibrium. In Lecture 14 we will talk about spin valve devices where magnetic contacts drive upspin and downspin states out of equilibrium faster than spin-flip processes can restore equilibrium.

In every case there are groups of states A, B etc that are driven out of equilibrium by smart contacts that can discriminate between them.

More and more of such examples can be expected in the coming years, as we learn to control current flow not just with gate electrodes that control the electrostatic potential, but with subtle contacting schemes that engineer the electrochemical potential(s). Many believe that nature does just that in designing many biological 'devices', but that is a different story. In the context of man-made devices there are many possibilities. Perhaps we will figure out how to contact s-orbitals differently from p-orbitals, or one valley differently from another valley, leading to fundamentally different devices.

But this requires a basic change in approach. Traditionally the work of device design has been divided neatly between two groups of specialists: physicists and material scientists who innovate new materials using atomistic theory and device engineers who worry about contacts and related issues using macroscopic theory. Future "solar cells" that seek to function effectively at the microscopic level may well require an approach that integrates materials and contacts at the atomistic level. Perhaps then we will be able to create devices that rival the marvels of nature like photosynthesis.

II. Old Topics in New Light

Thermoelectricity

Conductance measurements ordinarily do not tell us anything about the nature of the conduction process inside the conductor. If we connect the terminals of a battery across any conductor, electron current flows out of the negative terminal back to its positive terminal. Since this is true of all conductors, it clearly does not tell us anything about the conductor itself.

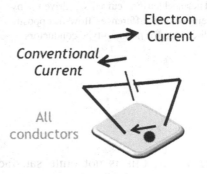

On the other hand, thermoelectricity, that is, electricity driven by a temperature difference, is an example of an effect that does. A very simple experiment is to look at the current between a hot probe and a cold probe (Fig.10.1).

For an n-type conductor (see Fig.5.1) the direction of the external current will be consistent with what we expect if electrons travel from the hot to the cold probe inside the conductor, but for a p-type conductor (see Fig.5.2) the direction is reversed, consistent with electrons traveling from the cold to the hot probe. Why?

It is often said that p-type conductors show the opposite effect because the carriers have the opposite sign. As we discussed in Lecture 5, p-type conductors involve the flow of electrons near the top of a band of energies and it is convenient to keep track of the empty states above μ rather than the filled states below μ. These empty states are called holes and since they represent the absence of an electron behave like positively charged entities.

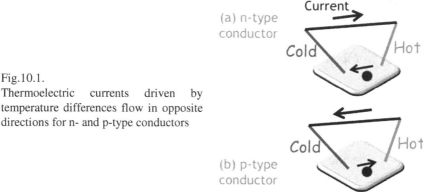

Fig.10.1.
Thermoelectric currents driven by temperature differences flow in opposite directions for n- and p-type conductors

However, this is not quite satisfactory since what moves is really an electron with a negative charge. "Holes" are at best a conceptual convenience and effects observed in a laboratory should not depend on subjective conveniences.

Fig.10.2. In n-type conductors the electrochemical potential is located near the bottom of a band of energies, while in p-type conductors it is located near the top. In n-conductors $D(E)$ increases with increasing E, while in p-conductors it decreases with increasing E.

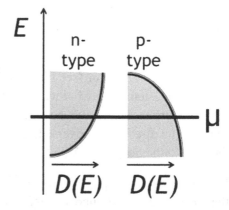

As we will see in this Lecture the difference between n- and p-conductors requires no new principles or assumptions beyond what we have already discussed, namely that the current is driven by the difference between f_1 and f_2. The essential difference between n- and p-conductors is that while one has a density of states $D(E)$ that increases with energy E, the other has a $D(E)$ decreasing with E.

Later in Lecture 13 we will discuss another important phenomenon called the Hall effect which changes sign for n-type and p-type conductors and this too is commonly blamed on negative and positive charges. This effect, however, has a totally different origin related to the negative mass $(m=p/v)$ associated with $E(p)$ relations in p-conductors that point downwards. By contrast the thermoelectric effect does not require a conductor to even have a $E(p)$ relation. Even small molecules show sensible thermoelectric effects (Baheti et al. 2008).

The basic idea is easy to see starting from our old expression for the current obtained in Lecture 3:

$$I = \frac{1}{q} \int_{-\infty}^{+\infty} dE\, G(E) \left(f_1(E) - f_2(E) \right)$$

(10.1, same as Eq.(3.3))

So far the difference in f_1 and f_2 has been driven by difference in electrochemical potentials μ_1 and μ_2. But it could just as well be driven by a temperature difference, since in general

$$f_1(E) = \frac{1}{\exp\left(\dfrac{E - \mu_1}{kT_1}\right) + 1}$$

$$f_2(E) = \frac{1}{\exp\left(\dfrac{E - \mu_2}{kT_2}\right) + 1}$$

and

(10.2)

But why would such a current reverse directions for an n-type and a p-type conductor?

To see this, consider two contacts with the same electrochemical potential μ, but with different temperatures as shown in Fig. 10.3.

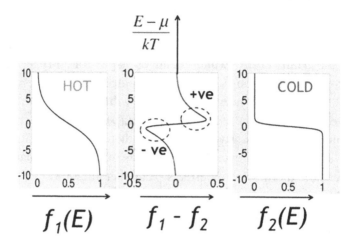

Fig.10.3. Two contacts with the same μ, but different temperatures: $f_1 - f_2$ is positive for $E>\mu$, and negative for $E<\mu$.

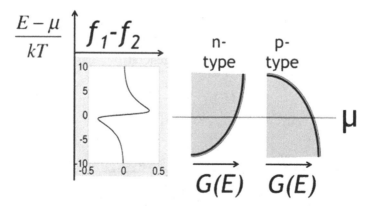

Fig.10.4 For n-type channels, the current for $E>\mu$ dominates that for $E<\mu$, while for p-type channels the current for $E<\mu$ dominates that for $E>\mu$. Consequently, electrons flow from hot to cold across an n-type channel, but from cold to hot in a p-type channel.

The key point is that the difference between $f_1(E)$ and $f_2(E)$ has a different sign for energies E greater than μ and for energies less than μ (see Fig.10.3).

In an n-type channel, the conductance $G(E)$ is an increasing function of energy, so that the net current is dominated by states with energy $E>\mu$ and thus flows from 1 to 2, that is from hot to cold (Fig.10.4). But in a p-type channel it is the opposite. The conductance $G(E)$ is a decreasing function of energy, so that the net current is dominated by states with energy $E<\mu$ and thus flows from 2 to 1, that is from cold to hot.

10.1.Seebeck Coefficient

We can use Eq.(10.1) directly to calculate currents without making any approximations. But it is often convenient to use a Taylor series expansion like we did earlier (Eq.(2.5)) to obtain results that are reasonably accurate for low "bias".

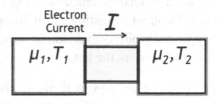

We could write approximately from Eq.(10.1)

$$I = G (V_1 - V_2) + G_S (T_1 - T_2) \qquad (10.3)$$

where we have defined V_1, V_2 as μ_1/q and μ_2/q. The conductance is given by

$$G = \int_{-\infty}^{+\infty} dE \, G(E) \left(\frac{\partial f_0}{\partial \mu} \right)$$

$$= \int_{-\infty}^{+\infty} dE \left(-\frac{\partial f_0}{\partial E} \right) G(E) \qquad (10.4a)$$

as we have seen before in Section 2.4.

The new coefficient G_S that we have introduced is given by

$$G_S = \frac{1}{q} \int\limits_{-\infty}^{+\infty} dE\, G(E) \left(\frac{\partial f_0}{\partial T} \right)$$

$$= \int\limits_{-\infty}^{+\infty} dE \left(-\frac{\partial f_0}{\partial E} \right) \frac{E - \mu_0}{qT} G(E)$$

$$\text{(10.4b)}$$

This last step, relating the derivatives with respect to T and with respect to E, requires a little algebra (see Appendix A).

One point regarding the notation: I should really use a different symbol for the averaged conductance G (which we have not used elsewhere in these lectures) to distinguish it from the energy-dependent conductance $G(E)$. To avoid confusion, in this Lecture I will try to write $G(E)$ whenever I mean the latter.

Eq.(10.4b) expresses mathematically the basic point we just discussed. Energies E greater and less than μ_0, contribute with opposite signs to the thermoelectric coefficient, G_S. It is clear that if we wanted to design a material with the best Seebeck coefficient, S we would try to choose a material with all its density of states on one side of μ_0 since anything on the other side contributes with an opposite sign and brings it down.

We can visualize Eq.(10.3) as shown in Fig.10.5, where the short circuit current is given by

$$I_{sc} = G_S (T_1 - T_2) \qquad (10.5)$$

Experimentally what is often measured is the open circuit voltage

$$V_{oc} = -\frac{I_{sc}}{G} = -\frac{G_S}{G} (T_1 - T_2) \qquad (10.6)$$

Note that we are using I and V for electron current and electron voltage μ/q whose sign is opposite that of the conventional current and voltage.

For n-type conductors, for example, G_S is positive, so that Eq.(10.6) tells us that V_{oc} is negative if $T_1 > T_2$. This means that the contact with the higher temperature has a negative electron voltage and hence a positive conventional voltage. By convention this is defined as a negative Seebeck coefficient.

$$S \equiv \frac{V_{oc}}{T_1 - T_2} = -\frac{G_S}{G} \tag{10.7}$$

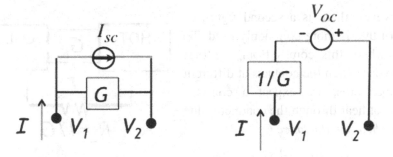

Fig.10.5. Circuit representations of Eq.(10.3).

10.2. Thermoelectric Figures of Merit

The practical importance of thermoelectric effects arise from the possibility of converting waste heat into electricity and from this point of view the important figure of merit is the amount of power that could be generated from a given $T_1 - T_2$. What load resistor R_L will maximize the power delivered to it (Fig.10.6)? A standard theorem in circuit theory says (this is not too hard to prove for yourself) that the answer is a "matched load" for which R_L equal to $1/G$:

$$P_{max} = V_{oc}^2 G/4 = S^2 G (T_1 - T_2)^2/4 \tag{10.8}$$

The quantity $S^2 G$ is known as the power factor and is one of the standard figures of merit for thermoelectric materials.

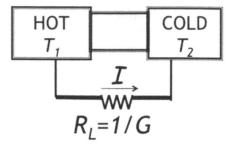

Fig.10.6. A thermoelectric generator can convert a temperature difference into an electrical output.

However, there is a second figure of merit that is more commonly used. To see where this comes from, we first note that when the contacts at different temperatures, we expect a constant flow of heat through the conductor due to its *heat conductance* G_K

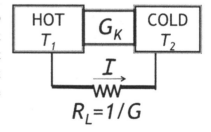

$$G_K \left(T_1 - T_2 \right)$$

which has to be supplied by the source that maintains the temperature difference. Actually this is not quite right, it only gives the heat flow under open circuit conditions and ignores a component that depends on I. But this is good enough for our purpose which is simply to provide an intuitive feeling for where the standard thermoelectric figure of merit comes from.

The ratio of the maximum generated power to the power that is supplied by the external source is a good measure of the efficiency of the thermoelectric material in converting heat to electricity and can be written as

$$\frac{P_{\text{max}}}{G_K(T_1-T_2)} = \underbrace{\frac{S^2GT}{G_K}}_{} \frac{T_1-T_2}{4T}$$

$$\equiv ZT \qquad (10.9)$$

where T is the average temperature $(T_1+T_2)/2$. The standard figure of merit for thermoelectric materials, called its *ZT product*, is proportional to the ratio of S^2G to G_K:

$$ZT \equiv \frac{S^2GT}{G_K} = \frac{S^2\sigma T}{\kappa} \qquad (10.10)$$

where κ is the thermal conductivity related to the thermal conductance G_K by the same geometric factor A/L connecting the corresponding electrical quantities G and σ. Indeed the Ohm's law for heat conduction (known as Fourier's law) also needs the same correction for interface resistance namely the replacement of L with $L+\lambda$.

However, while the electrical conductivity arises solely from charged particles like electrons, the thermal conductivity also includes a contribution from phonons which describes the vibrations of the atoms comprising the solid lattice. Ordinarily it is the phonon component that dominates the thermal conductivity and we will discuss it briefly in the next Lecture. For the moment let us talk about the heat carried by electrons, something we have not discussed so far at all.

10.3. Heat Current

We have discussed the thermoelectric currents in a material with any arbitrary conductance function $G(E)$. The nice thing about the elastic resistor is that channels at different energies all conduct in parallel, so that we can think of one energy at a time and add them up at the end. Consider a small energy range located between E and $E+dE$, either above or below the electrochemical potentials $\mu_{1,2}$ as shown in Fig.10.7.

As we discussed in Section 10.1, these two channels will make contributions with opposite signs to the Seebeck effect.

It has been known for a long time that the Seebeck effect is associated with a Peltier effect. The connection can be easily understood as follows. Earlier in Lecture 3 we saw that for an elastic resistor the associated Joule heat I^2R is dissipated in the contacts (see Fig.3.3). But if we consider the n-type or p-type channels in Fig.10.7 apparent that unlike Fig.3.3, both contacts do not get heated.

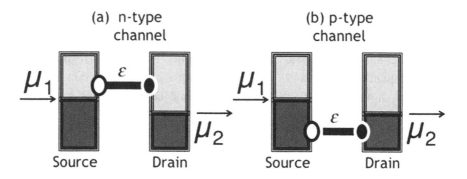

Fig.10.7. A one-level elastic resistor having just one level with E= ε, (a) above or (b) below the electrochemical potentials $\mu_{1,2}$,

Fig.10.8 is essentially the same as Fig.3.3 except that we have shown the heat absorbed from the surroundings rather than the heat dissipated. For n-type conductors the heat absorbed is positive at the source, negative at the drain, indicating that the source is cooled and the drain is heated. For p-type conductors it is exactly the opposite.

This is the essence of the Peltier effect that forms the basis for practical thermoelectric refrigerators. Note that the sign of the Peltier coefficient like that of the Seebeck coefficient is related to the sign of $E-\mu$ and not the sign of q.

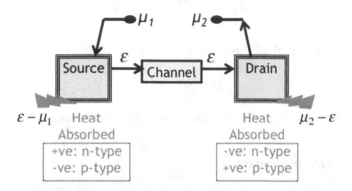

Fig.10.8. Same as Fig.3.3 but showing the heat absorbed at each contact. For n-type conductors the heat absorbed is positive at the source, negative at the drain showing that the electrons COOL the source and HEAT the drain. For p-type conductors it is exactly the opposite.

To write the heat current carried by electrons, we can simply extend what we wrote for the ordinary current earlier:

$$I = \frac{1}{q} \int_{-\infty}^{+\infty} dE\, G(E) \left(f_1(E) - f_2(E) \right) \qquad \text{(same as Eq.(3.3))}$$

Noting that an electron with energy E carrying a charge $-q$ also extracts an energy E-μ_1 from the source and dumps an energy E-μ_2 in the drain, we can write the heat currents I_{Q1} and I_{Q2} *extracted* from the source and drain respectively as

$$I_{Q1} = \frac{1}{q} \int_{-\infty}^{+\infty} dE\, \frac{E - \mu_1}{q}\, G(E) \left(f_1(E) - f_2(E) \right)$$

(10.11a)

$$I_{Q2} = \frac{1}{q} \int_{-\infty}^{+\infty} dE\, \frac{\mu_2 - E}{q}\, G(E) \left(f_1(E) - f_2(E) \right)$$

(10.11b)

The energy extracted from the external source per unit time is given by

$$I_E \;=\; \frac{\mu_1 - \mu_2}{q} I \;=\; V I \qquad\qquad (10.11c)$$

so that the sum of all three energy currents is zero:

$$I_{Q1} + I_{Q2} + I_E \;=\; 0$$

as we would expect due to overall energy conservation.

10.3.1. Linear response

Just as we linearized the current equation (Eq.(3.3)) to obtain an expression for the current in terms of voltage and temperature differences (Eqs.(10.4)), we can linearize the heat current equation to obtain

$$I_Q \;=\; G_P \, (V_1 - V_2) \;+\; G_Q \, (T_1 - T_2) \qquad\qquad (10.12)$$

where

$$G_P \;=\; \int\limits_{-\infty}^{+\infty} dE \left(-\frac{\partial f_0}{\partial E} \right) \frac{E - \mu_0}{q} G(E) \qquad\qquad (10.13a)$$

$$G_Q \;=\; \int\limits_{-\infty}^{+\infty} dE \left(-\frac{\partial f_0}{\partial E} \right) \frac{(E - \mu_0)^2}{q^2 \, T} G(E) \qquad\qquad (10.13b)$$

These are the standard expressions for the thermoelectric coefficients due to electrons which are usually obtained from the Boltzmann equation.

I should mention that the quantity G_Q we have obtained is not the thermal conductance G_K that is normally used in the ZT expression cited earlier (Eq.(10.10)). One reason is what we have stated earlier, namely that G_K also has a phonon component that we have not yet discussed. But there is another totally different reason.

The quantity G_K is defined as the heat conductance under electrical open circuit conditions ($I=0$):

$$G_K = \left(\frac{\partial I_Q}{\partial (T_1 - T_2)} \right)_{I=0}$$

while it can be seen from Eq.(10.12) that G_Q is the heat conductance under electrical short circuit conditions ($V=0$):

$$G_Q = \left(\frac{\partial I_Q}{\partial (T_1 - T_2)} \right)_{V_1 = V_2}$$

However, we can rewrite Eqs.(10.3) and (10.12) in a form that gives us the open circuit coefficients (as noted earlier, V and I represent the electron voltage μ/q and the electron current, which are opposite in sign to the conventional voltage and current)

$$(V_1 - V_2) = \frac{1}{G} I - \overbrace{\frac{G_S}{G}}^{\substack{S, \\ Seebeck}} (T_1 - T_2) \tag{10.14a}$$

$$I_Q = \underbrace{\frac{G_P}{G}}_{\substack{Peltier \\ -\Pi}} I + \underbrace{\left(G_Q - \frac{G_P G_S}{G} \right)}_{\substack{Heat\ conductance \\ G_K}} (T_1 - T_2) \tag{10.14b}$$

We have indicated the coefficients that are normally measured experimentally and are named after the experimentalists who discovered them. Eqs.(10.3) and (10.12), on the other hand, come more naturally in theoretical models because of our Taylor's series expansion and it is important to be aware of the difference.

Incidentally, if we use the expressions in Eqs.(10.13a) and (10.4b), the Peltier and Seebeck coefficients in Eq.(10.14) obey the Kelvin relation

$$\Pi = TS \tag{10.15}$$

which is a special case of the fundamental Onsager relations that the linear coefficients are required to obey (Lecture 15).

10.4. "Delta Function" Thermoelectric

It is instructive to look at a so-called "delta function" thermoelectric, which is a hypothetical material with a narrow conductance function located at energy ε with a width $\Delta\varepsilon$ that is much less than kT.

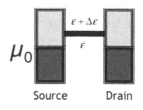

Source   Drain

It is straightforward to obtain the thermoelectric coefficients of this delta function thermoelectric formally starting from the general relations we have obtained in this Lecture, reproduced below for convenience:

$$G = \int_{-\infty}^{+\infty} dE \left(-\frac{\partial f_0}{\partial E} \right) G(E)$$

(same as 10.4a)

$$G_S = \int_{-\infty}^{+\infty} dE \left(-\frac{\partial f_0}{\partial E} \right) \frac{E - \mu_0}{qT} G(E)$$

(same as 10.4b)

$$G_P = \int_{-\infty}^{+\infty} dE \left(-\frac{\partial f_0}{\partial E} \right) \frac{E - \mu_0}{q} G(E)$$

(same as 10.13a)

$$G_Q = \int_{-\infty}^{+\infty} dE \left(-\frac{\partial f_0}{\partial E} \right) \frac{(E - \mu_0)^2}{q^2 T} G(E)$$

(same as 10.13b)

We argue that factors like $E - \mu_0$ can be pulled out of the integrals assuming they are almost constant over the very narrow energy range where $G(E)$ is non-zero. This gives

$$G = G(\varepsilon)\,\Delta\varepsilon\left(-\frac{\partial f_0}{\partial E}\right)_{E=\varepsilon} \qquad (10.16a)$$

$$G_S = \frac{\varepsilon-\mu_0}{qT}\,G \qquad (10.16b)$$

$$G_P = \frac{\varepsilon-\mu_0}{q}\,G \qquad (10.16c)$$

$$G_Q = \frac{(\varepsilon-\mu_0)^2}{q^2 T}\,G \qquad (10.16d)$$

From Eq.(10.14) we obtain the coefficients for the **delta function thermoelectric**:

$$S = -\frac{G_S}{G} = -\frac{\varepsilon-\mu_0}{qT} \qquad (10.17a)$$

$$\Pi = -\frac{G_P}{G} = -\frac{\varepsilon-\mu_0}{q} \qquad (10.17b)$$

$$G_K = G_Q - \frac{G_P G_S}{G} = 0 \qquad (10.17c)$$

Let us now see how we can understand these results from intuitive arguments without any formal calculations. The Seebeck coefficient in Eq.(10.17a) is the open circuit voltage required to maintain zero current. Since the channel conducts only at a single energy $E = \varepsilon$, in order for no current to flow, the Fermi functions at this energy must be equal:

$$f_1(\varepsilon) = f_2(\varepsilon) \rightarrow \frac{\varepsilon-\mu_1}{kT_1} = \frac{\varepsilon-\mu_2}{kT_2}$$

Hence

$$\frac{\varepsilon-\mu_1}{kT_1} = \frac{\varepsilon-\mu_2}{kT_2} = \frac{(\varepsilon-\mu_1)-(\varepsilon-\mu_2)}{k(T_1-T_2)} = -\frac{\mu_1-\mu_2}{k(T_1-T_2)}$$

Noting that the Seebeck coefficient is defined as

$$S \equiv \frac{(\mu_1-\mu_2)/q}{T_1-T_2} \quad \text{(with I = 0)}$$

we obtain $\quad S = -\dfrac{\varepsilon-\mu_1}{qT_1} = -\dfrac{\varepsilon-\mu_2}{qT_2} \approx -\dfrac{\varepsilon-\mu_0}{qT}$

in agreement with the result in Eq.(10.17a).

The expression in Eq.(10.17b) for the Peltier coefficient too can be understood in simple terms by arguing that every electron carries a charge $-q$ and a heat $\varepsilon-\mu_0$, and hence the ratio of the heat current to the charge current must be $(\varepsilon-\mu_0)/-q$.

That brings us to the zero heat conductance in Eq.(10.17c) which tells us that the heat current is zero under open circuit conditions, that is when the regular charge current is zero. This seems quite reasonable. After all if there is no electrical current, how can there be a heat current? But if this were the whole story, then no thermoelectric would have any heat conductance, and not just delta function thermoelectrics.

The full story can be understood by considering a two-channel thermoelectric with a temperature difference. Under open circuit conditions, there is a voltage between the two contacts with $\mu_1 < \mu_2$. Although the total current is zero, the individual currents in each level are non-zero. They are equal and opposite, thereby canceling each other out. But

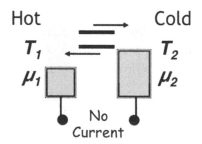

the corresponding energy currents do not cancel, since the channel with higher energy carries more energy. Zero charge current thus does not guarantee zero heat current, except for a delta function thermoelectric with its sharply peaked $G(E)$.

Since the delta function thermoelectric has zero heat conductance, the ZT product (see Eq.(10.10)) should be very large and it would be seem that is what an ideal thermoelectric should look like. However, as we mentioned earlier, even if the electronic heat conductance were zero, we would still have the phonon contribution which would prevent the ZT-product from getting too large. We will talk briefly about this aspect in the next Lecture.

10.4.1. Optimizing Power Factor

Let us end this Lecture by discussing what factors might maximize the power factor S^2G (see Eq.(10.8)) for a thermoelectric. If getting the highest Seebeck coefficient S were our sole objective then it is apparent from Eq.(10.17b) that we should choose our energy ε as far from μ_0 as possible. But that would make the conductance G from Eq.(10.17a) unacceptably low, because the factor $(-\partial f_0 / \partial E)$ dies out quickly as the energy E moves away from μ_0.

From Eq.(10.17a) and (10.16a) we have

$$S^2G = G(\varepsilon)\,\Delta\varepsilon \left(\frac{\varepsilon - \mu_0}{qT}\right)^2 \left(-\frac{\partial f_0}{\partial E}\right)_{E=\varepsilon}$$

$$= G(\varepsilon)\frac{\Delta\varepsilon}{kT}\left(\frac{k}{q}\right)^2 \underbrace{x^2 \frac{e^x}{(e^x+1)^2}}_{\equiv\ F(x)},\quad where \quad x \equiv \frac{\varepsilon - \mu_0}{kT}$$

$$(10.18)$$

Lessons from Nanoelectronics

Fig.10.9. Plot of

$$F(x) \equiv \frac{x^2 e^x}{(e^x + 1)^2}$$

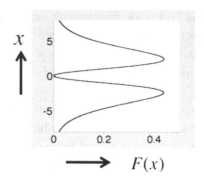

It is apparent from Fig.10.9 that the function $F(x)$ has a maximum around $x \sim 2$, suggesting that ideally we should place our level approximately $2kT$ above or below the electrochemical potential μ_0.

The corresponding Seebeck coefficient is approximately

$$S \approx 2\frac{k}{q} \qquad (10.19a)$$

$$S^2 G \approx 0.5\left(\frac{k}{q}\right)^2 G(\varepsilon)\frac{\Delta\varepsilon}{kT} \qquad (10.19b)$$

The best thermoelectrics typically have Seebeck coefficients that are not too far from the *2 (k/q) = 170 μV/K* expected from Eq.(10.19a). They are usually designed to place μ_0 a little below the bottom of the band so that the product of $G(E)$ and $(-\partial f_0 / \partial E)$ looks like a "delta function" around ε a little above the bottom of the band as shown in the sketch.

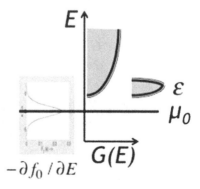

The problem is that the corresponding values of conductance are not as large as they could possibly be if μ_0 were located higher up in the band as sketched here. This would be characteristic of metals.

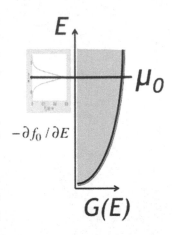

But metals show little promise as thermoelectric materials, because their Seebeck coefficients are far less than k/q, since the electrochemical potential lies in the middle of a band of states and there are nearly as many states above μ_0 as there are below μ_0.

For this reason the field of thermoelectric materials is dominated by semiconductors which show the highest power factors. However, the power factor determines only the numerator of the ZT product in Eq.(10.10). As we mentioned earlier the heat conductance in the denominator is dominated by phonon transport involving a physics that is very different from the electronic transport properties that this book is largely about. In the next lecture we will digress briefly to talk about phonon transport.

Lecture 11

Phonon Transport

11.1. Phonon Heat Current
11.2. Thermal Conductivity

We have seen earlier that the electrical conductivity is given by (Eq.(4.14))

$$\sigma = \frac{q^2}{h}\left(\frac{M\lambda}{A}\right)$$

(11.1)

where the number of channels per unit area M/A and the mean free path λ are evaluated in an energy window ~ a few kT around μ_0. The degeneracy factor g (Section 5.4) due to spins and valleys is assumed to be included in M.

In this Lecture we will extend the transport theory for electrons to handle something totally different, namely phonons and obtain a similar expression for the thermal conductivity due to phonons

$$\kappa_{ph} = \frac{\pi^2}{3}\frac{k^2 T}{h}\left(\frac{M\lambda}{A}\right)_{ph}$$

(11.2)

where the number of channels per unit area M/A and the mean free path λ are evaluated in a frequency window $\hbar\omega$ ~ a few kT. There is a degeneracy factor of $g=3$ due to three polarizations that is assumed to be included in M.

Our purpose in this Lecture is two-fold. The first is to provide an interesting perspective in the hunt for high-ZT thermoelectrics. We have

145

seen in Lecture 10 that with careful design it is possible to achieve a Seebeck coefficient ~ $2k/q$ while maximizing the numerator in Eq.(10.10). We can write

$$ZT \approx 4\frac{k^2 T}{q^2}\frac{\sigma}{\kappa + \kappa_{ph}} \approx 4\frac{k^2 T}{q^2}\frac{\sigma}{\kappa_{ph}}$$

if we assume that the thermoelectric has been designed to provide a Seebeck coefficient $S \sim 2k/q$ and the heat conductivity is dominated by phonons. Using Eqs.(11.1) and (11.2) we have

$$ZT \approx \frac{M\lambda/A}{(M\lambda/A)_{ph}}$$

(11.3)

where we have dropped a factor of $12/\pi^2 \sim 1$ since it is just an approximate number anyway. This is an interesting expression suggesting that once a material has been optimized to provide a respectable Seebeck coefficient (S), the ZT product we obtain simply reflects the ratio of $M\lambda/A$ for electrons and phonons.

As we discussed at the end of the last Lecture, the process of achieving a high S usually puts us in a regime with a low M/A for electrons. M/A for phonons on the other hand is often much higher ~ $(1 \text{ nm})^{-2}$ at room temperature, so that the ratio of M/A's in Eq.(11.3) is ~ *0.1* or less. But electrons tend to have a longer mean free path, resulting in a $ZT \sim 1$ for the best thermoelectrics. The most promising approach for improving ZT at this time seems to be to try to suppress the mean free path for phonons without hurting the electrons (the so-called "electron crystal, phonon glass").

The highest ZT is on the order of 1 to 3 and it has been that way for long time. Experts say that an increase of ZT to 4 - 10 would have a major impact on its practical applications and researchers hope that nanostructured materials might enable this increase. Whether they are right, only future experiments can tell, but it is clearly of interest to understand the principles that govern *ZT* in nanoscale materials and we hope this Lecture will contribute to this understanding.

But my real objective is to demonstrate the power of the elastic resistor approach that allows us not only to obtain linear transport coefficients for electrons easily, but also extend the results to a totally different entity (the phonons) with relative ease.

11.1. Phonon Heat Current

As we mentioned earlier the thermal conductance of solids has a significant phonon component in addition to the electronic component we just talked about. I will not go into this in any depth. My purpose is simply to show how easily our elastic transport model is extended to something totally different.

The atoms comprising the solid lattice are often pictured as an array of masses connected by springs as sketched here. The vibrational frequencies of such a system are described by a dispersion relation $\omega(\beta)$ analogous to the $E(k)$ relation that describes electron waves, with β playing the role of k, and $\hbar\omega$ playing the role of E.

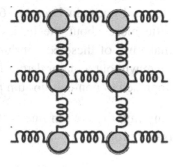

The key difference with electrons is that unlike electrons, there is no exclusion principle. Millions of phonons can be packed into a single channel creating a sound wave that we can even hear, if the frequency is low enough. One consequence of this lack of a exclusion principle is that the equilibrium distribution of phonons is given by a Bose function

$$n(\omega) \;\equiv\; \frac{1}{\exp\!\left(\dfrac{\hbar\omega}{kT}\right) - 1}$$

(11.4)

instead of the Fermi function for electrons introduced in Lecture 2:

$$f(E) \equiv \frac{1}{\exp\left(\dfrac{E-\mu}{kT}\right)+1}$$

(same as Eq.(2.2))

The difference with Eq.(11.4) is just the +1 instead of the -1 in the denominator, which restricts $f(E)$ to values between 0 and 1, unlike the $n(\omega)$ in Eq.(11.4). The other difference is the absence of an electrochemical potential μ in Eq.(11.4) which is attributed to the lack of conservation of phonons. Unlike electrons, they can appear and disappear as long as other entities are around to take care of energy conservation.

These results are of course not meant to be obvious, but they represent basic results from equilibrium statistical mechanics that are discussed in standard texts. In Lecture 16 on the Second Law, we will try to say a little more about the basics of equilibrium statistical mechanics. We make use of these equilibrium results but we cannot really do justice to them without a major detour from our main objective of presenting a new approach to non-equilibrium problems.

Anyway, the bottom line is that our result for the charge current carried by electrons

$$I = \frac{q}{h}\int_{-\infty}^{+\infty} dE \left(\frac{M\lambda}{L+\lambda}\right)\left(f_1(E)-f_2(E)\right)$$

can be modified to represent the heat current due to phonons

$$I_Q = \frac{1}{h}\int_0^{+\infty} d(\hbar\omega)\left(\frac{M\lambda}{L+\lambda}\right)_{ph}\hbar\omega\left(n_1(\omega)-n_2(\omega)\right) \qquad (11.5)$$

simply by replacing the charge with the energy: $q \to \hbar\omega$

and the Fermi functions with the Bose functions:

$$n_1(\omega) = \frac{1}{\exp(\hbar\omega/kT_1) - 1}$$

$$n_2(\omega) = \frac{1}{\exp(\hbar\omega/kT_2) - 1}$$

and changing the lower integration limit to zero.

Again we can linearize Eq.(11.5) to write (see Appendix A)

$$I_Q \approx G_K (T_1 - T_2) \tag{11.6}$$

where the thermal conductance due to phonons can be written as

$$G_K = \frac{k^2 T}{h} \int\limits_0^{+\infty} dx \left(\frac{M\lambda}{L+\lambda}\right)_{ph} \frac{x^2 e^x}{(e^x - 1)^2}, \quad x \equiv \frac{\hbar\omega}{kT} \tag{11.7}$$

Note that just as the elastic resistor model for electrons ignores effects due to the inelastic scattering between energy channels, this model for phonons ignores effects due to the so-called "anharmonic interactions" that cause phonons to convert from one frequency to another. While ballistic electron devices have been widely studied for nearly two decades, much less is known about ballistic phonon devices.

11.1.1. Ballistic Phonon Current

Before moving on let us take a brief detour to point out that the ballistic conductance due to phonons is well-known though in a slightly different form, similar to the Stefan-Boltzmann law for photons.

From Eq.(11.7) we can write the ballistic heat conductance as

$$[G_K]_{ballistic} = \frac{k^2 T}{h} \int\limits_0^{\infty} dx\, M_{ph} \frac{x^2 e^x}{(e^x - 1)^2} \tag{11.8}$$

To evaluate this expression we need to evaluate the number of modes which is related to the number of wavelengths that fit into the cross-section, as we discussed for electrons (see Eq.(5.17))

$$M_{ph} = \frac{\pi A}{(wavelength)^2} \underbrace{3}_{\substack{number\ of \\ polarizations}}$$

but we have a degeneracy factor of 3 for the three allowed polarizations. Noting that for phonons (c_s : acoustic wave velocity)

$$wavelength = \frac{c_s}{\omega / 2\pi}$$

we have

$$M_{ph} = \frac{3\omega^2 A}{4\pi c_s^2} = \frac{3k^2 T^2 A}{4\pi \hbar^2 c_s^2} x^2$$

From Eq.(11.8),

$$[G_\kappa]_{ballistic} = \frac{3k^4 T^3}{8\pi^2 \hbar^3 c_s^2} \underbrace{\int_0^\infty dx \frac{x^4 e^x}{(e^x - 1)^2}}_{= 4\pi^4 / 15} = \frac{\pi^2 k^4 T^3}{10 \hbar^3 c_s^2}$$

Making use of this expression we can write the ballistic heat current from Eq.(11.6) as

$$[I_Q]_{ballistic} = \frac{\pi^2 k^4 T^3}{10 \hbar^3 c_s^2} \Delta T$$

However, the ballistic current is usually written in a different form making use of the relation $T^3 \Delta T = \Delta(T^4 / 4)$:

$$[I_Q]_{ballistic} = \frac{\pi^2 k^4}{40 \hbar^3 c_s^2} \Delta T^4 = \frac{\pi^2 k^4}{40 \hbar^3 c_s^2} \left(T_1^4 - T_2^4\right) \tag{11.9}$$

The corresponding result for photons is known as the Stefan-Boltzmann relation for which the numerical factor differs by a factor of *2/3* because the number of polarizations is *2* instead of *3*. But this is just a detour. Let us get back to diffusive phonon transport.

11.2. Thermal Conductivity

Returning to Eq.(11.7) for the thermal conductance due to phonons, we could define the thermal conductivity

$$\kappa \;=\; \frac{k^2 T}{h} \int_{0}^{+\infty} dx \left(\frac{M\lambda}{A}\right)_{ph} \frac{x^2 e^x}{(e^x - 1)^2}$$

(11.10)

such that

$$G_\kappa = \frac{\kappa A}{\lambda + L}$$

Note the similarity with the electrical conductivity due to electrons:

$$\sigma \;=\; \frac{q^2}{h} \int_{-\infty}^{+\infty} dx \left(\frac{M\lambda}{L+\lambda}\right) \frac{e^x}{(e^x + 1)^2}$$

The function

$$F_T(x) \;\equiv\; \frac{e^x}{(e^x + 1)^2}$$

appearing in all electronic transport coefficients is different from the function

$$\frac{3}{\pi^2} \frac{x^2 e^x}{(e^x - 1)^2}$$

appearing in Eq.(11.10) but they are of similar shape as shown. The factor $3/\pi^2$ is needed to make the area under the curve equal to one, as it is for the broadening function $F_T(x)$ for electrons (see Eq.(2.4)).

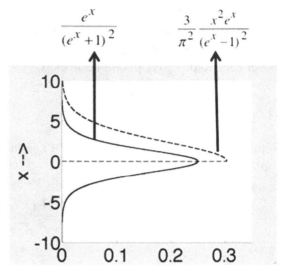

Fig.11.1. Broadening function for phonons compared to that for electrons, $F_T(x)$. These are the window functions defined by Jeong et al. (2011), see Eqs.(7e,f).

So we can think of electrical and thermal conductance at least qualitatively in the same way. Just as the electrical conductivity is given by the product

$$\underbrace{\frac{q^2}{h}}_{38\ \mu S}\left(\frac{M\lambda}{A}\right)$$

the thermal conductivity is given by

$$\underbrace{\frac{\pi^2}{3}\frac{k^2T}{h}}_{284\ pW/K}\left(\frac{M\lambda}{A}\right)_{ph}$$

The factor $\pi^2/3$ is just the inverse of the $3/\pi^2$ needed to normalize the phonon broadening function.

We mentioned at the end of the last Lecture that M/A for electrons tends to be rather small for good thermoelectric materials whose electrochemical potential μ lies within a kT of the bottom of the band. One way to get around this is to use materials where the entire electronic band of energies is a few kT wide, which is unusual. Unfortunately for the phonon band this condition is common, giving an average M/A close to the maximum value.

The most popular thermoelectric material Bi_2Te_3 appears to have a phonon bandwidth much less than kT, thus making the average value of M/A for phonons relatively small. The phonon mean free path is also relatively small, helping raise ZT. For example, $M/A \sim 4e17/m^2$, $\lambda \sim 15$ nm gives $\kappa \sim 2$ $W/m/K$, numbers that are approximately representative of Bi_2Te_3.

The possible role of nanostructuring in engineering a better thermoelectric is still a developing story whose ending is not clear. At this time all we can do is to present a different viewpoint that may help us see some new options. And that is what we have tried to do here.

Lecture 12

Measuring Electrochemical Potentials

12.1. The Landauer Formulas
12.2. Büttiker Formula

Electrochemical potentials have played an important role in our discussion, starting from Lecture 2 where I stressed that electron flow is driven by the difference in the electrochemical potentials μ_1 and μ_2 in the two contacts. However, talking about electrochemical potentials inside the channel, as we did later in Chapter 6 when discussing the diffusion equation, often raises eyebrows. This is because an electrochemical potential of μ implies that the occupation of all available states are described by the corresponding Fermi function (Eq.(2.2))

$$f(E) = \frac{1}{1 + \exp((E - \mu)/kT)}$$

This is approximately true of large contacts which always remain close to equilibrium, but not necessarily true of small conductors even for small applied voltages. As we saw in Lecture 6, it was important to introduce two separate electrochemical potentials μ^+ and μ^- in order to understand the interface resistance that is the key feature of the new Ohm's law (Eq.(4.2)).

Non-equilibrium electrochemical potentials of this type can be very useful in understanding current flow and is widely used by device engineers. It is common to use two different potentials (often called quasi-Fermi levels) for conduction and valence bands and in Lecture 14 we will talk about different potentials for upspin and downspin electrons.

Indeed in Lecture 9 we even argued that controlling such potentials with creatively designed "smart" contacts could lead to unique devices.

In spite of the obvious utility of the concept, many experts are uneasy about invoking non-equilibrium electrochemical potentials inside nanoscale devices which they view as ill-defined concepts that cannot be measured. Instead they feel conceptually on solid ground by sticking to terminal descriptions in terms of the electrochemical potentials at the contacts.

In this Lecture I would like to address some of these issues related to non-equilibrium potentials and their measurability using a simple example which will also allow us to connect our discussion to the Landauer formulas and the Büttiker formula that form the centerpiece of the transmission formalism widely used in mesoscopic physics.

So far we have talked about normal resistors with uniformly distributed scatterers characterized by a mean free path. Instead, following Landauer, let us consider an otherwise ballistic channel with a single localized defect that lets a fraction T of all the incident electrons proceed along the original direction, while the rest $1-T$ get turned around. (see Fig.12.1).

We could follow our arguments from Lecture 6 to obtain the spatial variation of the potentials μ^+ and μ^- across the scatterer, and use it to deduce the resistance of the scatterer. But experts are often uneasy about non-equilibrium potentials and one way to bypass these questions is to consider a four-terminal measurement (Fig.12.2) using two additional voltage probes that draw negligible current, to measure the voltage drop across the defect.

We will show that if the voltage probes are identical and weakly coupled (non-invasive) then this four-terminal conductance G_{4t} is given by

$$G_{4t} \;=\; \frac{I}{(\mu_{1*} - \mu_{2*})/q} \;=\; M\frac{q^2}{h}\frac{T}{1-T}$$

$$(12.1)$$

M being the number of channels or modes in the conductor discussed in Lecture 4. But if we were to determine the conductance using the actual voltage applied to the current-carrying terminals we would obtain a lower conductance:

$$G_{2t} \;=\; \frac{I}{(\mu_1 - \mu_2)/q} \;=\; M \frac{q^2}{h} T \tag{12.2}$$

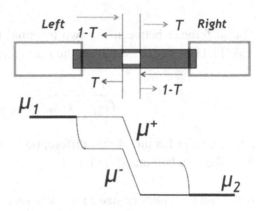

Fig.12.1.
Potential variation across a defect.

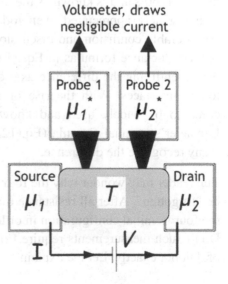

Fig.12.2: Four-terminal measurement of conductance of an otherwise ballistic one-dimensional conductor having a single "defect" in the middle, through which electrons have a probability T of transmitting.

Fig.12.3: The two-terminal resistance can
be viewed as the four-terminal resistance
in series with the interface resistance.

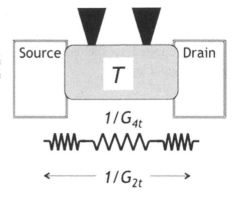

The difference between the two-terminal (Eq.(12.2)) and four-terminal
(Eq.(12.1)) resistances reflects the same *interface resistance*

$$\frac{1}{G_{2t}} - \frac{1}{G_{4t}} = \frac{h}{q^2 M}$$

introduced in Lecture 4 that differentiates the new Ohm's law (Eq.(1.4))
from the standard one (Eq.(1.1a)).

Although the interface resistance was recognized for metallic resistors in
the late 1960's and is known as the *Sharvin resistance*, its ubiquitous role
is not widely appreciated even today. In the early 1980's there was
considerable confusion and discussion about the difference between the
two conductance formulas in Eqs.(12.1) and (12.2) and *Imry* is credited
with identifying the difference as a quantized Sharvin resistance related
to the interfaces. With the rise of mesoscopic physics, Eq.(12.2) has
come to be widely used and known as the Landauer formula while
Landauer's original formula (Eq.(12.1)) is relatively forgotten, and not
many recognize the difference.

The reader may wonder why the four-terminal Landauer formula came to
be "forgotten." After all resistance measurements are commonly made in
the four terminal configuration in order to exclude any contact resistance.
Don't such measurements require Eq.(12.1) for their interpretation? Sort
of, but not exactly. Let me explain.

One problem in the early days of mesoscopic physics was that the voltage probes were strongly coupled to the main conductor and behaved like "additional defects" whose effect could not simply be ignored. In order to interpret real experiments using four-terminal configurations, Büttiker (see Büttiker 1988) found an elegant solution by writing the current I_m at terminal m of a multi-terminal conductor in terms of the terminal potentials μ_n:

$$I_m = (1/q) \sum_n G_{mn} (\mu_m - \mu_n)$$

$$(12.3)$$

where $G_{m,n}$ is the conductance determined by the transmission $T_{m,n}$ between terminals m and n. With just two terminals, Büttiker's formula reduces to

$$I_1 = (1/q) \, G_{12}(\mu_1 - \mu_2) = -I_2$$

which is the same as the two-terminal Landauer formula (Eq.(12.2)) if we identify G_{12} as $(q^2/h)M$. But the power of Eq.(12.3) lies in its ability to provide a quantitative basis for the analysis of multi-terminal structures like the one in Fig.12.2.

Knowing G_{mn}, if we knew all the potentials μ_m, we could use Eq.(12.3) to calculate the currents I_m at all the terminals. Of course for the voltage probes 1* and 2* we do not know the voltages they will float to and so we do not know μ_{1*} or μ_{2*}, to start with. But we do know the currents I_{1*} and I_{2*}, each of which must be zero, since the high impedance voltmeter draws negligible current. The point is that if we know either μ_m or I_m at each terminal m we can solve Eq.(12.3) to obtain whatever we do not know.

In this Lecture we will look at a specific problem, namely the voltage drop across a defect (Fig.12.1) and show that with weakly coupled non-invasive probes the Büttiker formula indeed gives the same answers as we get by looking directly at the electrochemical potentials μ^+ and μ^- inside the conductor.

This is reassuring because the approach due to Büttiker deals directly with measurable terminal quantities and so appears conceptually on more comfortable ground. The development of scanning probe microscopy (SPM) has made it possible to use nanoscale tunneling contacts as voltage probes whose effect is indeed negligible. Measurements using such "non-invasive" probes do provide experimental support for the four-terminal Landauer formula, but there is a subtlety involved.

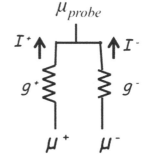

What a voltage probe measures is some weighted average of the two potentials μ^+ and μ^- , the exact weighting depending on the construction of the probes. We could model it by associating conductances g^+ and g^- with the transmission of electrons from the + and the - streams into the probes respectively.

Setting the net probe current to zero we can write

$$g^+ (\mu^+ - \mu_{probe}) + g^- (\mu^- - \mu_{probe}) = 0$$

$$\mu_{probe} = \underbrace{\frac{g^+}{g^+ + g^-}}_{\alpha} \mu^+ + \underbrace{\frac{g^-}{g^+ + g^-}}_{1-\alpha} \mu^-$$

so that   (12.4)

For atomic scale probes that are much smaller than an electron wavelength we expect the two conductances to be similar so that the weighting factor $\alpha \sim 50\%$, so that the probe measures the average potential

$$\mu_{probe} = (\mu^+ + \mu^-)/2$$

For larger probes, however, it is possible for a voltage probe to have a pronounced bias for one stream or the other leading to a weighting factor α different from 50%. If this weighting happens to be different for the two probes 1* and 2*, it could change the measured resistance from that predicted by Eq.(12.1).

Indeed, experimental measurements have even shown *negative resistance*, something that cannot be understood in terms of Eq.(12.1). However, some of this is due to quantum interference effects that make the simple semiclassical description in terms of $\mu^{\pm}$ inadequate as we will see in Lecture 20. However, one could use a more sophisticated version of Eq.(12.4) (Lecture 22) or use the Büttiker formula, with the conductances $G_{m,n}$ calculated from an appropriate quantum transport model.

The bottom line is that if we know the correct internal state of the conductor in terms of a set of non-equilibrium electrochemical potentials, we can predict what a specific non-invasive voltage probe will measure and the result should match what the Büttiker formula predicts. The reverse, however, is not true. Knowing what a specific probe will measure, we cannot deduce the internal state of the conductor. But if we have measurements from multiple probes we could back out the internal state as we will see in Lecture 14 when we discuss non-equilibrium spin potentials.

With that rather long "introduction" let us now look at the two Landauer formulas (Eqs.(12.1), (12.2)) and the Büttiker formula (Eq.(12.3)) in a little more detail.

12.1. The Landauer formulas (Eqs.(12.1), (12.2))

Getting back to the problem of finding the potential variation across a defect in an otherwise ballistic conductor (Fig.12.1), we start by relating the outgoing currents to the incoming currents as follows

$$I^+ (Right) \;=\; T I^+ (Left) \;+\; (1-T) I^- (Right)$$

$$I^- (Left) \;=\; (1-T) I^+ (Left) \;+\; T I^- (Right)$$

We can then convert the currents to occupation factors (see Eqs.(6.13))

$$f^+ (Right) \;=\; T f^+ (Left) + (1-T) f^- (Right)$$

$$f^- (Left) \;=\; (1-T) f^+ (Left) + T f^- (Right)$$

and then to potentials using the same argument as in Lecture 6 (see discussion following Eq.(6.11b)):

$$\mu^+ (Right) \;=\; T \mu^+ (Left) \;+\; (1-T) \mu^- (Right)$$
$$\qquad\qquad =\; T \mu_1 \;+\; (1-T) \mu_2 \qquad\qquad (12.5a)$$

$$\mu^- (Left) \;=\; (1-T) \mu^+ (Left) \;+\; T \mu^- (Right)$$
$$\qquad\qquad =\; (1-T) \mu_1 \;+\; T \mu_2 \qquad\qquad (12.5b)$$

The algebra can be simplified by choosing the potential for one of the contacts as zero and the other as one. The actual potential differences can then be obtained by multiplying by the actual $\mu_1 - \mu_2 = qV$.

Fig.12.4. Spatial profile of μ^+ and μ^- across a scatterer normalized to an overall potential difference of one. The actual potential differences can be obtained by multiplying by the actual $\mu_1 - \mu_2 = qV$.

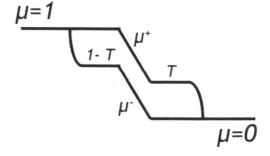

Eqs.(12.5a,b) then give us the picture shown in Fig.12.4 leading to

$$\mu^+ - \mu^- = T(\mu_1 - \mu_2)$$

as long as both μ^+ and μ^- are evaluated at the same location on the left or on the right of the scatterer. Using

$$I = \frac{q}{h} M(\mu^+ - \mu^-)$$

for the current we obtain the standard Landauer formula (Eq.(12.2)).

To obtain the first Landauer formula we find the drop in either μ^+ or μ^- across the scatterer:

$$\mu^+(Left) - \mu^+(Right) = (1-T)(\mu_1 - \mu_2) \tag{12.6a}$$

$$\mu^-(Left) - \mu^-(Right) = (1-T)(\mu_1 - \mu_2) \tag{12.6b}$$

and then divide the current by it to obtain the result stated in Eq.(12.1):

$$G_{4t} = \frac{q^2}{h} M \frac{T}{1-T}$$

Note, however, that we are looking at the electrochemical potentials inside the conductor. How does this relate to the voltage measured by non-invasive voltage probes implemented using scanning tunneling probes (STP)? Assuming that the probe measures the average of μ^+ and μ^- we obtain the plot shown in Fig.12.5 from Fig.12.4.

What if the probe measures a weighted average of μ^+ and μ^- with some α (see Eq.(12.4)) other than 50%? As long as α is the same for both probes, the drop across the scatterer would still be given by

$$\mu_{probe}(Left) - \mu_{probe}(Right) = (1-T)(\mu_1 - \mu_2) \tag{12.6c}$$

thus leading to the same Landauer formula (Eq.(12.1)). But if the weighting factor α were different for the two probes then the result would not match Eq.(12.1). As an extreme example if α were zero on the left and one on the right,

$$\mu_{probe}(Left) - \mu_{probe}(Right) = (1-2T)(\mu_1 - \mu_2)$$

leading to a negative resistance for $T > 0.5$.

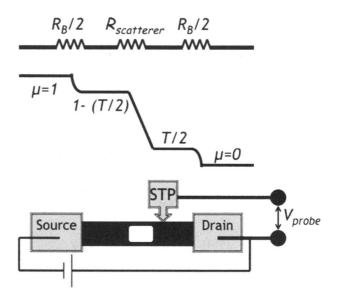

Fig.12.5. A scanning tunneling probe (STP) measures the average electrochemical potential $(\mu^+ + \mu^-)/2$.

Clearly the concept of non-equilibrium potentials μ^+ and μ^- should be used wisely with caution. But it does lead to intuitive understandable results. The potential drops across the defect but not across the ballistic regions, suggesting that the defect represents a resistance given by Eq.(12.1). Note, however, that we are still talking about elastic resistors. We have an IR drop in the voltage, but no corresponding I^2R in Joule dissipation. All dissipation is still in the contacts.

12.2. Büttiker Formula (Eq.(12.3))

Eq.(12.3) deals directly with the experimentally measured terminal quantities bypassing any questions regarding the internal variables. The point we wish to stress is the general applicability of this result irrespective of whether the resistor is elastic or not. Indeed, as we will see we can obtain it invoking very little beyond linear circuit theory.

We start by defining a multi-terminal conductance

$$G_{m,n} \equiv -\frac{\partial I_m}{\partial(\mu_n/q)}, \quad m \neq n$$

(12.7a)

$$G_{m,m} \equiv +\frac{\partial I_m}{\partial(\mu_m/q)}$$

(12.7b)

Why do we have a negative sign for $m \neq n$, but not for $m = n$? The motivation can be appreciated by looking at a representative multi-terminal structure (Fig.12.6).

An increase in μ_1 leads to an incoming or positive current at terminal 1, but leads to **outgoing** or negative currents at the other terminals. The signs in Eq.(12.7a,b) are included to make the coefficients come out positive as we intuitively expect a conductance to be.

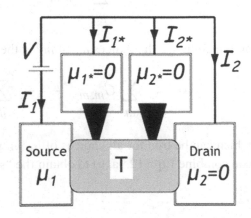

Fig.12.6. Thought experiment based on the four-terminal measurement set-up in Fig.12.1.

In terms of these conductance coefficients, we can write the current as

$$I_m = G_{m,m} \frac{\mu_m}{q} - \sum_{n \neq m} G_{m,n} \frac{\mu_n}{q} \qquad (12.8)$$

The conductance coefficients must obey two important "sum rules" in order to meet two important conditions.

Firstly, the currents predicted by Eq.(12.8) must all be zero if all the μ's are equal, since there should be no external currents at equilibrium. This requires that

$$G_{m,m} = \sum_{n \neq m} G_{m,n} \qquad (12.9a)$$

Secondly, for any choice of μ's, the currents I_m must add up to zero. This requires that

$$G_{m,m} = \sum_{n \neq m} G_{n,m} \qquad (12.9b)$$

but it takes a little algebra to see this from Eq.(12.8). First we sum over all m

$$\sum_m I_m = 0 = \sum_m G_{m,m} \frac{\mu_m}{q} - \sum_m \sum_{n \neq m} G_{m,n} \frac{\mu_n}{q}$$

and interchange the indices n and m for the second term to write

$$0 = \sum_m G_{m,m} \frac{\mu_m}{q} - \sum_m \sum_{n \neq m} G_{n,m} \frac{\mu_m}{q}$$

which can be true for all choices of μ_m only if Eq.(12.9b) is satisfied. We can combine Eqs.(12.9a, b) to obtain the "sum rule" succinctly:

$$G_{m,m} \;=\; \sum_{n \neq m} G_{m,n} \;=\; \sum_{n \neq m} G_{n,m}$$

$$(12.10)$$

Making use of the sum rule (Eq.(12.10)) we can re-write the first term in Eq.(12.8) to obtain Eq.(12.3):

$$I_m \;=\; (1/q) \sum_n G_{m,n} (\mu_m - \mu_n)$$

(same as Eq.(12.3))

Note that it is not necessary to restrict the summation to $n \neq m$, since the term with $n=m$ is zero anyway. An alternate form that is sometimes useful is to write

$$I_m \;=\; \sum_n g_{m,n} \frac{\mu_n}{q}$$

$$(12.11)$$

where the response coefficients defined as

$$g_{m,n} \;\equiv\; -G_{m,n}, \quad m \neq n$$

$$(12.12a)$$

$$g_{m,m} \;\equiv\; G_{m,m}$$

$$(12.12b)$$

The sum rule in Eq.(12.10) can be rewritten in term of this new response coefficient:

$$\sum_n g_{m,n} \;=\; \sum_n g_{n,m} \;=\; 0$$

$$(12.13)$$

12.2.1. Application

In Section 12.1 we analyzed the potential profile across a single scatterer with transmission probability T. Based on this discussion (Fig.12.5) we would expect that two non-invasive probes inserted before and after the scatterer should float to potentials *1-(T/2)* and *T/2* as indicated in Fig.12.7. But will Büttiker's approach get us the same result?

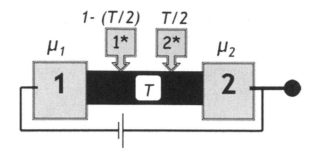

Fig.12.7. Based on Fig.12.5, we expect that two non-invasive probes inserted before and after a scatterer with transmission probability T to float to potentials 1-$(T/2)$ and $T/2$ respectively.

We start from Eq.(12.11) noting that we have four currents and four potentials, labeled 1, 2, 1* and 2*:

$$
\begin{Bmatrix} I_1 \\ I_2 \\ I_{1*} \\ I_{2*} \end{Bmatrix} = \frac{Mq}{h} \begin{bmatrix} A & B \\ C & D \end{bmatrix} \begin{Bmatrix} \mu_1 \\ \mu_2 \\ \mu_{1*} \\ \mu_{2*} \end{Bmatrix}
\tag{12.14}
$$

where A,B,C and D are each (2x2) matrices.

Since
$$
\begin{Bmatrix} I_{1*} \\ I_{2*} \end{Bmatrix} = \begin{Bmatrix} 0 \\ 0 \end{Bmatrix}
$$

we have
$$
\begin{Bmatrix} \mu_{1*} \\ \mu_{2*} \end{Bmatrix} = -D^{-1} C \begin{Bmatrix} \mu_1 \\ \mu_2 \end{Bmatrix}
\tag{12.15}
$$

Now we can write [C] and [D] in the form

$$
\begin{bmatrix} C & D \end{bmatrix} = \begin{bmatrix} -t_1 & -t_2 & r & 0 \\ -t_2' & -t_1' & 0 & r' \end{bmatrix}
\tag{12.16}
$$

where the elements t_1, t_2, t_1' and t_2' of the matrix [C] can be visualized as the probability that an electron transmit from 1 to 1*, 2 to 1*, 2 to 2* and 1 to 2* respectively as sketched in Fig.12.8.

We have assumed that both probes 1* and 2* are weakly coupled so that any direct transmission between them can be ignored. The sum rule in Eq.(12.13) then requires that

$$r = t_1 + t_2$$

and
$$r' = t_2' + t_1' \qquad (12.17)$$

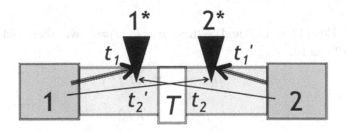

Fig.12.8. The elements t_1, t_2, t_1' and t_2' of the matrix [C] can be visualized as the probability that an electron transmit from 1 to 1*, 2 to 1*, 2 to 2* and 1 to 2* respectively.

Using Eqs.(12.15)-(12.17) we have

$$\mu_{1*} = \frac{t_1}{t_1+t_2}\mu_1 + \frac{t_2}{t_1+t_2}\mu_2 \qquad (12.18a)$$

and
$$\mu_{2*} = \frac{t_2'}{t_1'+t_2'}\mu_1 + \frac{t_1'}{t_1'+t_2'}\mu_2 \qquad (12.18b)$$

So far we have kept things general, making no assumptions other than that of weakly coupled probes. Now we note that for our problem (Fig.12.8), t_1 can be written as

$$t_1 \;\; = \;\; \tau \;\; + \;\; (1-T)\tau \tag{12.19a}$$

since an electron from 1 has a probability of τ to get into probe 1*
directly plus a probability of $1-T$ times τ to get reflected from the
scatterer and then get into probe 1*.

Similarly t_2 can be written as

$$t_2 \;\; = \;\; T\,\tau \tag{12.19b}$$

since an electron from 2 has to cross the scatterer (probability T) and
then enter the weakly coupled probe 1* (probability τ) . Similarly we
can argue $t_1 = t_1'$, $t_2 = t_2'$.

Using Eqs.(12.19a,b) and setting $\mu_1=1$, $\mu_2=0$, we then obtain from
Eqs.(12.18a,b),

$$\mu_{1*} \;\; = \;\; 1-(T/2)$$

$$\mu_{2*} \;\; = \;\; T/2$$

in agreement with what we expected from the last section (Fig.12.7). As
mentioned earlier, this is reassuring since the Büttiker formula deals only
with terminal quantities, bypassing the subtleties of non-equilibrium
electrochemical potentials.

However, the real strength of Eq.(12.3) lies in its model-independent
generality. It should be valid in the linear response regime for all
conductors, simple and complex, large and small. The conductances G_{mn}
in Eq.(12.3) can be calculated from a microscopic transport model like
the Boltzmann equation introduced in Lecture 7 or the quantum transport
model discussed in Part three of these lectures. Sometimes they can even
be guessed and as long as we are careful about not violating the sum
rules we should get reasonable results.

12.2.2. Is Eq.(12.3) obvious?

Some might argue that Eq.(12.3) is not really telling us much. After all, we can always view any structure as a network of effective resistors like the one shown in Fig.12.9 for three terminals? Wouldn't the standard circuit equations applied to this network give us Eq.(12.3)?

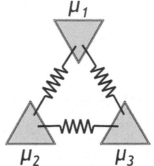

Fig.12.9. The Büttiker formula (Eq.(12.3)) can be visualized as a network of resistors, only if the conductances are reciprocal, that is, if $G_{mn} = G_{nm}$.

The answer is "yes" if we consider only normal circuits for which electrons transmit just as easily from m to n as from n to m so that

$$G_{m\leftarrow n} \;=\; G_{n\leftarrow m}$$

where we have added the arrows in the subscripts to denote the standard convention for the direction of electron transfer. Eq.(12.3), however, goes far beyond such normal circuits and was intended to handle conductors in the presence of magnetic fields for which

$$G_{m\leftarrow n} \;\neq\; G_{n\leftarrow m}$$

For such conductors, Eq.(12.3) is not so easy to justify. Indeed if we were to reverse the subscripts m and n in Eq.(12.3) to write

$$I_m = \; (1/q) \sum_n G_{nm} (\mu_m - \mu_n)$$

WRONG!

it would not even be correct. Its predictions would be different from those of Eq.(12.3) for multi-terminal non-reciprocal circuits of the type we will discuss in the next Lecture.

Lecture 13

Hall effect

13.1. Why n- and p- Conductors Are Different
13.2. Spatial Profile of Electrochemical Potential
13.3. Measuring the Potential
13.4. Non –Reciprocal Circuits

Let me briefly explain what the Hall effect is about. Consider a two-dimensional conductor (see Fig.13.1) carrying current, subject to a perpendicular magnetic field along the y-direction which exerts a force on the electrons perpendicular to its velocity.

$$\vec{F} = \frac{d\vec{p}}{dt} = -q\vec{v} \times \vec{B}$$

(13.1)

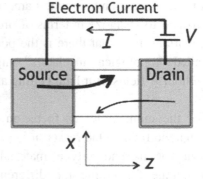

Fig.13.1.
A magnetic field B in the y-direction makes electrons from the source veer "up"wards.

This would cause an electron from the source to veer "up"wards and an electron from the drain to veer "down"wards as shown. Since there are more electrons headed from source to drain, we expect electrons to pile up on the top side causing a voltage V_H to develop in the x-direction transverse to current flow (see Fig.13.2).

Fig.13.2. Basic structure with two voltage probes whose potential difference measure the Hall voltage, $qV_H = \mu_{1*} - \mu_{2*}$.

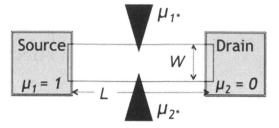

The Hall effect has always been important since its discovery around 1880, and has acquired a renewed importance since the discovery of the quantum Hall effect in 1980 at high magnetic fields. In these lectures we have seen the conductance quantum q^2/h appear repeatedly and it is very common in the context of nanoelectronics and mesoscopic physics. But the quantum Hall effect was probably the first experimental observation where it played a clear identifiable role and the degree of precision is so fantastic that the National Bureau of Standards uses it as the resistance standard. We will talk briefly about it later at the end of this lecture.

I will also talk about an interesting property that magnetic (*B*) fields introduce into any circuit, namely that of non-reciprocity, making it difficult to visualize in terms of ordinary resistances. This is particularly important, now that there is the possibility that a new class of materials called "topological insulators" might show such non-reciprocity even without B-fields. But I am getting ahead of myself.

For the moment let us focus on the conventional Hall effect at low magnetic fields. One reason it is particularly important is that it changes sign for n- and p-type materials, thus providing an experimental technique for telling the difference. Like the thermoelectric current discussed in Lecture 10, this too is commonly explained by invoking "holes" as the positive charge carriers in p-type materials. Once again, this is not satisfactory because it is really the electrons that move even in p-type conductors. Both n-type and p-type conductors have negative charge carriers.

For the thermoelectric effect we saw that its sign is determined by the slope of the density of states $D(E)$, that is whether it is an increasing or a decreasing function of the energy E. By contrast, the sign of the Hall effect is determined by the sign of the effective mass defined as the ratio of the momentum p to the velocity dE/dp (see Lecture 5). As a result although the magnetic force (see Eq.(13.1)) is the same for both n- and p-type conductors, giving the same dp/dt, the resulting dv/dt has opposite signs. This makes electrons in p-conductors veer in the opposite direction giving rise to a Hall voltage of the opposite sign.

Clearly this requires the existence of an $E(p)$ relation underlying the density of states function. Perhaps it is for this reason that amorphous semiconductors which lack a well-defined $E(p)$ often show strange results for the Hall effect and yet show reasonable thermoelectric effect.

The simple theory of the Hall effect given in freshman physics texts goes like this. First the current is written as

$$I = q\,(N/L)\,v_d$$

(13.2a)

with the drift velocity given by the product of the mobility and the electric field in the z-direction:

$$v_d = \bar{\mu}\,(V/L)$$

(13.2b)

These two relations are normally combined to yield the Drude formula (see Eq.(5.1)) that we discussed in Lecture 5

$$\frac{I}{V} = \underbrace{q\frac{N}{WL}\bar{\mu}}_{\sigma}\frac{W}{L}$$

(13.2c)

For the Hall effect, it is argued that an electric field V_H/W must appear in the x-direction to offset the magnetic force

$$V_H / W = v_d B \qquad (13.3)$$

Combining Eq.(13.3) with Eq.(13.2a) one obtains the standard expression for the Hall resistance

$$R_H = \frac{V_H}{I} = \frac{B}{q(N/LW)} \qquad (13.4)$$

One reason the Hall effect is widely used is that Eq.(13.4) allows us to determine the electron density N/LW from the slope of the Hall resistance versus B-field curve.

This looks like a straightforward transparent theory for a well-established effect. What more could we add to it? The main concern we have about this derivation is the same concern that we voiced regarding the Drude formula, namely that if electric field were indeed what drives currents then all electrons should feel its effect.

Indeed Eq.(13.4) for the Hall resistance conveys the impression that the Hall effect depends on the total electron density N/LW over all energies. But we believe this is not correct. Like the other transport coefficients we have discussed, the Hall resistance too is a "Fermi surface property" that depends only on the electrons in an energy window ~ a few kT around $E=\mu_0$ and not on the total number of electrons obtained by integrating over energy.

We will show that the Hall resistance for a single energy channel of an elastic resistor is given by

$$R_H(E) = \frac{2B\,LW}{q\,D(E)v(E)p(E)} \qquad (13.5a)$$

which can be averaged over an energy window of a few kT around $E=\mu_0$ using our standard broadening function:

$$\frac{1}{R_H} = \int\limits_{-\infty}^{+\infty} dE \left(-\frac{\partial f_0}{\partial E} \right) \frac{1}{R_H(E)}$$

(13.5b)

Note that in general we should integrate the conductance $1/R_H$ rather than the resistance R_H since different energy channels all have the same voltage so that they conduct "in parallel" as circuit theorists would put it.

Eq.(13.5) can be reduced to the standard result (Eq.(13.4)) by making use of the single band relation obtained in Lecture 5

$$D(E)v(E)p(E) = N(E).d \quad \text{(same as Eq.(5.4))}$$

with $d= 2$ for a two-dimensional conductor and relating the average of $N(E)$ to the total number of electrons as we did in Section 5.3. But if the single band relation (Eq.(5.7)) is not applicable one should use the expression in Eq.(13.5) rather than Eq.(13.4).

In any case, Eq.(13.5b) shows that the effect really does not involve electrons at all energies. One reason this point causes some confusion is the existence of equilibrium currents inside the sample in the presence of a magnetic field which involve all electrons at all energies.

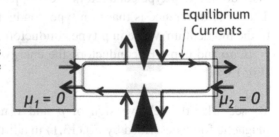

Fig.13.3. Equilibrium currents can exist in any conductor in the presence of a magnetic field.

However, these are dissipationless currents of the type that exist even if we put a hydrogen atom in a magnetic field and have nothing to do with the transport coefficients we are talking about. In any transport model it is important to eliminate these non Fermi surface currents. A similar issue arises with respect to spin currents even without any magnetic fields (see Rashba 2003, Lecture 22).

Getting back to the problem of determining the Hall voltage, as we saw in the last Lecture there are two approaches: (a) calculate the non-equilibrium electrochemical potential inside the conductor or (b) treat it as a four terminal structure using the Büttiker equation. We will discuss both approaches sequentially in Sections 13.2 and 13.3 respectively after we have briefly discussed the dynamics of electrons in a *B*-field in Section 13.1.

13.1. Why n- and p- Conductors Are Different

Fig.13.4
The Hall resistance changes sign for n- and p-type conductors and is inversely proportional to $N(E)$, at $E=\mu_0$.

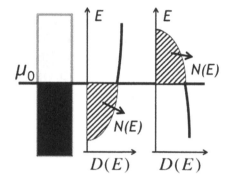

Why do n- and p-type conductors show opposite signs of the Hall effect? The basic difference is that in n-type conductors, the velocity is parallel to the momentum, while in p-type conductors, it is anti-parallel because $v = dE/dp$ and in p-type conductors, the energy decreases with increasing p (Fig.13.4).

To see why the relative sign of p and v matters, let us consider the magnetic force described by Eq.(13.1) in a little more detail.

For any isotropic $E(p)$ relation, the velocity and momentum are collinear (parallel or anti-parallel) pointing, say at an angle θ to the z-axis, so that

$$\vec{p} = p\cos\theta\,\hat{z} + p\sin\theta\,\hat{x}$$

$$\vec{v} = v\cos\theta\,\hat{z} + v\sin\theta\,\hat{x}$$

Inserting into Eq.(13.1) we obtain

$$\frac{d\theta}{dt} = \frac{qvB}{p} \qquad (13.6)$$

showing that the angle θ increases linearly with
time so that the velocity and momentum vectors
rotate uniformly in the z-x plane.

But the sense of rotation is opposite for n- and p-type conductors because
the ratio p/v has opposite signs. This is the ratio we defined as mass (see
Eq.(5.6)) and is constant for parabolic dispersion (Eq.(5.7a)).

$$\omega_c = \left| \frac{qvB}{p} \right|_{E=\mu_0} = \left| \frac{qB}{m} \right|_{E=\mu_0} \qquad (13.7)$$

But for linear dispersion (Eq.(5.7b)) the mass increases with energy, so
that the cyclotron frequency decreases with increased carrier density, as
is observed in graphene.

The magnetic field tries to make the electron would go round and round
in a circle with an angular frequency ω_c. However, it gets scattered after
a mean free time τ, so that if $\omega_c \tau \ll 1$ the electron never really gets to
complete a full rotation. This is the low field regime where the Hall
resistance in given by Eq.(13.5), while the high field regime
characterized by $\omega_c \tau \gg 1$ leads to the quantum Hall effect mentioned
earlier.

Let us now discuss our first approach to the problem of determining the
Hall resistance (Eq.(13.5)) based on looking at the non-equilibrium
electrochemical potentials inside the conductor.

13.2. Spatial Profile of Electrochemical Potential

As I mentioned earlier, the textbook derivation of the Hall resistance
(Eq.(13.4)) looks fairly straightforward, but we are attempting to provide

a different expression (Eq.(13.5)) motivated by the same reasons that prompted us to describe an alternative expression for the conductivity back in Lecture 5.

In our elastic resistor model, the role of the drift velocity in the text-book derivation is played by the potential separation

$$\delta\mu = \mu^+ - \mu^-$$

between drainbound and sourcebound states, so that instead of Eq.(13.2a)) we have from Lecture 6 (see Eq.(6.16))

$$I(E) \;=\; \frac{q}{h} M(E)\left(-\frac{\partial f_0}{\partial E}\right)\delta\mu$$

(13.8a)

with
$$\frac{M(E)}{h} \;=\; \frac{D(E)v(E)}{\pi L}$$

(13.8b)

where we have used the result for 2D conductors from Eq.(4.13).

Instead of Eq.(13.2b), we have the potential separation $\delta\mu$ related to the applied voltage by (see Eq.(9.1))

$$\delta\mu \;=\; \frac{qV\lambda}{L+\lambda} \;\cong\; q\lambda\frac{V}{L}$$

(13.8c)

Just as Eqs.(13.2a,b) yield the Drude expression for the conductivity, similarly Eqs.(13.8a,b) can be combined to yield the more general expression that we discussed in Lecture 5 (see Eq.(5.15)).

For the Hall effect we need a replacement for Eq.(13.3)

$$V_H / W \;=\; v_d B$$

which we will show is given by

$$\frac{V_H}{W} = \frac{2}{\pi}\frac{\delta\mu}{p}B \qquad (13.9)$$

Eq.(13.9) together with Eqs.(13.8a,b) gives us the result for Hall resistance stated earlier in Eq.(13.5).

Unfortunately we do not have a one-line argument for Eq.(13.9) like the one used for Eq.(13.3). Instead I need to describe a two-page argument using the BTE discussed in Lecture 7.

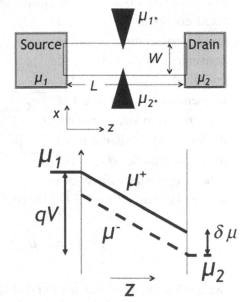

Fig.13.5.
Spatial variation of $\mu^{\pm}$ along z.

Back in Lecture 6 we obtained a solution for a subset of this problem based on a solution of

$$v_z\frac{\partial\mu}{\partial z} = -\frac{\mu-\mu_0}{\tau} \qquad (13.10)$$

and obtained the solutions for the electrochemical potentials μ^{+} and μ^{-} sketched above in Fig.13.5.

The solutions could be written in the form

$$\mu(z,\theta) \;=\; \bar{\mu}(z) + \frac{2}{\pi}\,\delta\mu\cos\theta$$

(13.11)

where we have separated out a z-dependent part $\bar{\mu}(z)$ from the momentum-dependent part at a specific location, z. The latter needs a little discussion.

Since we are discussing an elastic resistor for which electrons have a fixed energy E and hence a fixed momentum p, it is convenient to use cylindrical coordinates for the momentum $(p,\ \theta)$ instead of $(p_x\ ,\ p_z)$. Suppose we were dealing only with electrons at a fixed angle θ (or the exact opposite direction) then making use of Eq.(13.8c) we could write

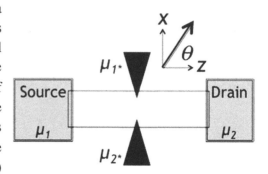

$$\mu(z) \;=\; \bar{\mu}(z) + \frac{qV}{L}\,v\tau\cos\theta$$

(13.12a)

noting that the mean free path in this case is simply

$$2v_z\tau = 2v\tau\cos\theta\,.$$

Comparing with Eq.(13.11), we have

$$\delta\mu \;\approx\; \frac{qV}{L}\frac{\pi}{2}\,v\tau$$

(13.12b)

The question is how we expect the solution in Eq.(13.11) to change when we turn on the magnetic field so that it exerts a force on the electrons. For this we could use a linearized version of the BTE like Eq.(7.17), but retaining both z- and x- components since we have a two-dimensional problem

$$v_x \frac{\partial \mu}{\partial x} + v_z \frac{\partial \mu}{\partial z} + F_x \frac{\partial \mu}{\partial p_x} + F_z \frac{\partial \mu}{\partial p_z} = -\frac{\mu - \mu_0}{\tau} \quad (13.13)$$

Note that Eq.(13.10) is a "subset" of this equation which includes three extra terms. The last two coming from the magnetic force (Eq.(13.1)) can be written as

$$F_x \frac{\partial \mu}{\partial p_x} + F_z \frac{\partial \mu}{\partial p_z} = \vec{F} \cdot \vec{\nabla}_p \mu = \frac{F_\theta}{p} \frac{\partial \mu}{\partial \theta} + F_r \frac{\partial \mu}{\partial p}$$

The force due to a magnetic field has no radial component, only a θ component:

$$F_r = 0 , \quad F_\theta = -qvB$$

This is because the velocity is purely radial and so when we take a cross-product with a magnetic field in the z-direction, we get a vector that is purely in the θ-direction. This allows us to rewrite Eq.(13.13) in the form

$$v_x \frac{\partial \mu}{\partial x} + v_z \frac{\partial \mu}{\partial z} - \frac{qvB}{p} \frac{\partial \mu}{\partial \theta} = -\frac{\mu - \mu_0}{\tau} \quad (13.14)$$

Noting that our solution in Eq.(13.11) satisfies Eq.(13.10), it is easy to check that if we add an extra term varying only with x to it, the resulting expression

$$\mu(z, \theta, x) = \bar{\mu}(z) + \frac{2}{\pi} \delta\mu \cos\theta - \frac{2}{\pi} \frac{\delta\mu}{p} qBx \quad (13.15)$$

will satisfy Eq.(13.14). From this solution we obtain Eq.(13.9) by writing

$$-qV_H = \mu(x = W) - \mu(x = 0) = -\frac{2}{\pi} \frac{\delta\mu}{p} qBW$$

13.3. Measuring the Potential

Let us now look at how we could calculate the Hall voltage using the Büttiker equation for a four-terminal conductor with two current probes and two voltage probes (Fig.13.6) not unlike the one we discussed in the last Lecture. But the two probes are now on two sides of the conductor and would normally not register any potential difference. But when an applied potential causes electrons to flow from left to right, the applied B-field causes them to veer upwards or downwards making μ_{1*} different from μ_{2*} and we would like to calculate the resulting voltage for a given current.

The basic result from the last Lecture still holds:

$$\mu_{1*} = \frac{t_1}{t_1+t_2}\mu_1 + \frac{t_2}{t_1+t_2}\mu_2 \qquad \text{(same as (12.18a))}$$

and

$$\mu_{2*} = \frac{t_2'}{t_1'+t_2'}\mu_1 + \frac{t_1'}{t_1'+t_2'}\mu_2 \qquad \text{(same as (12.18b))}$$

as long as we interpret the various probabilities according to Fig.13.6.

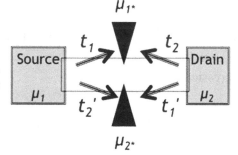

Fig.13.6. We can use the results from Eq.(12.18) as long as we visualize the different probabilities as shown.

These probabilities can be calculated numerically using either semiclassical or numerical models, but I do not have a simple analytical argument showing that this indeed yields a Hall voltage

$$-qV_H \;=\; \mu_{1*} - \mu_{2*}$$

in agreement with our basic result in Eq.(13.9).

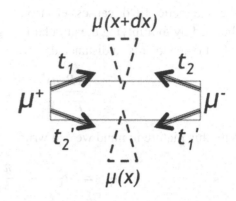

To obtain Eq.(13.9) we could apply our basic result in Eqs.(12.18a,b) to a thin slice of the conductor less than a mean free path long somewhere in the middle of a long channel with electrons from the left having a potential μ^+ and those from the right having a potential μ^-.

We can then use Eqs.(12.18a,b) to write down the potential on the upper and lower edges which we could treat as conceptual probes 1* and 2*:

$$\mu(x+dx) \;=\; \frac{t_1}{t_1+t_2}\,\mu^+ + \frac{t_2}{t_1+t_2}\,\mu^- \tag{13.16a}$$

and

$$\mu(x) \;=\; \frac{t_2{}'}{t_1{}'+t_2{}'}\,\mu^+ + \frac{t_1{}'}{t_1{}'+t_2{}'}\,\mu^- \tag{13.16b}$$

To write down the probabilities appearing in Eqs.(13.16a,b) we note that without a magnetic field all electrons with a velocity lying within the range of angles

$$0 \;<\; \theta \;<\; +\pi/2$$

will go upwards while those with a velocity in the range

$$-\pi/2 \;<\; \theta \;<\; 0$$

will go downwards.

The magnetic field causes electrons to bend upwards by an angle obtained by multiplying the angular rate given by Eq.(13.6) with the time dx/v it takes to cross a distance dx:

$$\frac{qvB}{p}\frac{dx}{v} = \frac{qB\,dx}{p}$$

With this picture in mind we can write

$$t_1 = t_1' \sim \frac{\pi}{2} + \frac{qB}{p}\,dx$$

(13.17a)

$$t_2 = t_2' \sim \frac{\pi}{2} - \frac{qB}{p}\,dx$$ (13.17b)

and use it in Eqs.(13.16a,b) to obtain

$$\mu(x+dx) - \mu(x) = \frac{2qB\,dx}{\pi p}\left(\mu^+ - \mu^-\right)$$

so that
$$\frac{d\mu}{dx} = \frac{2qB}{\pi p}\,\delta\mu$$

in agreement with Eq.(13.15) and hence with Eq.(13.9).

13.3.1. Edge states:

Some of the most illuminating use of the Büttiker approach is in the quantum Hall regime where the B-fields are so high that no electron from the source ever gets to probe 2*, and no electron from the drain gets to 1*. This makes

$$\frac{V_H}{(\mu_1 - \mu_2)/q} = \frac{t_1 - t_2'}{t_1 + t_2'} = 1$$

since $t_2 = t_2' = 0$ so that the Hall voltage becomes equal to the applied voltage making the Hall resistance equal to the ordinary two-terminal resistance.

Interestingly, in this regime this resistance is given by

$$R = \frac{h}{q^2 i} \tag{13.18}$$

where i is an integer to a fantastic degree of precision, making this a resistance standard used by the National Bureau of Standards. It is as if we have an unbelievably perfect ballistic conductor whose only resistance is the interface resistance. Since these conductors are often hundreds of micrometers long, this perfect ballisticity is amazing and was recognized with a Nobel prize in 1985 (von Klitzing K. et al. 1980)

In these lectures we have seen the conductance quantum q^2/h appear repeatedly and it is very common in the context of nanoelectronics and mesoscopic physics. But the quantum Hall effect was probably the first experimental observation where it played a clear identifiable role.

The simplest picture that helps understand it is the so-called "skipping orbits" (Fig.13.7) that result in a "divided" electronic highway with drainbound electrons so well-separated from the sourcebound electrons that backscattering is extremely unlikely, resulting in an incredibly ballistic conductor.

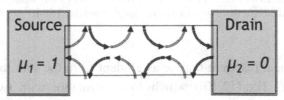

Fig.13.7. Skipping orbits in high B-fields leads to a "divided highway" with drainbound electrons on one side and sourcebound electrons on the other.

This simple picture, however, is a little too simple. It does not for example tell us the significance of the integer i in Eq.(13.18) which requires some input from quantum mechanics, as we will see in Lecture 21.

13.4. Non-Reciprocal Circuits

This may be a good place to raise an interesting property of conductors with non-reciprocal transmission of the type expected from edge states. Consider the structure shown in Fig.13.8 with a B-field that makes an electron coming in from contact 2 veer towards contact 1, but makes an electron coming from contact 1 veer away from contact 2.

Is $G_{1,2} \neq G_{2,1}$?

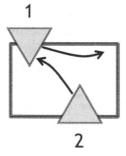 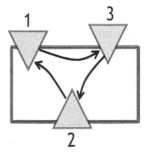

Fig.13.8. A magnetic field makes an electron coming in from contact 2 veer towards contact 1, but makes an electron coming from contact 1 veer away from contact 2. Is $G_{1,2} \neq G_{2,1}$? Yes, if there are more than two terminals, but not in a two-terminal circuit.

Not in the linear response regime as evident from the sum rule discussed in Lecture 12 (Eq.(12.10)) which for a structure with two terminals requires that

$$G_{1,1} = G_{1,2} = G_{2,1}.$$

However, there is no such requirement for a structure with more than two terminals. For example with three terminals, Eq.(12.10) tells us that

$$G_{1,1} \;=\; G_{1,2} + G_{1,3} \;=\; G_{2,1} + G_{3,1}$$

which does not require $G_{1,2}$ to equal $G_{2,1}$. The effects of such non-reciprocal transmission have been observed clearly with "edge states" in the quantum Hall regime.

This idea of "edge states" providing unidirectional ballistic channels over macroscopic distances is a very remarkable effect, but it has so far been restricted to low temperatures and high B-fields making it not too relevant from an applied point of view. That may change with the advent of new materials like "topological insulators" which show edge states even without B-fields.

But can we have non-reciprocal transmission without magnetic fields? In general the conductance matrix (which is proportional to the transmission matrix) obeys the Onsager reciprocity relation

$$G_{n,m}(+B) \;=\; G_{m,n}(-B) \tag{13.19}$$

requiring the current at n due to a voltage at m to equal the current at m due to a voltage at n with any magnetic field reversed. This is one of the deep principles of statistical mechanics which is usually proved for large conductors starting from the Kubo formula (Lecture 15).

Doesn't this Onsager relation require the conductance to be reciprocal

$$G_{m,n} \;=\; G_{n,m}$$

when $B=0$? The answer is yes if the structure does not include magnetic materials. Otherwise we need to reverse not just the external magnetic field but the internal magnetization too.

$$G_{n,m}(+B,+M) \;=\; G_{m,n}(-B,-M) \tag{13.20}$$

For example if one contact is magnetic, Onsager relations would require the $G_{1,2}$ in structure (a) to equal $G_{2,1}$ in structure (b) with the contact magnetization reversed as sketched above. ***But that does not mean $G_{1,2}$ equals $G_{2,1}$ in the same structure,*** (a) or (b).

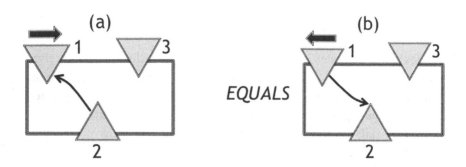

And so based on our current understanding a "topological insulator" which is a non-magnetic material could not show non-reciprocal conductances at zero magnetic field with ordinary contacts, but might do so if magnetic contacts were used. But this is an evolving story whose ending is not yet clear.

What has become very clear in the last twenty years is the operation of magnetic contacts, and that is what we will discuss next.

Lecture 14

Spin valve

14.1. Mode Mismatch and Interface Resistance
14.2. Spin Potentials
14.3. Spin-Torque
14.4. Polarizers and Analyzers

One of the major developments in the last two decades is the spin valve, a device with two magnetic contacts (Fig.14.1) If they are magnetized in the same direction (parallel configuration, P) the resulting resistance is lower than if they are magnetized in opposite directions (anti-parallel configuration, AP). Since its first demonstration in 1988, it rapidly found application as a "reading" device to sense the information stored in a magnetic memory and the discovery was recognized with a Nobel prize in 2007.

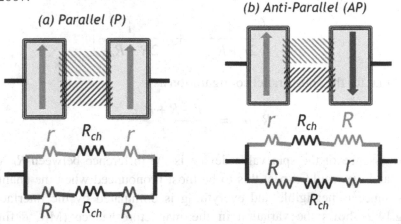

Fig.14.1. Spin valve: (a) Parallel (P) configuration. (b) Anti-Parallel (AP) configuration.

So far we have only mentioned spin as part of a "degeneracy factor, g" (Section 5.4), the idea being that electronic states always come in pairs, one corresponding to each spin. We could call these "up" and "down" or "left" and "right" or even "red" and "blue" as we have done in Fig.14.1. Note that the two spins are not spatially separated even though we have separated the red and the blue channel for clarity. Ordinarily the two channels are identical and we can calculate the conductance due to one and remember to multiply by two.

But in spin valve devices the contacts are magnets that treat the two spin channels differently and the operation of a spin valve can be understood in fairly simple terms if we postulate that each spin channel has a different interface resistance with the magnet depending on whether it is parallel (majority spin) or anti-parallel (minority spin) to the magnetization.

If we assume the interface resistance for majority spins to be r and for minority spins to be R $(r < R)$ we can draw simple circuit representations for the P and AP configurations as shown, with R_{ch} representing the channel resistance. Elementary circuit theory then gives us the resistance for the parallel configuration as

$$R_P = \left(\frac{1}{2r + R_{ch}} + \frac{1}{2R + R_{ch}} \right)^{-1}$$

and that for the anti-parallel configuration as

$$R_{AP} = \frac{r + R + R_{ch}}{2}$$

The essence of the spin valve device is the difference between R_P and R_{AP} and we would expect this to be most pronounced when the channel resistance is negligible and everything is dominated by the interfaces. Fig.14.2 shows the variation in the magnetoresistance (MR, defined below) as a function of the channel resistance (per spin) R_{ch} normalized to $r+R$ assuming $P=0.5$. Note how the MR dies out quickly once the normalized R_{ch} increases beyond say ~5.

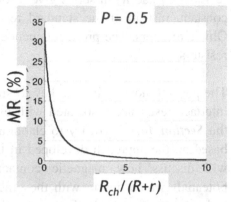

Fig.14.2. Variation in magneto-resistance (MR) as a function of the normalized channel resistance.

If we set $R_{ch} = 0$, we obtain a simple result for the maximum MR

$$MR \equiv \frac{R_{AP}}{R_P} - 1 \;=\; \frac{(R-r)^2}{4rR}, \quad \text{if } R_{ch} = 0 \qquad (14.1)$$

which can be written in terms of the polarization:

$$P \;\equiv\; \frac{R-r}{R+r} \qquad (14.2)$$

$$MR \;=\; \frac{P^2}{1-P^2}, \quad \text{if } R_{ch} = 0 \qquad (14.3)$$

I should mention here that the expression commonly seen in the literature has an extra factor of 2

$$MR \;=\; \frac{2P^2}{1-P^2} \qquad (14.4)$$

which is applicable to magnetic tunnel junctions (MTJ's) that use short tunnel junctions as channels instead of the metallic channels we have been discussing. We get this extra factor of 2, if we assume that two resistors R_1 and R_2 in series give a total resistance of KR_1R_2, K being a constant, instead of the standard result $R_1 + R_2$ expected of ordinary Ohmic resistors. The product dependence captures the physics of tunnel resistors.

The point to note is that the key to spin valve operation is the different interface resistance associated with each spin for magnetic contacts. In the *Section 14.1* I will try to elaborate on the origin of this difference based on the approach developed in this book. Next (*Section 14.2*) we will discuss how magnetic contacts lead to nonequilibrium spin potentials which ties in with the other important theme we have been trying to stress, namely that nanoscale device design of the future will need to emphasize the control of electrochemical potentials through creative design of contacts (Lecture 9).

While spin valves showed us how to use magnets to inject spins and control spin potentials, later researchers have shown how to use non-equilibrium spins to turn nanoscale magnets thus integrating spintronics and magnetics into a single and very active area of research with exciting possibilities that are yet to be explored.

In *Section 14.3* we will try to give the reader a feeling for this intriguing interplay of spins and magnets. Finally in *Section 14.4* we will try to illustrate the interesting dichotomy presented by spins, where some aspects can be understood in semiclassical terms, while others require a quantum viewpoint which we will take up in part three (Lecture 22).

14.1. Mode Mismatch and Interface Resistance

The original spin valve devices used metallic channels like copper and have developed rapidly since the original experiments in 1988. For many practical applications they have now been largely replaced by MTJ's that use insulating oxides as channels, due to the much larger values of MR that have been achieved.

By contrast all efforts to use semiconductors as the channel material proved singularly unsuccessful till around the year 2000, when the cause for poor MR was identified as the high R_{ch} compared to $R+r$ and this led to the idea of deliberately increasing the interface resistances by inserting barrier layers as sketched in Fig.14.3 (See for example, review by Schmidt 2005).

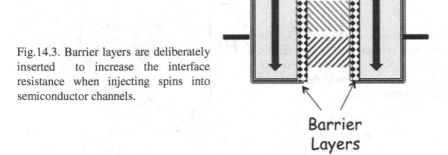

Fig.14.3. Barrier layers are deliberately inserted to increase the interface resistance when injecting spins into semiconductor channels.

Barrier Layers

Inserting a barrier layer is now a standard technique used by all experimentalists trying to inject spins into a semiconducting channel and so the "facts" seem quite clear. But why does it work?

The common explanation is that the barriers increase the interface resistances R, r thereby reducing the ratio $R_{ch}/(R+r)$ and increasing MR in accordance with Fig.14.2. However, it seems to us that if this were the whole story we should be able to increase the MR by reducing R_{ch} using a channel that is short enough to be ballistic (Length, $L \ll$ mean free path, λ). But experimentally it seems clear that making the channel short does not help.

Fig.14.4. shows a sketch of the number of channels $M(E)$ (or the density of states, $D(E)$) for a magnetic contact and a non-magnetic channel plotted separately for upspin and downspin electrons to the right and to the left respectively.

In the non-magnetic channel, the two are identical, but in the magnetic contact the minority spin band is typically shifted up in energy making the number of modes at E=μ smaller for the minority spin M_{dn} than for the majority spin M_{up}. What are the interface resistances?

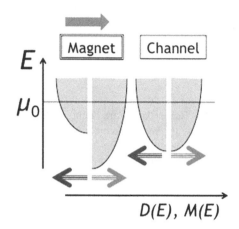

Fig.14.4. Sketch of D(E) or M(E) for magnetic contact and non-magnetic channel plotted separately for upspin and downspin electrons to the right and to the left respectively.

We will show shortly (Section 14.1.1) that the interface resistance at an interface between two materials with different numbers of modes ($M_1 > M_2$) is given by

$$R_{int} = \frac{h}{2q^2}\left(\frac{1}{M_2} - \frac{1}{M_1}\right)$$ (14.5)

If M_1 is much much greater than M_2, then

$$R_{int} \approx \frac{h}{2q^2 M_2}$$

which is the result we have discussed earlier (Fig.6.2) corresponding to **good contacts** ($M_1 >> M_2$).

Typically for a metallic channel the number of modes in the channel has a value intermediate between the two. Ideally

$$M_{up} >> M >> M_{dn}$$

so that the magnet provides a good contact for the majority spin but not for the minority spin:

$$r = \frac{h}{2q^2}\left(\frac{1}{M} - \frac{1}{M_{up}}\right) \approx \frac{h}{2q^2 M}$$

$$R = \frac{h}{2q^2}\left(\frac{1}{M_{dn}} - \frac{1}{M}\right) \approx \frac{h}{2q^2 M_{dn}}$$

Metallic channel
$M_{up} > M,\ M_{dn} < M$

Semiconducting channel
$M_{up},\ M_{dn} \gg M$

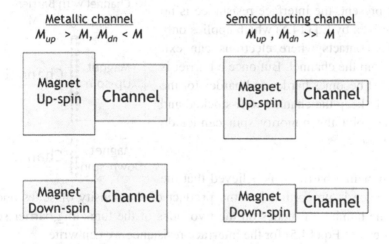

With semiconducting channels, on the other hand the number of modes in the channel is typically the smallest

$$M_{up} > M_{dn} \gg M$$

so that

$$r = R = \frac{h}{2q^2 M}$$

and the polarization P (see Eq.(14.2)) is zero.

In other words, it seems to us that the problem with spin injection into semiconducting channels is not just related to a high R_{ch} which could be eliminated with short ballistic channels, but is also related to the loss of distinction in the interface resistances R and r for the two spins. With 10 modes in a channel, it really does not matter whether the contact has 100 modes (minority spin) or 1000 modes (majority spin). Both are equally effective in keeping the channel well stocked with electrons.

Why does it help to insert a barrier? With a barrier present, the interface resistance is no longer given by Eq.(14.5) which applies only to good contacts where electrons can exit easily from the channel. But once a barrier is inserted it becomes harder and harder for the contact to keep the channel well-stocked and at some point the minority spin cannot do that any more.

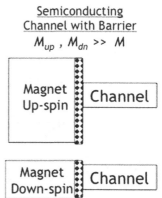

Semiconducting
Channel with Barrier
M_{up} , M_{dn} >> M

For a tunneling barrier it is believed that the conductance is proportional to the product of the density of states and hence the number of modes on the two sides of the tunneling barrier so that instead of Eq.(14.5) for the interface resistance we can write

$$\frac{1}{R_{int}} = K M_1 M_2 \qquad (14.6)$$

K being a constant. This seems reasonable if an electron from any mode on the left can transition into any mode on the right, but the exact conditions under which Eq.(14.5) changes to Eq.(14.6) need further discussion which we will not get into. We simply note that if we accept Eq.(14.6) we obtain

$$\frac{1}{r} = K M_{up} M , \quad \frac{1}{R} = K M_{dn} M$$

so that the polarization P (Eq.(14.2)) can now be sizeable irrespective of the number of modes M in the channel:

$$P = \frac{M_{up} - M_{dn}}{M_{up} + M_{dn}}$$

Of course the interface resistance values are larger than those for Ohmic interfaces described by Eq.(14.5).

14.1.1. Interface Resistance Due to Mode Mismatch

Let me briefly explain where Eq.(14.5) for the interface resistance comes from. Consider an interface between two channels with different mode numbers $M_1 > M_2$ with large contacts (effectively infinite number of modes) at both ends as shown in Fig.14.5.

Fig.14.5. Interface between two channels with mode numbers M_1 and M_2 with large contacts (infinite number of modes) at either end.

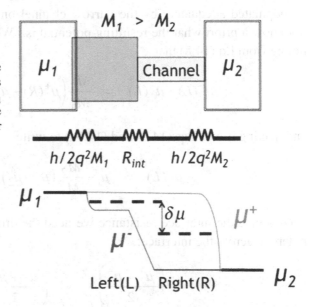

Consider the electrochemical potentials μ^+, μ^- for right-moving and left-moving electrons respectively. As we discussed in Lecture 6, the contacts impose the boundary conditions (L: Left, R: Right)

$$\mu^+(L) \;=\; \mu_1 \text{ and } \mu^-(R) \;=\; \mu_2 \qquad (14.7)$$

The current I is the same on the left and on the right and is given by

$$I \;=\; \frac{q}{h} M_1\!\left(\mu^+ - \mu^-\right)_L \;=\; \frac{q}{h} M_2\!\left(\mu^+ - \mu^-\right)_R \qquad (14.8)$$

The electrons flow freely across the interface, so that the right-moving lanes in the narrow channel on the right remain in equilibrium with the right-moving electrons on the left:

$$\mu^+(R) \;=\; \mu_1 \qquad (14.9a)$$

The left-moving lanes in the wide channel on the left, however, cannot be populated adequately by the narrow channel on the right and we do not know a priori what the resulting potential is. We can determine it by noting from Eq.(14.8) that

$$\mu^+(L) - \mu^-(L) \;=\; \frac{M_2}{M_1}\!\left(\mu^+(R) - \mu^-(R)\right)$$

and making use of Eqs.(14.7) and (14.9a) to write

$$\mu^-(L) \;=\; \mu_1 - \frac{M_2}{M_1}\!\left(\mu_1 - \mu_2\right) \qquad (14.9b)$$

To evaluate the interface resistance we need the drop $\delta\mu$ in the average potential across the interface:

$$\delta\mu \;=\; \left(\frac{\mu^+ + \mu^-}{2}\right)_L \;-\; \left(\frac{\mu^+ + \mu^-}{2}\right)_R$$

Making use of Eqs.(14.7) and (14.9)

$$\delta\mu = \frac{1}{2}\left(1-\frac{M_2}{M_1}\right)(\mu_1-\mu_2)$$

$$I = \frac{q}{h}M_2(\mu_1-\mu_2)$$

so that we obtain the result stated earlier in Eq.(14.5):

$$R_{int} \equiv \frac{\delta\mu/q}{I} = \frac{h}{2q^2}\left(\frac{1}{M_2}-\frac{1}{M_1}\right)$$

14.2. Spin Potentials

The difference in the interface resistance between a magnet and the up and downspin channels allows us to create "spin potentials" electrically inside a non-magnetic conductor, a phenomenon that is at the heart of the field of spintronics. This has been possible in metallic conductors like copper ever since its first demonstration in 1985, but has now become possible in semiconductors as well, once the idea of deliberately introducing a barrier layer was demonstrated.

The concept of "spin potentials" is easy to appreciate considering a simple structure having just one magnetic contact, as shown in Fig.14.6a. If no spin was involved we would expect the electrochemical potential to vary spatially as sketched in Fig.14.6b. We could obtain a quantitative plot by solving the diffusion equation (See Eqs.(6.1a,b))

$$\frac{dI}{dz} = 0,$$

$$I = -\frac{\sigma A}{q}\frac{d\mu}{dz} \qquad \text{(same as Eqs.(6.1a,b))}$$

subject to the appropriate boundary conditions on $\mu(z)$ at the contacts. Now because the interface resistance is different for the two spins we would expect different drops across the magnet-channel interface for them and so when we solve Eqs.(6.1a,b) for the upspins we will get a different profile from that for the downspins as sketched roughly in Fig.14.6c.

We expect the electrochemical potentials for the two spins to separate around the magnetic contact but they are eventually brought back down to a common level by spin-flip processes, that continually try to restore local equilibrium by equalizing μ_{up} and μ_{dn}.

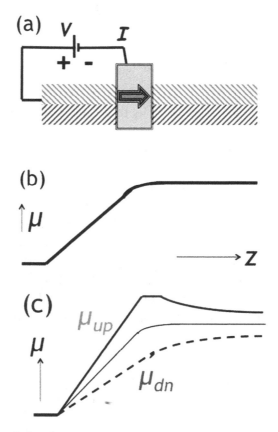

Fig.14.6. (a) Structure designed to cause separation of spin potentials in a channel using magnetic contacts. (b) Expected potential variation for non-magnetic contacts. (c) Magnetic contacts cause a separation of potentials for up and down spins.

Quantitative plots can be obtained by writing separate diffusion equations for up and down spins

$$I_{up} = -\frac{\sigma A}{2q}\frac{d\mu_{up}}{dz} \tag{14.10a}$$

$$I_{dn} = -\frac{\sigma A}{2q}\frac{d\mu_{dn}}{dz} \tag{14.10b}$$

We are using half the total conductivity σ for each of the up and down spin channels. It is the interface that discriminates between them, and this is reflected in a different interface resistance and hence a different boundary condition at the magnetic contact (Eqs.(6.4a,b)).

Noting that spin-flip processes convert upspin currents into downspin currents and viceversa, so that

$$\frac{d I_{up}}{dz} = -\frac{d I_{dn}}{dz} = -K\left(\mu_{up} - \mu_{dn}\right) \tag{14.11}$$

where K is a constant proportional to the strength of spin-flip processes that seek to equalize μ_{up} and μ_{dn}.

Combining Eqs.(14.11) with (14.10a,b) we obtain

$$\frac{d^2\mu_{up}}{dz^2} = \frac{\mu_{up} - \mu_{dn}}{2\lambda_{sf}^2} = -\frac{d^2\mu_{dn}}{dz^2} \tag{14.12}$$

where

$$\lambda_{sf} = \frac{1}{2}\sqrt{\frac{\sigma A}{qK}}$$

can be interpreted as a characteristic length that provides a measure of the distance over which spin-related information is preserved. It varies widely from tens of nanometers to hundreds of micrometers depending on the material and the temperature.

Eq.(14.12), known as the Valet-Fert equation, was originally obtained from the Boltzmann equation discussed in Lecture 7 and is widely used to analyze spin diffusion problems. We could use the upspin and downspin potentials to define charge and spin potentials

$$\mu \equiv (\mu_{up} + \mu_{dn})/2 \quad,$$

$$\mu_s \equiv \mu_{up} - \mu_{dn} \tag{14.13a}$$

Similarly the upspin and downspin currents in Eq.(14.10) can be used to define charge and spin currents:

$$I = I_{up} + I_{dn}$$

$$I_s = I_{up} - I_{dn} \tag{14.13b}$$

It is straightforward to show that the charge components obey the normal equations (Eqs.(6.1a,b)), while the spin component is affected by the spinflip length λ_{sf}.

$$\frac{d^2\mu_s}{dz^2} = \frac{\mu_s}{\lambda_{sf}^2} \tag{14.14}$$

Can we measure the spin voltage inside the channel? The answer is yes, not just along the current path, but also outside the path as shown in Fig.14.7. The latter is often referred to as a non-local spin voltage and is "routinely" measured in spin transport experiments.

The spin voltage is measured by looking at the change in the potential at the output probe when it is switched from parallel to anti-parallel relative to the injecting probe. We will show that this spin voltage is given by

$$V_s \equiv \frac{\mu_P - \mu_{AP}}{q} = P_1 P_2 \, I \, R_S \, e^{-L/\lambda_{sf}} \tag{14.15}$$

where P_1 and P_2 are the polarizations of the injecting and detecting contacts and

$$R_S = \lambda_{sf} / \sigma A \qquad (14.16)$$

These results in a more general form are discussed in Takahashi and Maekawa (2003). Here I just want to give the reader a feeling for the physics by going through the case when both contacts have high resistance due to tunneling barriers. Unfortunately even this simplest case requires a relatively extended discussion.

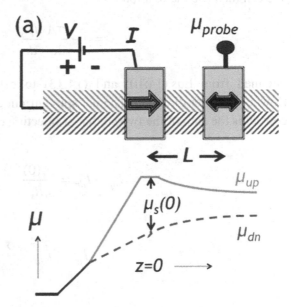

Fig.14.7. The spin voltage can be measured even outside the current path as shown.

14.2.1. Non-local spin voltage, Eq.(14.15)

Obtaining Eq.(14.15) involves two steps. Step 1 is to show that the spin potential at the injecting probe is given by

$$\mu_s(0) = P_1 q I R_s \qquad (14.17a)$$

Step 2 is to show that the difference between the output potentials for the parallel and anti-parallel configurations is given by

$$\mu_P - \mu_{AP} = P_2 \mu_s(0) e^{-L/\lambda_{sf}} \tag{14.17b}$$

Combining the two we obtain the stated result in Eq.(14.15).

Step 1 starts by noting that the spin potential obeys Eq.(14.14) which requires it to decay exponentially away from the injecting probe in either direction as sketched in Fig.14.8.

$$\mu_s = \mu_s(0) e^{-|z|/\lambda_{sf}} \tag{14.18}$$

We can then use the relation

$$I_s = -\frac{\sigma A}{2q}\frac{d\mu_s}{dz} \tag{14.19}$$

obtained from Eqs.(14.10) and (14.13) to calculate the spin current flowing in either direction. The net spin current drawn from the injecting contact is the sum of the two so that at injecting contact we should have

$$I_{up} - I_{dn} = \frac{\mu_s(0)}{qR_s} \tag{14.20}$$

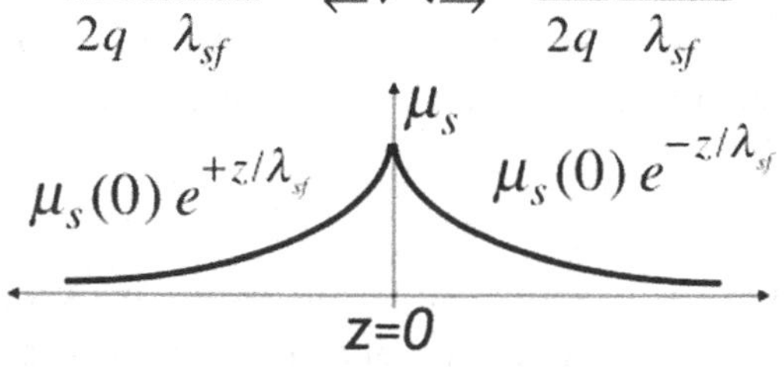

$$I_{up} - I_{dn} = \frac{\sigma A}{q}\frac{\mu_s(0)}{\lambda_{sf}} = \frac{\mu_s(0)}{qR_s}$$

Fig.14.8. Calculating the net spin current at the injecting probe in the structure of Fig.14.7.

$$\frac{\sigma A}{2q}\frac{\mu_s(0)}{\lambda_{sf}} \qquad \frac{\sigma A}{2q}\frac{\mu_s(0)}{\lambda_{sf}}$$

$$\mu_s(0) e^{+z/\lambda_{sf}} \qquad \mu_s \qquad \mu_s(0) e^{-z/\lambda_{sf}}$$

$$z=0$$

Consider now the flow of current at the injecting contact modeling it in terms of two interfacial conductances g_{up}, g_{dn} for up and down spins respectively.

Simple circuit theory yields

$$\frac{\mu_s(0)}{q} \equiv \frac{\mu_{up} - \mu_{dn}}{q} = \frac{I_{dn}}{g_{dn}} - \frac{I_{up}}{g_{up}}$$

which can be rewritten in the form

$$\frac{\mu_s(0)}{q} = \frac{g_{up} + g_{dn}}{2 g_{up} g_{dn}} \left(P_1 I - (I_{up} - I_{dn}) \right)$$

so that using Eq.(14.20)

$$\left(I_{up} - I_{dn} \right) R_s = \frac{g_{up} + g_{dn}}{2 g_{up} g_{dn}} \left(P_1 I - (I_{up} - I_{dn}) \right) \tag{14.21}$$

where P_1 is the polarization of the injecting probe defined as

$$P_1 \equiv \frac{g_{up} - g_{dn}}{g_{up} + g_{dn}} \tag{14.22}$$

Now the resistance R_s (Eq.(14.16)) represents the resistance of a section of the channel of length equal to the spin-flip length λ_{sf} which is usually much smaller than the interface resistances $1/g_{up}$ or $1/g_{dn}$ which are relatively large due to the use of barriers to enhance the polarization. Under these conditions we can set the right hand side to zero to obtain

$$I_{up} - I_{dn} = P_1 I$$

which combined with Eq.(14.20) yields our desired result in Eq.(14.17a).

To obtain Eq.(14.17b) (Step 2), we start by noting that the potential at the detecting probe is given by

$$\mu_s(L) \;=\; \mu_s(0)\,e^{-L/\lambda_{sf}} \qquad\qquad (14.22)$$

To find the potential registered by the detecting probe we adopt a circuit model similar to that for the injecting probe. Note also the similarity with the model used to obtain Eq.(12.4), but with the role positive and negative going electrons replaced by up and down spins.

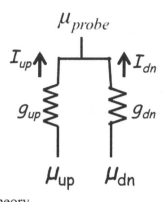

Setting the current equal to zero as we did earlier in Lecture 12 in a different context (see Eq.(12.4)), we have from simple circuit theory

$$I = 0 = g_{up}(\mu_{up} - \mu_{probe}) + g_{dn}(\mu_{dn} - \mu_{probe})$$

so that

$$\mu_{probe} = \frac{g_{up}\mu_{up} + g_{dn}\mu_{dn}}{g_{up} + g_{dn}}$$

We can use Eq.(14.13a) to write μ_{up} and μ_{dn} in terms of μ and μ_s

$$\mu_{up} = \mu + \frac{\mu_s}{2}$$

$$\mu_{dn} = \mu - \frac{\mu_s}{2}$$

and use these relations to write

$$\mu_{probe} = \mu + \frac{P_2\mu_s}{2}$$

where P_2 is the polarization of the detecting probe defined as before (Eq.(14.2)) in terms of the interface conductances.

With the probe in anti-parallel configuration, the potential is given by the same expression but with g_{up} and g_{dn} interchanged, that is with P_2 replaced by $-P_2$:

$$\mu_{AP} = \mu - \frac{P_2\mu_s}{2}$$

Hence
$$\mu_P - \mu_{AP} = P_2\mu_s(L) \tag{14.23}$$

Combining Eqs.(14.23) with (14.22) we obtain the result stated in Eq.(14.15).

14.3. Spin-Torque

We have seen how the spin-specific interface resistances associated with magnetic contacts have led to the creation of spin potentials that have been measured experimentally and give rise to large magnetic field dependent resistances that are used routinely to read information stored in magnets.

Another important development has been the demonstration of "spin-torque" which allows spin currents to turn magnets provided they are not more than a few atomic layers in thickness. The basic experimental fact is summarized in Fig.14.9 showing a spin valve structure with a fixed magnet on the left pointing down and a nanomagnet on the right free to point up or down. A negative voltage on the fixed magnet creates a large down spin potential

$$\mu_s \equiv \mu_{up} - \mu_{dn} < 0$$

that exerts a "spin-torque" on the nanomagnet which if it is sufficiently large can make it turn from up to down. Next if we reverse the polarity of the voltage, the positive voltage on the fixed magnet pulls out down spins from the channel reversing the spin potential

$$\mu_s \equiv \mu_{up} - \mu_{dn} > 0$$

Again if this potential is large enough it can turn the nanomagnet back to the up configuration. This effect is now experimentally well-established and it seems likely that it will soon be used to **write** information into nanomagnets, just as magnetoresistance phenomena are used to **read** information from them.

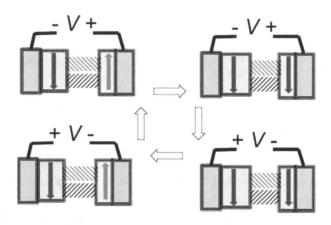

Fig.14.9. Spin valve structure with a fixed magnet on the left pointing down, and a nanomagnet on the right free to point up or down.

These two demonstrations, namely that magnets can create excess spins and excess spins can turn magnets, have combined spintronics and nanomagnetics into a single field that will require its practitioners to acquire an understanding of both spin transport and nanomagnet dynamics (Fig.14.10).

Fig.14.10. Theoretical models need to combine spin transport with nanomagnet dynamics.

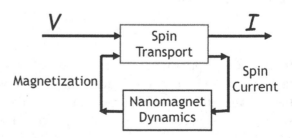

These Lectures are largely about transport, which in this context represents the top half, namely spin transport. We have discussed the spin diffusion equation in the last Section and will discuss the full quantum version later in Lecture 22. These spin transport models let us calculate the spin current given the magnetization.

To complete the story we need a model for the nanomagnet dynamics which will tell us the magnetization of a magnet given the spin current that is incident on it. This model is based on the Landau-Lifshitz-Gilbert (LLG) equation and in the rest of this Section let me try to say a few brief words about it.

The first point to note is that roughly speaking (we will discuss some subtleties in Part III) every electron is like an elementary magnet with a magnetic moment given by the Bohr magneton

$$\mu_B \equiv q\hbar/2m = 9.27e-24\ A-m^2 \qquad (14.24)$$

roughly what we would get if a current of 10 µA were circulating in a square loop with dimensions 1 nm x 1 nm, or say a current of 1 mA in a 0.1 nm x 0.1 nm loop.

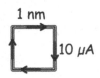

This was established back in the 1920's by the celebrated experiment due to Stern and Gerlach. More correctly the electron magnetic moment is given by

$$\mu_{el} = \frac{g_s}{2}\mu_B$$

g_s being the "g-factor" which is approximately equal to 2 for electrons in vacuum but could be significantly different in solids, just as the effective mass of electrons in solids can differ from that in vacuum. We will not worry about this "detail" and assume $g_s = 2$ for the following discussion.

If each electron is like a magnet then why are all materials not magnetic? Because usually the electrons are all paired with every up magnet balanced by a corresponding down magnet. It is only in magnetic materials like iron that internal interactions make a large number of electrons line up in the same direction giving rise to a macroscopic magnetization whose **magnitude** is given by

$$M_s = \mu_B \frac{N_s}{\Omega} \qquad (14.25)$$

N_s being the number of spins in a volume Ω.

The magnitude of the magnetization of a magnet is fixed but its **direction** denoted by the unit vector $\hat{m}$ can change when a magnetic field $\vec{H}$ are applied. The dynamics of $\hat{m}$ is described by the LLG equation

$$(1+\alpha^2)\frac{d\hat{m}}{dt} = \underbrace{-\gamma\mu_0\left(\hat{m}\times\vec{H}\right)}_{Dynamics} - \underbrace{\alpha\gamma\mu_0\left(\hat{m}\times\hat{m}\times\vec{H}\right)}_{Dissipation} \quad (14.26)$$

where γ is the "gyromagnetic ratio" given by

$$\gamma \equiv \frac{q}{m} = \frac{2\mu_B}{\hbar}$$

and μ_0 is the permeability of vacuum which may not appear explicitly in much of the literature since it is common to use cgs units rather than the SI units we are using.

As indicated in Eq.(14.26) the first term on the right represents dynamical processes while the second term represents "frictional" processes, α being known as the damping coefficient, typically ~ 0.01.

To get some insight, let us see how we can use this equation to understand a basic experimental fact, namely that a magnet has an "easy axis" (assumed to be along z). An external magnetic field H_{ext} can be used to change its magnetization between -1 and + 1 if it exceeds a critical field H_K, as sketched in Fig.14.11. With the magnetic field pointing along the z-direction

$$\vec{H} = H\,\hat{z}$$

Eq.(14.26) has the form (dropping the term α^2 which is usually $<< 1$)

$$\frac{d\hat{m}}{dt} = -\gamma\mu_0 H\left(\hat{m}\times\hat{z}\right) - \alpha\gamma\mu_0 H\left(\hat{m}\times\hat{m}\times\hat{z}\right)$$

so that taking its dot product with the unit vector $\hat{z}$ we have

$$\frac{dm_z}{dt} = \left(1-m_z^2\right)\alpha\gamma\mu_0 H \qquad (14.27)$$

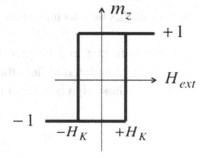

Fig.14.11. A magnet has an "easy axis" (assumed to be along z). An external magnetic field H_{ext} can be used to change its magnetization between -1 and + 1 if it exceeds a critical field H_K.

Clearly the two steady states predicted by this equation are

$$m_z = +1 \quad and \quad m_z = -1$$

since either choice makes dm$_z$/dt = 0. But are they stable? To answer this question let us assume a small deviation δ from +1

$$m_z = +1-\delta$$

so that Eq.(14.27) reduces to

$$-\frac{d}{dt}\delta \approx (2\alpha\gamma\mu_0 H)\delta$$

showing that such deviations will die out if H is positive. Similarly if we assume

$$m_z = -1 + \delta$$

so that Eq.(14.27) reduces to

$$\frac{d}{dt}\delta \approx (2\alpha\gamma\mu_0 H)\delta$$

showing that such deviations will die out if H is negative. In short,

$$m_z = +1 \quad is\ stable \quad if\ H > 0 \qquad (14.28a)$$

$$m_z = -1 \quad is\ stable \quad if\ H < 0 \qquad (14.28b)$$

How does this help us understand Fig.14.11?

First we note that in a magnet, the total field H consists of an external field H_{ext} and an internal field that each electron feels due to all the other electrons, whose sign is determined by m_z:

$$H = H_{ext} + \underbrace{H_K m_z}_{int ernal\ field} \qquad (14.29)$$

Taking this into account the stability conditions in Eq.(14.28) translate into

$$m_z = +1 \quad is\ stable \quad if\ H_{ext} > -H_K \qquad (14.30a)$$

$$m_z = -1 \quad is\ stable \quad if\ H_{ext} < H_K \qquad (14.30b)$$

which is exactly what Fig.14.11 indicates.

How do we describe switching with a spin current, $\vec{I}_s$? We have to add an extra term to the right hand side of Eq.(14.26)

$$\frac{d\hat{m}}{dt} \;=\; \left(\cdots From\ Eq.(14.26)\cdots\right) \;-\; \left(\hat{m}\times\hat{m}\times\frac{\vec{I}_s}{qN_s}\right) \quad (14.31)$$

proportional to the spin current per spin (N_s is the number of spins comprising the magnet). Why is the additional term

$$\hat{m}\times\hat{m}\times\frac{\vec{I}_s}{qN_s} \quad and\ not\ just \quad \frac{\vec{I}_s}{qN_s}?$$

What the operation $\hat{m}\times\hat{m}$ does to any vector $\vec{V}$ is to subtract out any component of $\vec{V}$ along $\hat{m}$ as described by the following identity:

$$-\hat{m}\times\hat{m}\times\vec{V} \;=\; \vec{V}-(\hat{m}.\vec{V})\hat{m}$$

Hence

$$-\hat{m}\times\hat{m}\times\frac{\vec{I}_s}{qN_s} \;=\; component\ of\ \frac{\vec{I}_s}{qN_s}\ perpendicular\ to\ \hat{m}$$

which is justified by noting that the magnitude of the magnetization of a magnet does not change, only its direction. To ensure this, the right hand side of the LLG equation must be perpendicular to the direction of $\hat{m}$. Actually there is an additional term due to the spin current

$$\alpha\hat{m}\times\frac{\vec{I}_s}{qN_s}$$

which is also perpendicular to $\hat{m}$ but we will ignore it since α is typically very small.

Starting from Eq.(14.31) we can project along the easy axis ($\hat{z}$) to obtain

$$\frac{dm_z}{dt} = \left(1 - m_z^2\right)\left(\alpha\gamma\mu_0 H_K m_z + \frac{I_s}{qN_s}\right) \quad\quad (14.32)$$

We can argue exactly as we argued with Eq.(14.27) that the critical spin current needed to switch the magnet is given by

$$\left(\frac{I_s}{qN_s}\right)_{critical} = \alpha\gamma\mu_0 H_K \quad\quad (14.33)$$

This relation has a simple physical interpretation, noting from our earlier discussion leading to Eq.(14.28) that the right-hand side of Eq.(14.33) is ~ inverse of the time constant τ for switching. Eq.(14.33) tells us that the critical current needed is such that the total number of spins delivered by the spin current, $I_s\tau/q$, is approximately equal to the number of spins N_s comprising the magnet.

The critical current itself is proportional to the product of the number of spins and the inverse time constant ~ N_s/τ. Making use of Eq.(14.25) we have

$$\left(I_s\right)_{critical} = \frac{4q\alpha}{\hbar}\underbrace{\left(\frac{1}{2}\mu_0 H_K M_s\Omega\right)}_{Energy\ Barrier} \quad\quad (14.33)$$

The quantity appearing in parenthesis in Eq.(14.33) represents the energy barrier separating the two states of the magnet and in order for the magnet to exist stably in one of these, the barrier needs to be at least several tens of *kT*. Otherwise the magnet will keep flipping back and forth between the two states many times in time scales of interest.

For an energy barrier ~ *40 kT*, and α =0.01, Eq.(14.33) predicts a critical spin current of ~ 10 μA. The critical current is expected to be somewhat larger and experimentally values ~ 50-100 μA have been demonstrated.

14.4. Polarizers and Analyzers

Let me end this long Lecture with a few words pointing out a subtle aspect of spin that we have glossed over so far. This aspect can be appreciated by considering the non-local spin voltage measurement that we discussed earlier (see Fig.14.7) except that the probe magnet is neither parallel nor anti-parallel to the injecting probe, but instead makes some arbitrary angle θ with it (Fig.14.12)?

Fig.14.12. Same as Fig.14.7, except that the output probe is at an arbitrary angle to the injecting probe. What voltage would it measure?

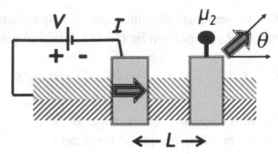

The answer can be stated quite simply:

$$\mu_2 = \mu + \frac{\vec{P}_2 \cdot \vec{\mu}_s}{2} \qquad (14.34)$$

where the polarization $\vec{P}_2$ points in the direction of the output magnet, while the spin potential $\vec{\mu}_s$ points in the direction of the spin polarization in the channel, which we assume to be in the direction of the injecting magnet. Earlier (see Eq.(14.23)) we considered two special cases when the angle between the two magnets was either zero (parallel, P) or 180 degrees (antiparallel, AP).

How do we understand the general result in Eq.(14.34)? For those unfamiliar with electron spin, the simplest analogy is

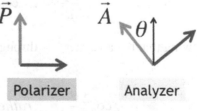

probably that of photon polarization. As we learn in freshman physics, a

polarizer-analyzer combination lets through a flux proportional to $\cos^2 \theta$
It is maximum when the two are parallel ($\theta = 0$ degrees), and a minimum
when the two are perpendicular ($\theta = 90$ degrees).

What about electrons? Suppose we have electrons that are all up, so that
from Eq.(14.13a)

$$\mu_s = \mu_{up} = 2\mu$$
$$(14.35a)$$

then as we rotate the direction of magnetization of the probe θ, the
measured voltage from Eq.(14.18) should change as

$$\frac{\mu_2}{\mu} = 1 + P_2 \cos\theta \qquad (14.35b)$$

As with photons, the voltage is a
maximum when the probe (analyzer)
is parallel to the electron polarization (
$\theta = 0$ degrees). But with electrons the
minimum occurs, not when the two
are perpendicular ($\theta = 90$ degrees) but
when the two are **antiparallel** ($\theta = 180$
degrees).

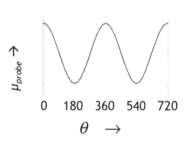

Indeed if we assume a perfect voltage probe having $P_2 = 1$, Eq.(14.35b)
reduces to

$$\frac{\mu_2}{\mu} = 1 + \cos\theta = 2\cos^2\frac{\theta}{2}$$

showing that the analyzer lets through a fraction of electrons proportional
to

$$\underbrace{\cos^2\frac{\theta}{2}}_{Electrons} \qquad rather \ than \qquad \underbrace{\cos^2\theta}_{Photons}$$

This basic difference between electrons and photons is apparent from the spin valve experiment that we started this Lecture with (see Fig.14.1). The current is a minimum, not when the two magnets are perpendicular, but when they are anti-parallel. Orthogonal directions are not represented by say z and x that are 90 degrees apart. Rather they are represented by up and down that are 180 degrees apart. And that is why a proper description of electron spin requires *spinors* rather than vectors as we will discuss later in Lecture 22.

One point that causes some confusion is the following. It seems that if we had electrons in the channel whose spin direction we did not know, we could measure it using a magnet. As we turn the magnet the measured voltage should go through maxima and minima as shown in Fig.4.10, and the direction corresponding to a maximum tells us the direction of the electron spin.

But doesn't quantum mechanics teach us that the spin of an electron cannot be exactly measured? Yes, but that is true if we had just one electron. Here we are talking of an "army" of electrons identically prepared by an injecting contact and what our magnet measures is the average over many many such electrons. This is not in violation of any basic principle.

Lecture 15

Kubo formula

15.1. Kubo Formula for an Elastic Resistor
15.2. Onsager Relations

In our discussion we have stressed the non-equilibrium nature of the problem of current flow requiring contacts with different electrochemical potentials (see Fig.2.4). Just as heat flow is driven by a difference in temperatures, current flow is driven by a difference in electrochemical potentials. Our basic current expression (see Eqs.(3.3), (3.4))

$$I \;=\; q \int\limits_{-\infty}^{+\infty} dE \, \frac{D(E)}{2t(E)} \, (f_1(E) - f_2(E))$$

(15.1)

is applicable to arbitrary voltages but so far we have focused largely on the low bias approximation (see Eq.(3.1))

$$G \;=\; q^2 \int\limits_{-\infty}^{+\infty} dE \left(-\frac{\partial f_0}{\partial E} \right) \frac{D(E)}{2t(E)}$$

(15.2)

Although we have obtained this result from the general non-equilibrium expression, it is interesting to note that the low bias conductance is really an *equilibrium property*. Indeed there is a fundamental theorem relating the low bias conductance for small voltages to the fluctuations in the current that occur at equilibrium when no voltage is applied. Let me explain.

Consider a conductor with no applied voltage (see Fig.15.1) so that both source and drain have the same electrochemical potential μ_0. There is of course no net current without an applied voltage, but even at equilibrium, every once in awhile, an electron crosses over from source to drain and on the average an equal number crosses over the other way from the drain to the source, so that

$$\left\langle I(t_0) \right\rangle_{eq} = 0$$

where the angular brackets $\langle \cdot \cdot \rangle$ denote either an "ensemble average" over many identical conductors or more straightforwardly a time average over the time t_0.

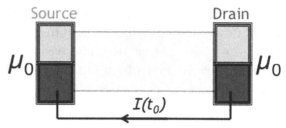

Fig.15.1.
At equilibrium both contacts have the same electrochemical potential μ_0. No net current flows, but there are equal currents I_0 from source to drain and back.

However, if we calculate the current correlation

$$C_I = \int\limits_{-\infty}^{+\infty} d\tau \ \left\langle I(t_0 + \tau) \, I(t_0) \right\rangle_{eq}$$

(15.3)

we get a non-zero value even at equilibrium, and *the Kubo formula* relates this quantity to the low bias conductance :

$$G = \frac{C_I}{2kT} = \frac{1}{2kT} \int\limits_{-\infty}^{+\infty} d\tau \ \left\langle I(t_0 + \tau) \, I(t_0) \right\rangle_{eq}$$

(15.4)

This is a very powerful result because it allows one to calculate the conductance by evaluating the current correlations using the methods of equilibrium statistical mechanics, which are in general more well-developed than the methods of non-equilibrium statistical mechanics. Indeed before the advent of mesoscopic physics in the late 1980's, the Kubo formula was the only approach used to model quantum transport. However, its use is limited to linear response. In these Lectures (Part three) we will stress the Non-Equilibrium Green's Function (NEGF) method for quantum transport, which allows us to address the non-equilibrium problem head on for quantum transport, just as the BTE discussed in Lecture 7 does for semiclassical transport.

In this Lecture, however, my purpose is primarily to connect our discussion to this very powerful and widely used approach. The Kubo formula in principle applies to large conductors with inelastic scattering, though in practice it may be difficult to evaluate the effect of complicated inelastic processes on the current correlation. The usual approach is to evaluate transport in long conductors with a high frequency alternating voltage, for which electrons can slosh back and forth without ever encountering the source or drain contacts. One could then obtain the zero frequency conductivity by letting the sample size L tend to infinity *before* letting the frequency tend to zero (see for example, Chapter 5 of Doniach and Sondheimer (1973)).

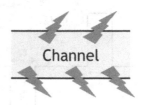

What we will do is something far simpler, namely look at the effect of contacts on the current correlations in an elastic resistor. We will show that applied to an elastic resistor the Kubo formula does lead to our old result (Eq.(15.2)) from Lecture 3. We will then discuss briefly how the Kubo formula leads to the Onsager relations mentioned in Lecture 13 (Eq.(13.19)).

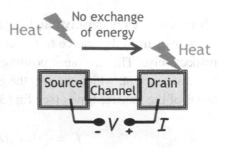

15.1. Kubo Formula for an Elastic Resistor

15.1.1. One-Level Resistor

In the spirit of the bottom-up approach, consider first the one-level resistor from Chapter 3 connected to two contacts with the same electrochemical potential μ_0 and hence the same Fermi function $f_0(E)$ (see Fig.15.2).

Fig.15.2.
At equilibrium with the same electrochemical potential in both contacts, there is no net current. But there are random pulses of current as electrons cross over in either direction.

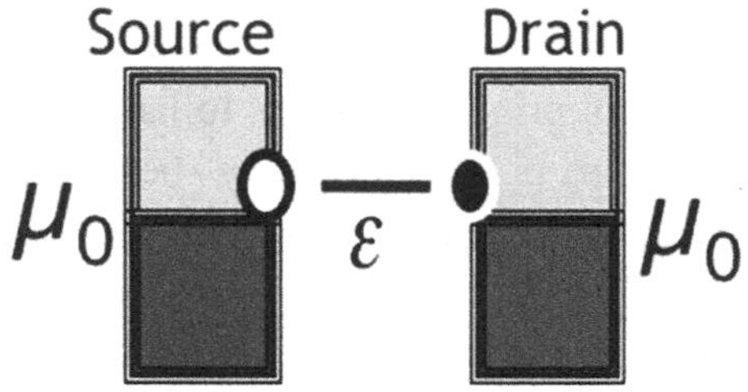

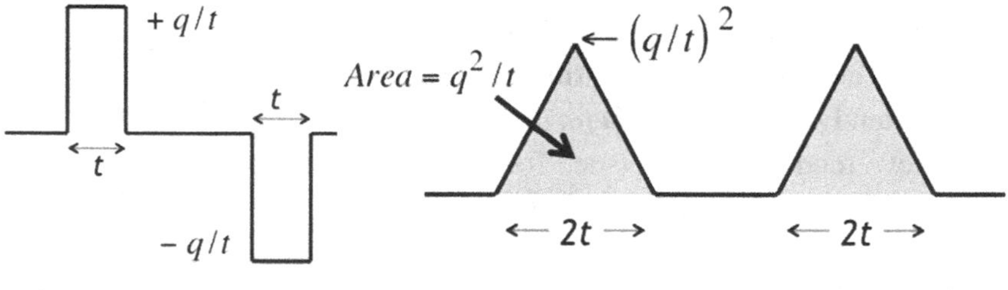

There are random positive and negative pulses of current as electrons cross over from the source to the drain and from the drain to the source respectively. The average positive current is equal to the average negative current, which we call the equilibrium current I_0 and write it in terms of the transfer time t (see Eq.(3.2))

$$I_0 = \frac{q}{t}\, f_0(\varepsilon)\left(1 - f_0(\varepsilon)\right)$$

(15.5a)

where the factor $f_0(\varepsilon)(1-f_0(\varepsilon))$ is the probability that an electron will be present at the source ready to transfer to the drain but no electron will be present at the drain ready to transfer back. The correlation is obtained by treating the transfer of each electron from the source to the drain as an independent stochastic process.

The integrand in Eq.(15.3) then looks like a sequence of triangular pulses as shown each having an area of q^2/t, so that

$$C_I = 2\frac{q^2}{t} f_0(\varepsilon)(1-f_0(\varepsilon))$$

$$(15.5b)$$

where the additional factor of 2 comes from the fact that I_0 only counts the positive pulses, while both positive and negative pulses contribute additively to C_I.

14.1.2 Elastic Resistor

We will now show that the Kubo formula (Eq.(15.4)) applied to an elastic resistor leads to the same conductance expression (Eq.(15.2)) that we obtained earlier. Generalizing our one-level results from Eqs.(15.5) to an elastic resistor with an arbitrary density of states, $D(E)$ as before we have

$$I_0 = q \int_{-\infty}^{+\infty} dE \, \frac{D(E)}{2t(E)} f_0(E)(1-f_0(E))$$

$$(15.6a)$$

$$C_I = 2q^2 \int_{-\infty}^{+\infty} dE \, \frac{D(E)}{2t(E)} f_0(E)(1-f_0(E))$$

$$(15.6b)$$

Note that $C_I = 2qI_0$. Making use of Eq.(15.4) we have for the conductance

$$G = \frac{C_I}{2kT} = q^2 \int\limits_{-\infty}^{+\infty} dE \, \frac{f_0(E)(1-f_0(E))}{kT} \, \frac{D(E)}{2t(E)}$$

$$(15.7)$$

which is the same as our expression in Eq.(15.2), noting that

$$\left(-\frac{\partial f_0}{\partial E}\right) = \frac{f_0(E)\,(1-f_0(E))}{kT}$$

$$(15.8)$$

In summary, the Kubo formula (Eq.(15.4)) applied to an elastic resistor leads to the result (Eq.(15.2)) we obtained in Lecture 3 from elementary arguments. Interestingly, the identity in Eq.(15.8) is key to this equivalence, since our elementary arguments lead to a conductance proportional to

$$\frac{f_1 - f_2}{\mu_1 - \mu_2} \cong -\frac{\partial f_0}{\partial E}$$

while the current correlations in the Kubo formula lead to

$$\frac{f_0\,(1-f_0)}{kT}$$

Note how the current correlation requires us to invoke the exclusion principle for the $1\text{-}f_0$ factor, but the elementary argument does not. For phonons (Lecture 11) the elementary arguments lead to (see Eq.(11.8) for the Bose function, n)

$$\frac{n_1 - n_2}{\hbar\omega} \cong -\frac{\partial n}{\partial(\hbar\omega)} = \frac{n(1+n)}{kT}$$

$$(15.9)$$

and agreement with the corresponding Kubo formula would require a $1\text{+}n$ factor instead of the $1\text{-}f$ factor for electrons. We will talk a little more about Fermi and Bose functions in Lecture 16, but the point here is that the theory of noise is more intricate than the theory for the average current that we will focus on in these Lectures. However, I should mention that there is at present an extensive body of work on subtle correlation effects in elastic resistors some of which have been experimentally observed (see for example, Büttiker 2009).

15.2. Onsager Relations

A very important application of the Kubo formula is as a starting point for a very fundamental result like the Onsager relations mentioned in Lecture 13 (Eq.(13.17)).

$$G_{n,m}(+B) \ = \ G_{m,n}(-B) \tag{15.10}$$

requiring the current at n due to a voltage at m to be equal to the current at m due to a voltage at n with any magnetic field reversed.

This is usually proved starting from the multiterminal version of the Kubo formula

$$G_{m,n} \ = \ \frac{1}{2kT} \int\limits_{-\infty}^{+\infty} d\tau \ \left\langle I_m(t_0+\tau) I_n(t_0) \right\rangle_{eq} \tag{15.11}$$

involving the correlation between the currents at two different terminals.

Consider a three terminal structure with a magnetic field $(B > 0)$ that makes electrons entering contact 1 bend towards 2, those entering 2 bend towards 3 and those entering 3 bend towards 1.

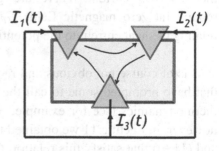

We would expect the correlation

$$\left\langle I_2(t_0+\tau) I_1(t_0) \right\rangle_{eq}$$

to look something like this sketch with the correlation extending further for positive τ.

This is because electrons go from 1
to 2, and so the current I_1 at time t_0
is strongly correlated to the current
I_2 at a later time ($\tau > 0$), but not to
the current at an earlier time.

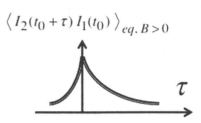

If we reverse the magnetic field ($B < 0$), it is argued that the trajectories
of electrons are reversed, so that

$$\left\langle I_1(t_0 + \tau)\, I_2(t_0) \right\rangle_{eq,\, B<0}$$
$$= \left\langle I_2(t_0 + \tau)\, I_1(t_0) \right\rangle_{eq,\, B>0} \qquad (15.12)$$

This is the key argument. If we accept this, the Onsager relation
(Eq.(15.10)) follows readily from the Kubo formula (Eq.(15.11)).

What we have discussed here is really the simplest of the Onsager
relations for the generalized transport coefficients relating generalized
forces to fluxes. For example, in Lecture 10 we discussed additional
coefficients like G_S (see Eq.(10.3)) relating a temperature difference to
the electrical current. There are generalized Onsager relations that
require (at zero magnetic field) $G_P = T\, G_S$, G_P being the coefficient
relating the heat current to the potential difference (see Eq.(10.12)).

This is of course not obvious and requires deep and profound arguments
that have prompted some to call the Onsager relations the fourth law of
thermodynamics (see for example, Yourgrau et al. 1966). Interestingly,
however, in Lecture 11 we obtained transport coefficients (see Eqs.(11.5)
and (11.6)) that satisfy this relation $G_P = T\, G_S$ straightforwardly without
any profound or subtle arguments. We could cite this as one more
example of the power and simplicity of the elastic resistor that comes
from disentangling mechanics from thermodynamics.

Lecture 16

Second law

16.1. Asymmetry of Absorption and Emission
16.2. Entropy
16.3. Law of Equilibrium
16.4. Fock Space States
16.5. Alternative Expression for Entropy

Back in Lecture 10, when discussing the heat current carried by electrons we drew a picture (Fig.10.8) showing the flow of electrons and heat in an elastic resistor consisting of a channel with two contacts (source and drain) with a voltage applied across it (Fig.16.1a). Fig.16.1b shows a slightly generalized version of the same picture that will be useful for the present discussion.

Fig.16.1b shows an elastic channel receiving N_1, N_2 electrons with contacts 1 and 2, held at potentials μ_1 and μ_2 respectively. Of course both N_1 and N_2 cannot be positive. If N_1 electrons enter the channel from one contact an equal number must leave from the other contact so that

$$N_1 + N_2 = 0 \qquad (16.1a)$$

For generality I have also shown an exchange of energy E_0 (but not electrons) with the surroundings at temperature T_0, possibly by the emission and absorption of phonons and/or photons. This exchange is absent in elastic resistors.

The principle of energy conservation requires that the total energy entering the channel is zero

$$E_1 + E_2 + E_0 = 0 \qquad (16.1b)$$

229

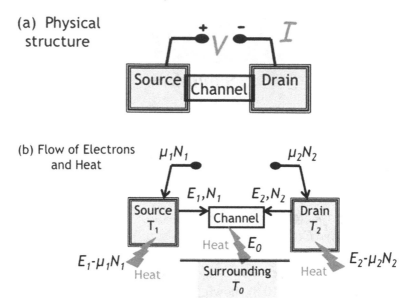

Fig.16.1. The flow of electrons and heat in the specific physical structure we have been discussing can be depicted in general terms as shown. For an elastic resistor, $E_0 = 0$.

This could be called an example of the **first law** of thermodynamics. However, there is yet another principle

$$\frac{E_1 - \mu_1 N_1}{T_1} + \frac{E_2 - \mu_2 N_2}{T_2} + \frac{E_0}{T_0} \leq 0 \qquad (16.2)$$

known as the **second law** of thermodynamics. Unlike the first law, the second law involves an inequality. While most people are comfortable with the first law or the principle of energy conservation, the second law still continues to excite debate and controversy.

And yet in some ways the second law embodies ideas that we know from experience. Suppose for example we assume all contacts to be at the same temperature ($T_2 = T_1 = T_0$). In this case Eq.(16.2) simply says that the total heat absorbed from the surroundings

$$(E_1 - \mu_1 N_1) \ + \ (E_2 - \mu_2 N_2) \ + \ E_0 \ \leq \ 0 \qquad (16.3a)$$

Making use of Eq.(16.1b), this implies

$$\mu_1 N_1 \ + \ \mu_2 N_2 \ \geq \ 0 \qquad (16.3b)$$

The total energy exchanged in the process $E_1 + E_2 + E_0$ has two parts: One that came from the thermal energy of the surroundings and the other that came from the battery. Eq. (16.3a) tells us that the former must be negative, and Eq.(16.3b) tells us that the latter must be positive. In other words, **we can take energy from a battery and dissipate it as heat, but we cannot take heat from the surroundings and charge up our battery.**

This should come as no surprise to anybody. After all if we could use heat from our surroundings to charge a battery (perhaps even run a car!) then there would be no energy problem. But the point to note is that this is not prohibited by the first law since energy would still be conserved. It is the second law that makes a distinction between the energy stored in a battery and the thermal energy in our surroundings. The first is easily converted into the second, but not the other way around because thermal energy is distributed among many degrees of freedom. We can take energy from one degree of freedom and distribute it among many degrees of freedom, but we cannot take energy from many degrees of freedom and concentrate it all in one.

This intuitive feeling is quantified and generalized by the second law (Eq.(16.2)) based on solid experimental evidence. For example if we have multiple "contacts" at different temperatures then it is possible to take heat from the hotter contact, dump a part of it in the colder contact, use the difference to charge up a battery and still be compliance with the second law.

Are all the things we have discussed so far in compliance with the second law? The answer is yes. For the elastic resistor $E_0 = 0$, and we can write the second law from Eq.(16.1b) in the form

$$\frac{\varepsilon - \mu_1}{T_1} N_1 \;+\; \frac{\varepsilon - \mu_2}{T_2} N_2 \;\le\; 0$$

where we have written $E_1 = \varepsilon N_1$ and $E_2 = \varepsilon N_2$, assuming that each electron entering and exiting the channel has an energy of ε. Making use of Eq.(16.1a) this means that

$$\left(\frac{\varepsilon - \mu_1}{T_1} \;-\; \frac{\varepsilon - \mu_2}{T_2} \right) N_1 \;\le\; 0$$

Our description of the elastic resistor always meets this condition, since the flow of electrons is determined by $f_1 - f_2$, as we saw in Lecture 3. N_1 is positive indicating electron flow from source to drain if

$$f_1(\varepsilon) \;>\; f_2(\varepsilon)$$

that is, if
$$\frac{1}{1 + e^{(\varepsilon - \mu_1)/kT}} \;>\; \frac{1}{1 + e^{(\varepsilon - \mu_2)/kT}}$$

$$\frac{\varepsilon - \mu_1}{T_1} \;<\; \frac{\varepsilon - \mu_2}{T_2}$$

Similarly we can show that N_1 is negative if

$$\frac{\varepsilon - \mu_1}{T_1} \;>\; \frac{\varepsilon - \mu_2}{T_2}$$

In either case we have

$$\left(\frac{\varepsilon - \mu_1}{T_1} \;-\; \frac{\varepsilon - \mu_2}{T_2} \right) N_1 \;\le\; 0$$

thus ensuring that the second law is satisfied.

But what if we wish to go beyond the elastic resistor and include energy exchange within the channel. What would we need to ensure that we are complying with the second law?

16.1. Asymmetry of Absorption and Emission

The answer is that our model needs to ensure that for all processes involving the exchange of electrons with a contact held at a potential μ and temperature T, the probability of *absorbing E,N* be related to the probability of *emitting E,N* by the relation

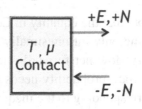

$$\frac{P(+E,+N)}{P(-E,-N)} = e^{-(E-\mu N)/kT} \qquad (16.4a)$$

If only energy is exchanged, but not electrons, then the relation is modified to

$$\frac{P(+E)}{P(-E)} = e^{-E/kT} \qquad (16.4b)$$

To see how this relation (Eq.(16.4)) ensures compliance with the second law (Eq.(16.2)), consider the process depicted in Fig.16.1 involving energy and/or electron exchange with three different "contacts". Such a process should have a likelihood proportional to

$$P(E_1, N_1)\, P(E_2, N_2)\, P(E_0)$$

while the likelihood of the reverse process will be proportional to

$$P(-E_1, -N_1)\, P(-E_2, -N_2)\, P(-E_0)$$

In order for the former to dominate their ratio must exceed one:

$$\frac{P(+E_1, +N_1)\, P(+E_2, +N_2)\, P(+E_0)}{P(-E_1, -N_1)\, P(-E_2, -N_2)\, P(-E_0)} \geq 1$$

If all processes obey the relations stated in Eqs.(16.4a,b), we have

$$\exp\left(-\frac{E_1 - \mu_1 N_1}{kT_1}\right) \exp\left(-\frac{E_2 - \mu_2 N_2}{kT_2}\right) \exp\left(-\frac{E_0}{kT_0}\right) \geq 1$$

which leads to the second law stated in Eq.(16.2), noting that *exp (- x)* is greater than one, only if *x* is less than zero.

Note that the equality in Eq.(16.2) corresponds to the forward probability being only infinitesimally larger than the reverse probability, implying a very slow net forward rate. To make the "reaction" progress faster, the forward probability needs to exceed the reverse probability significantly, corresponding to the inequality in Eq.(16.2).

So how do we make sure our model meets the requirement in Eq.(16.4)? Consider for example a conductor with one inelastic scatterer in the middle separating a region having an energy level at E_1 from another having a level at E_2. Electrons flow from contact 1 to 2 by a process of emission whose probability is given by

$$D_{2\leftarrow 1}\ f_1(E_1)(1-f_2(E_2))$$

while the flow from 2 to 1 requires an absorption process with probability

$$D_{1\leftarrow 2}\ f_2(E_2)(1-f_1(E_1))$$

Since one process involves emission while the other involves absorption, the rates should obey the requirement imposed by Eq.(16.4b):

$$\frac{D_{2\leftarrow 1}}{D_{1\leftarrow 2}}\ =\ e^{(E_1-E_2)/kT_0} \qquad (16.5)$$

as we had stated earlier in Lecture 9 in a different context (Eq.(9.7)). T_0 is the temperature of the surroundings with which electrons exchange energy.

The current in such an inelastic resistor would be given by an expression of the form (suppressing the arguments E_1, E_2 for clarity)

$$I \sim D_{2 \leftarrow 1} f_1(1 - f_2) - D_{1 \leftarrow 2} f_2(1 - f_1)$$

which reduces to the familiar form for elastic resistors

$$I \sim (f_1 - f_2)$$

only if $$D_{2 \leftarrow 1} = D_{1 \leftarrow 2}$$

corresponding to elastic scattering $E_2 = E_1$. Ordinary resistors have both elastic and inelastic scatterers intertwined and there is no simple expression relating the current to f_1, f_2.

The bottom line is that any model that includes energy exchange in the channel should make sure that absorption and emission rates are related by Eq.(16.5) if the surroundings are in equilibrium with a fixed temperature. Any transport theory, semiclassical or quantum needs to make sure it complies with this requirement to avoid violating the second law.

16.2. Entropy

The asymmetry of emission and absorption expressed by Eqs.(16.4) is actually quite familiar to everyone, indeed so familiar that we may not recognize it. We all know that if we take a hydrogen atom and place its lone electron in an excited (say 2p) state, it will promptly emit light and descend to the 1s state. But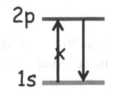
an electron placed in the 1s state will stay there forever. We justify it by saying that the electron "naturally" goes to its lowest energy state.

But there is really nothing natural about this. Any mechanical interaction (quantum or classical) that takes an electron from 2p to 1s will also take it from 1s to 2p. The natural descent of an electron to its lowest energy state is driven by a force that is not mechanical in nature. It is "entropic" in origin, as we will try to explain.

Basically it comes from a ***property of the surroundings*** expressed by Eq.(16.4) which tells us that it much harder to absorb anything from a reservoir, compared to emitting something into it. At zero temperature, a system can only emit and never absorb, and so an electron in state 2p can emit its way to the lowest energy state 1s, but an electron in state 1s can go nowhere.

This behavior is of course quite well-established and does not surprise anyone. But it embodies the key point that makes transport and especially quantum transport such a difficult subject in general. Any theoretical model has to include entropic processes in addition to the familiar mechanical forces.

So where does the preferential tendency to lose energy rather than gain energy from any "reservoir" come from? Eq.(16.4) can be understood by noting that when the electron loses energy the contact gains in energy so that the ratio of the rate of losing energy to the rate of gaining energy is equal to the ratio of the density of states at $E_0 + \varepsilon$ to that at E_0 (Fig.16.2):

$$\frac{P(-\varepsilon)}{P(+\varepsilon)} = \frac{W(E_0 + \varepsilon)}{W(E_0)}$$

Here $W(E)$ represents the number of states available at an energy range E in the contact which is related to its entropy by the Boltzmann relation

$$S = k \ln W \qquad (16.7)$$

so that $\qquad \dfrac{P(-\varepsilon)}{P(+\varepsilon)} = \exp\dfrac{S(E_0 + \varepsilon) - S(E_0)}{k} \qquad (16.8)$

Assuming that the energy exchanged ε is very small compared to that of the large contact, we can write

$$S(E_0 + \varepsilon) - S(E_0) \approx \varepsilon \left(\frac{dS}{dE}\right)_{E=E_0} = \frac{\varepsilon}{T}$$

with the temperature defined by the relation

$$\frac{1}{T} = \left(\frac{dS}{dE}\right)_{E=E_0}$$ (16.9)

This is of course a very profound result saying that regardless of the detailed construction of any particular reservoir, as long as it is in equilibrium, dS/dE can be identified as its temperature.

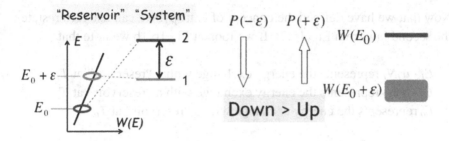

Fig.16.2. Electrons preferentially go down in energy because it means more energy for the "reservoir" with a higher density of states. It is as if the lower state has a far greater "weight" as indicated in the lower panel.

If we accept this, then Eq.(16.7) gives us the basic relation that governs the exchange of energy with any "reservoir" in equilibrium with a temperature T:

$$\frac{P(-\varepsilon)}{P(+\varepsilon)} = e^{\varepsilon/kT}$$

as we stated earlier (see Eq.(16.4b)).

If the emission of energy involves the emission of an electron which eventually leaves the contact with an energy μ, then ε should be replaced by $\varepsilon - \mu$, as indicated in Eq.(16.4a). The key idea is the same as what we introduced in Fig.10.8 when discussing thermoelectric effects, namely that when an electron is added to a reservoir with energy ε, an amount $\varepsilon - \mu$ is dissipated as heat, the remaining μ representing an increase in the energy of the contact due to the added electron. Indeed that is the definition of the electrochemical potential μ. Eventually the added electron leaves the contact as shown.

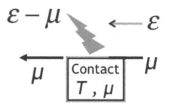

16.2.1. Total Entropy Always Increases

Now that we have defined the concept of entropy, we can use it to restate the second law from Eq.(16.2). If we look at Fig.16.1b we note that

$E_1 - \mu_1 N_1$ represents the energy exchange with a "reservoir" at T_1,
$E_2 - \mu_2 N_2$ represents the energy exchange with a "reservoir" at T_2,
E_0 represents the energy exchange with a "reservoir" at T_0.

Based on the definition of temperature in Eq.(16.9), we can write the corresponding changes in entropy $\Delta S_1, \Delta S_2, \Delta S_0$ as shown below

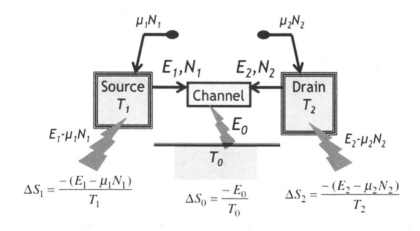

Note that these are exactly the same terms (except for the negative sign) appearing in Eq.(16.2), which can now be restated as

$$(\Delta S)_1 \; + \; (\Delta S)_2 \; + \; (\Delta S)_0 \; \geq \; 0 \qquad (16.10)$$

In other words, the second law requires the total change in entropy of all the reservoirs to be positive.

16.2.2. *Free energy always decreases*

At zero temperature, any system in coming to equilibrium with its surroundings, goes to its state having the lowest energy. This is because a reservoir at zero temperature will only allow the system to give up energy, but not to absorb any energy. Interestingly, at non –zero temperatures, one can define a quantity called the *free energy*

$$F \; = \; E - TS \qquad (16.11)$$

such that at equilibrium a system goes to its state with minimum free energy. At $T=0$, the free energy, F is the same as the total energy, E.

To see this, consider a system that can exchange energy with a reservoir such that the total energy is conserved.

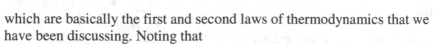

Using the subscript "R" for reservoir quantities we can write

$$\Delta E \; + \; \Delta E_R \; = \; 0 \qquad (16.12a)$$

$$\Delta S \; + \; \Delta S_R \; \geq \; 0 \qquad (16.12b)$$

which are basically the first and second laws of thermodynamics that we have been discussing. Noting that

$$\Delta S_R \; = \; \frac{\Delta E_R}{T}$$

we can combine Eqs.(16.12a,b) to write

$$\Delta F \equiv \; \Delta E \; - \; T\Delta S \; \leq \; 0 \qquad (16.13)$$

which tells us that all energy exchange processes permitted by the first and second laws will cause the free energy to decrease, so that the final equilibrium state will be one with minimum free energy.

16.3. Law of Equilibrium

The preferential tendency to lose energy rather than gain energy from any surrounding "reservoir" as expressed in Eq.(16.4) leads to a universal law stating that any system in equilibrium having states i with energy E_i and with N_i particles will occupy these states with probabilities

$$p_i = \frac{1}{Z} e^{-(E_i - \mu N_i)/kT} \qquad (16.14)$$

where Z is a constant chosen to ensure that all the probabilities add up to one.

To see this we note that all reservoirs in equilibrium have the property

$$\frac{P(+E,+N)}{P(-E,-N)} = e^{-(E-\mu N)/kT} \qquad \text{(same as 16.4a)}$$

Suppose we have a system with two states as shown exchanging energy and electrons with the surroundings. At equilibrium, we require upward transitions to balance downward transitions, so that

$$E = E_1 - E_2 ,$$
$$N = N_1 - N_2$$

"Fock space"

$$p_2 \, P(E,N) = p_1 \, P(-E,-N)$$

Making use of Eq.(16.4a), we have

$$\frac{p_1}{p_2} = \frac{P(+E,+N)}{P(-E,-N)} = e^{-((E_1-\mu N_1)-(E_2-\mu N_2))/kT}$$

It is straightforward to check that the probabilities given by Eq.(16.14) satisfy this requirement and hence represent an acceptable equilibrium solution.

How can we have a law of equilibrium so general that it can be applied to all systems irrespective of its details? Because as we noted earlier it comes from the *property of the surroundings* and not the system.

Eq.(16.14) represents the key principle or equilibrium statistical mechanics, Feynman (1965) called it the "summit". But it looks a little different from the two equilibrium distributions we introduced earlier, namely the Fermi function (Eq.(2.2)) and the Bose function (Eq.(11.8)).

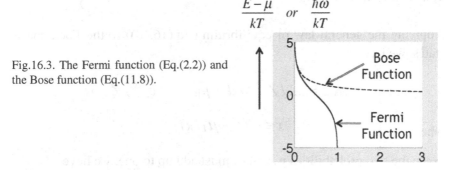

Fig.16.3. The Fermi function (Eq.(2.2)) and the Bose function (Eq.(11.8)).

$$\frac{E-\mu}{kT} \quad or \quad \frac{\hbar\omega}{kT}$$

Bose Function

Fermi Function

Fig.16.3 shows these two functions which look the same at high energies but deviate significantly at low energies. Electrons obey the exclusion principle and so the occupation $f(E)$ is restricted to values between 0 and 1. The Bose function is not limited between 0 and 1 since there is no exclusion principle.

Interestingly, however, both the Bose function and the Fermi function are special cases of the general law of equilibrium in Eq.(16.14). To see this, however, we need to introduce the concept of *Fock space* since the energy levels appearing in Eq.(16.14) do not represent the one-electron states we have been using throughout these Lectures. They represent the so-called Fock space states, a new concept that needs some discussion.

16.4. Fock space states

Consider a simple system with just one energy level, ε. In the one electron picture we think of electrons going in and out of this level. In the Fock space picture we think of the two possible states of the system, one corresponding to an empty state with energy $E=0$, and one corresponding to a filled state with energy $E = \varepsilon$ as shown.

When an electron comes in the system goes from the empty state *(0)* to the full state *(1)*, while if an electron leaves, the system goes from *1* to *0*.

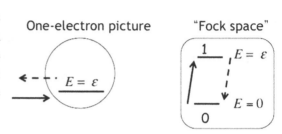

Applying the general law of equilibrium (Eq.(16.14)) to the Fock space states, we have

$$p_0 \;=\; 1/Z \quad and \quad p_1 \;=\; e^{-x}/Z$$

where
$$x \equiv (\varepsilon - \mu)/kT$$

Since the two probabilities p_0 and p_1 must add up to one, we have

$$Z = 1 + e^{-x}$$

$$p_0 \;=\; \frac{1}{e^{-x}+1} \;=\; 1 - f_0(\varepsilon)$$

$$p_1 \;=\; \frac{e^{-x}}{e^{-x}+1} \;=\; \frac{1}{e^{x}+1} \;=\; f_0(\varepsilon)$$

The probability of the system being in the full state, p_1 thus equals the Fermi function while the probability of the system being in the empty state, p_0 equals one minus the Fermi function, as we would expect.

16.4.1. Bose function

The Bose function too follows from Eq.(16.14), but we need to apply it to a system where the number of particles go from zero to infinity. Fock space states for electrons on the other hand are restricted to just zero or one because of the exclusion principle.

Eq.(16.14) then gives us the probability of the system being in the N-photon state as

$$p_N = \frac{e^{-Nx}}{Z}, \quad where \; x \equiv \frac{\hbar\omega}{kT}$$

To ensure that all probabilities add up to one, we have

$$Z = \sum_{N=0}^{\infty} e^{-Nx} = \frac{1}{1-e^{-x}}$$

so that the average number of photons is given by

$$n = \sum_{N=0}^{\infty} N\, p_N = \frac{1}{Z}\sum_{N=0}^{\infty} N e^{-Nx}$$

Noting that

$$\sum_{N=0}^{\infty} N e^{-Nx} = -\frac{d}{dx}\sum_{N=0}^{\infty} e^{-Nx} = -\frac{d}{dx}Z$$

we can show with a little algebra that

$$n = \frac{1}{e^{x}-1}$$

which is the Bose function stated earlier in Eq.(11.8).

The reason we have $E-\mu$ appearing in the Fermi function for electrons but not $\hbar\omega - \mu$ in the Bose function for photons or phonons is that the latter are not conserved. As we discussed in Section 16.2, when an electron enters the contact with energy E, it relaxes to an average energy of μ, and the energy dissipated is $E-\mu$. But when a photon or a phonon with energy $\hbar\omega$ is emitted or absorbed, the energy dissipated is just that.

(sidebar)

• • •

3 — $E = 3\hbar\omega$

2 — $E = 2\hbar\omega$

1 — $E = \hbar\omega$

0 — $E = 0$

"Fock space" for photons

However, there are conserved particles (not photons or phonons) that also obey Bose statistics, and the corresponding Bose function has $E-\mu$ and not just E.

16.4.2. Interacting electrons

The general law of equilibrium (Eq.(16.14)) not only gives us the Fermi and Bose functions but in principle can also describe the equilibrium state of complicated interacting systems, if we are able to calculate the appropriate Fock space energies. Suppose we have an interacting system with two one-electron levels corresponding to which we have four Fock space states as shown labeled 00, 01, 10 and 11. The 11 state with both levels occupied has an extra interaction energy U_0 as indicated.

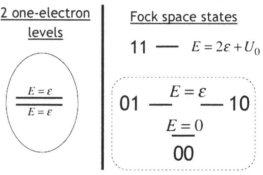

What is the average number of electrons if the system is in equilibrium with an electrochemical potential μ? Once again defining

$$x \equiv \frac{\varepsilon - \mu}{kT}, \quad \text{we have from Eq.(16.14)}$$

$$p_{00} = \frac{1}{Z}$$

$$p_{01} = p_{10} = \frac{e^{-x}}{Z}$$

$$p_{11} = \frac{e^{-2x}}{Z} e^{-U_0/kT}$$

The average number of electrons is given by

$$n = 0.p_{00} + 1.p_{01} + 1.p_{10} + 2.p_{11}$$

$$= \frac{2(e^{-x} + e^{-2x} e^{-U_0/kT})}{Z}$$

We could work out the details for arbitrary interaction energy U_0, but it is instructive to look at two limits. Firstly, the non-interacting limit with $U_0 \rightarrow 0$ for which

$$Z = 1 + 2e^{-x} + e^{-2x} = (1 + e^{-x})^2$$

so that with a little algebra we have

$$n = \frac{2}{1 + e^{(\varepsilon - \mu)/kT}}, \quad U_0 \rightarrow 0$$

equal to the Fermi function times two as we might expect since there are two non-interacting states.

The other limit is that of strongly interacting electrons for which $Z = 1 + 2e^{-x}$ so that

$$n = \frac{1}{1 + \frac{1}{2}e^{(\varepsilon - \mu)/kT}}, \quad U_0 \rightarrow \infty$$

a result that does not seem to follow in any simple way from the Fermi function. With g one-electron states present, it takes a little more work to show that the number is

$$n = \frac{1}{1 + \frac{1}{g}e^{(\varepsilon - \mu)/kT}}, \quad U_0 \rightarrow \infty$$

This result may be familiar to some readers in the context of counting electrons occupying localized states in a semiconductor.

Equilibrium statistical mechanics is a vast subject and we are of course barely scratching the surface. My purpose here is simply to give a reader a feel for the concept of Fock space states and how they relate to the one electron states we have generally been talking about.

This is important because the general law of equilibrium (Eq.(16.14)) and the closely related concept of entropy (Eq.(16.7)) are both formulated in terms of Fock space states. We have just seen how the law of equilibrium can be translated into one-electron terms for non-interacting systems. Next let us see how one does the same for entropy.

16.5.Alternative Expression for Entropy

Consider a system of independent localized spins, like magnetic impurities in the channel. At equilibrium, half the spins randomly point up and the other half point down. What is the associated entropy?

$$S = Nk \, \ell n \, 2$$

Eq.(16.7) defines the entropy S as $k \, \ell n W$, W being the total number of Fock space states accessible to the system. In the present problem we could argue that each spin has two possible states (up or down) so that a collection of N spins has a total of 2^N states:

$$W = 2^N \quad \rightarrow \quad S = k \, \ell n W = Nk \, \ell n 2 \qquad (16.15)$$

This is correct, but there is an alternative expression that can be used whenever we have a system composed of a large number of identical independent systems, like the N spin collection we are considering:

$$S = -Nk \sum_i \tilde{p}_i \, \ell n \, \tilde{p}_i \qquad (16.16)$$

where the $\tilde{p}_i$'s denote the probabilities of finding an individual system in its ith state. An individual spin, for example has a probability of *1/2* for being in either an up or a down state, so that from Eq.(16.16) we obtain

$$S = -Nk\left(\frac{1}{2}\ell n\frac{1}{2} + \frac{1}{2}\ell n\frac{1}{2}\right) = Nk\,\ell n\,2$$

exactly the same answer as before (Eq.(16.15)).

Eq.(16.16), however, is more versatile in the sense that we can use it easily even if the $\tilde{p}_i$'s happen to be say *1/4* and *3/4* rather than *1/2* for each. Besides it is remarkably similar to the expression for the Shannon entropy associated with the information content of a message composed of a string of *N* symbols each of which can take on different values *i* with probability $\tilde{p}_i$. In the next Lecture I will try to elaborate on this point further.

Let me end this Lecture simply by indicating how this new expression for entropy given in Eq.(16.16) is obtained from our old one that we used in Eq.(16.15). This is described in standard texts on statistical mechanics (see for example, Dill and Bromberg (2003)).

16.5.1. From Eq.(16.7) to Eq.(16.16)

Consider a very large number N of identical systems each with energy levels *{E$_i$}* occupied according to probabilities *{ $\tilde{p}_i$ }*, such that the number of these syatems in state *i* is given by

$$N_i = N\,\tilde{p}_i$$

$\{E_i, \tilde{p}_i\}$	$\{E_i, \tilde{p}_i\}$	$\{E_i, \tilde{p}_i\}$
$\{E_i, \tilde{p}_i\}$	$\{E_i, \tilde{p}_i\}$	$\{E_i, \tilde{p}_i\}$
$\{E_i, \tilde{p}_i\}$	$\{E_i, \tilde{p}_i\}$	$\{E_i, \tilde{p}_i\}$

The total number of ways in which we can have a particular set of {N_i} should equal W, so that from standard combinatorial arguments we can write

$$W = \frac{N!}{N_1! N_2! \quad \cdots}$$

Taking the logarithm and using Stirling's approximation for large n

$$\ell n \, n! \cong n \, \ell n \, n - n$$

we have
$$\ell n \, W = \ell n \, N! - \ell n \, N_1! - \ell n \, N_2! - \quad \cdots$$

$$= N \, \ell n \, N - N \tilde{p}_1 \, \ell n \, N \tilde{p}_1 - N \tilde{p}_2 \, \ell n \, N \tilde{p}_2 - \cdots$$

Making use of the condition that all the probabilities { $\tilde{p}_i$ } add up to one, we have

$$\ell n \, W = -N \left(\tilde{p}_1 \, \ell n \, \tilde{p}_1 + \tilde{p}_2 \, \ell n \, \tilde{p}_2 + \quad \cdots \right) = -N \sum_i \tilde{p}_i \, \ell n \, \tilde{p}_i$$

This gives us W in terms of the probabilities, thus connecting the two expressions for entropy in Eq.(16.7) and Eq.(16.16).

16.5.1. Equilibrium Distribution from Minimizing Free Energy

One last observation before we move on. In general the system could be in some arbitrary state (not necessarily the equilibrium state) where each energy level E_i is occupied with some probability $\tilde{p}_i$. However, we have argued that for the equilibrium state, the probabilities $\tilde{p}_i$ are given by

$$\left[\tilde{p}_i \right]_{equilibrium} = \frac{1}{Z} e^{-E_i / kT} \equiv p_i \quad (\text{see Eq.(16.14)})$$

where Z is a constant chosen to ensure that all the probabilities add up to one. We have also argued that the equilibrium state is characterized by a minimum in the free energy $F = E - TS$. Can we show that of all the possible choices for the probabilities { $\tilde{p}_i$ }, the equilibrium distribution

$\{p_i\}$ is the one that will minimize the free energy ?

Noting that the energy of an individual system is given by

$$E = \sum_i E_i \, \tilde{p}_i$$

and using S/N from Eq.(16.16) we can express the free energy as

$$F = \sum_i \tilde{p}_i \, (E_i + kT \ell n \, \tilde{p}_i) \qquad (16.17)$$

which can be minimized with respect to changes in $\{\tilde{p}_i\}$

$$dF = 0 = \sum_i d\tilde{p}_i \, (E_i + kT \ell n \, \tilde{p}_i + kT)$$

$$= \sum_i d\tilde{p}_i \, (E_i + kT \ell n \, \tilde{p}_i)$$

noting that the sum of all probabilities is fixed, so that

$$\sum_i d\tilde{p}_i = 0$$

We can now argue that in order to ensure that dF is zero for arbitrary choices of $d\tilde{p}_i$ we must have

$$E_i + kT \ell n \, \tilde{p}_i = 0$$

which gives us the equilibrium probabilities in Eq.(16.14).

Even if the system is not in equilibrium we can use Eq.(16.17) to calculate the free energy F of an out-of equilibrium system if we know the probabilities $\tilde{p}_i$. But the answer should be a number larger than the equilibrium value.

In the next Lecture I will argue that in principle we can build a device that will harness the excess free energy

$$\Delta F = F - F_{eq}$$

of an out-of-equilibrium system to do useful work.

The excess free energy has two parts:

$$\underbrace{\Delta F}_{\substack{excess\ free \\ energy}} = \underbrace{\Delta E}_{\substack{excess \\ energy}} - T \underbrace{\Delta S}_{inf\,ormation}$$

The first part represents real energy, but the second represents information that is being traded to convert energy from the surrounding reservoirs into work. Let us now talk abut this "fuel value of information."

Chapter 17

Fuel value of information

17.1. Information-Driven Battery
17.2. Fuel Value Comes From Knowledge
17.3. Landauer's Principle
17.4. Maxwell's Demon

A system in equilibrium contains no information, since the equilibrium state is independent of past history. Usually information is contained in systems that are stuck in some out of equilibrium state. We would like to argue that if we have such an out-of-equilibrium system, we can in principle construct a device that extracts an amount of energy less than or equal to

$$E_{available} = F - F_{eq} \qquad (17.1)$$

where F is the free energy of the out-of-equilibrium system and F_{eq} is the free energy of the system once it is restored to its equilibrium state. Let me explain where this comes from.

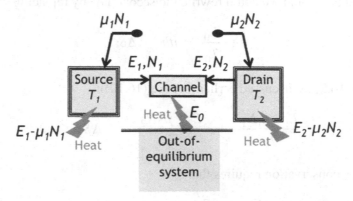

Fig.17.1. An out-of-equilibrium system can in principle used to construct a battery.

251

Consider the general scheme discussed in the last Lecture, but with both contacts at the same temperature T and with the electrons interacting with some metastable system. Since this system is stuck in an out-of-equilibrium state we cannot in general talk about its temperature.

For example a collection of independent spins in equilibrium would be randomly half up and half down at any temperature. So if we put them into an all-up state, as shown below, we cannot talk about the temperature of this system. But we could still use Eq.(16.16) to find its entropy, which would be zero.

Equilibrium state, $S = Nk \, \ell n \, 2$ *Out – of – equilibrium state,* $S = 0$

Fig.17.2. A collection of N independent spins in equilibrium would be randomly half up and half down, but could be put into an out-of-equilibrium state with all spins pointing up.

With this in mind we could rewrite the second law by replacing

$$\frac{E_0}{T_0} \quad with \quad -\Delta S$$

in Eq.(16.2) as discussed earlier (see Eq.(16.10)):

$$\frac{E_1 - \mu_1 N_1}{T_1} + \frac{E_2 - \mu_2 N_2}{T_2} - \Delta S \leq 0 \qquad (17.2a)$$

Energy conservation requires that

$$E_1 + E_2 = -E_0 \equiv \Delta E \qquad (17.2b)$$

where ΔE is the change in the energy of the metastable system.

Combining Eqs.(17.2a,b), assuming $T_1 = T_2 = T$, and making use of $N_1 + N_2 = 0$, we have

$$(\mu_1 - \mu_2)N_1 \geq \Delta E - T\Delta S = \Delta F \qquad (17.3)$$

Ordinarily, ΔF can only be positive, since a system in equilibrium is at its minimum free energy and all it can do is to increase its F. In that case, Eq.(17.3) requires that N_1 have the same sign as $\mu_1 - \mu_2$, that is, electrons flow from higher to lower electrochemical potential, as in any resistor.

But a system in an out-of-equilibrium state can relax to equilibrium with a corresponding decrease in free energy, so that ΔF is negative, and N_1 could have a sign opposite to that of $\mu_1 - \mu_2$, without violating Eq.(17.3). Electrons could then flow from lower to higher electrochemical potential, as they do inside a battery. The key point is that a *metastable non-equilibrium state can at least in principle be used to construct a battery*.

In a way this is not too different from the way real batteries work. Take the lithium ion battery for example. A charged battery is in a metastable state with excess Lithium ions intercalated in a carbon matrix at one electrode. As Lithium ions migrate out of the carbon electrode, electrons flow in the external circuit till the battery is discharged and the electrodes have reached the proper equilibrium state with the lowest free energy. The maximum energy that can be extracted is the change in the free energy.

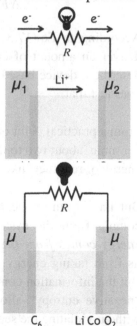

Usually the change in the free energy F comes largely from the change in the real energy E (recall that $F = E - TS$). That does not sound too surprising. If a system starts out with an energy E that is greater than its equilibrium energy E_0, then as it relaxes, it seems plausible that a cleverly designed device could capture the extra energy $E-E_0$ and deliver it as useful work.

What makes it a little more subtle, is that the extracted energy could come from the change in entropy as well. For example the system of localized spins shown in Fig.17.2 in going from the all-up state to its equilibrium state suffers no change in the actual energy, assuming that the energy is the same whether a spin points up or down. In this case the entire decrease in free energy comes from the increase in entropy:

$$\Delta E = 0$$
$$\Delta S = Nk\,\ell n\,2 \tag{17.5}$$
$$\Delta F = \Delta E - T\Delta S = -NkT\,\ell n\,2$$

According to Eq.(17.3) we should be able to build a device that will deliver an amount of energy equal to $NkT\,\ell n2$. In this Lecture I will describe a device based on the anti-parallel spin valve (Lecture 15) that does just that.

From a practical point of view, $NkT\,\ell n2$, amounts to about 2.5 kiloJoules per mole, about two to three orders of magnitude lower than the available energy of real fuels like coal or oil which comes largely from ΔE.

But the striking conceptual point is that the energy we extract is not coming from the system of spins whose energy is unchanged. *The energy comes from the surroundings.* Ordinarily the second law stops us from taking energy from our surroundings to perform useful work. But the information contained in the non-equilibrium state in the form of "negative entropy" allows us to extract energy from the surroundings without violating the second law.

From this point of view we could use the relation $F=E-TS$ to split up the right hand side of Eq.(17.1) into an actual energy and an info-energy that can be extracted from the surroundings by making use of the information available to us in the form of a deficit in entropy S relative to the equilibrium value S_{eq}:

$$E_{available} = \underbrace{(E - E_{eq})}_{Energy} + \underbrace{T\,(S_{eq} - S)}_{Info-Energy} \qquad (17.6)$$

For a set of independent localized spins in the all-up state, the available energy is composed entirely of info-energy: there is no change in the actual energy.

17.1. Information-Driven Battery

Let us see how we could design a device to extract the info-energy from a set of localized spins. Consider a perfect anti-parallel spin-valve device (Lecture 14) with a ferromagnetic source that only injects and extracts upspin electrons and a ferromagnetic drain that only injects and extracts downspin electrons from the channel (Fig.17.3).

These itinerant electrons interact with the localized spins through an exchange interaction of the form

$$u+D \iff U+d \qquad (17.7)$$

where u, d represent up and down channel electrons, while U, D represent up and down localized spins.

Ordinarily this "reaction" would be going equally in either direction. But by starting the localized spins off in a state with $U \gg D$, we make the reaction go predominantly from right to left and the resulting excess itinerant electrons u are extracted by one contact while the deficiency in d electrons is compensated by electrons entering the other contact. After some time, there are equal numbers of localized U and D spins and the reaction goes in either direction and no further energy can be extracted.

But what is the maximum energy that can be extracted as the localized spins are restored from their all up state to the equilibrium state? The answer is $NkT \ln 2$ equal to the change in the free energy of the localized spins as we have argued earlier.

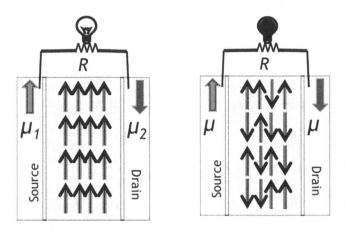

Fig.17.3.
An ***info-battery***: A perfect anti-parallel spin-valve device can be used to extract the excess free energy from a collection on N localized spins, all of which are initially up. Eventually the battery runs down when the spins have been randomized.

But let us see how we can get this result from a direct analysis of the device. Assuming that the interaction is weak we expect the upspin channel electrons (u) to be in equilibrium with contact 1 and the downspin channel electrons (d) to be in equilibrium with contact 2, so that

$$f_u(E) = \frac{1}{\exp\left(\dfrac{E - \mu_1}{kT}\right) + 1}$$

and

$$f_d(E) = \frac{1}{\exp\left(\dfrac{E - \mu_2}{kT}\right) + 1} \tag{17.8}$$

Assuming that the reaction

$$u + D \quad \Leftrightarrow \quad U + d \qquad \text{(same as Eq.(17.7))}$$

proceeds at a very slow pace so as to be nearly balanced, we can write

$$P_D \, f_u \, (1 - f_d) \;=\; P_U \, f_d \, (1 - f_u)$$

$$\frac{P_U}{P_D} \;=\; \frac{f_u}{1 - f_u} \, \frac{1 - f_d}{f_d} \;=\; e^{\Delta \mu / kT} \tag{17.9}$$

where

$$\Delta \mu \;\equiv\; \mu_1 - \mu_2$$

Here we assumed a particular potential $\mu_{1,2}$ and calculated the corresponding distribution of up and down localized spins. But we can reverse this argument and view the potential as arising from a particular distribution of spins.

$$\Delta \mu \;=\; kT \, \ell n \frac{P_U}{P_D} \tag{17.10}$$

Initially we have a larger potential difference corresponding to a preponderance of upspins (Fig.17.3, left), but eventually we end up with equal up and down spins (Fig.17.3, right) corresponding to $\mu_1 = \mu_2 = \mu$.

Looking at our basic reaction (Eq.(17.7)) we can see that everytime a D flips to an U, a u flips to a d which goes out through the drain. But when a U flips to a D, a d flips to a u which goes out through the source. So the net number of electrons transferred from the source to the drain equals half the change in the difference in the number of U and D spins:

$$n \, (Source \rightarrow Drain) = \; - \Delta N_U$$

We can write the energy extracted as the potential difference times the number of electrons transferred

$$E = - \int_{Initial}^{Final} \Delta\mu \ dN_U$$

Making use of Eq.(17.10) we can write

$$E = - NkT \int_{Initial}^{Final} \left(\ell n \, P_U - \ell n \, P_D \right) dP_U$$

Noting that
$$dP_U + dP_D = 0$$

and that
$$S = - Nk \left(P_U \, \ell n \, P_U + P_D \, \ell n \, P_D \right) \qquad \text{(Eq.(16.16))}$$

we can use a little algebra to rewrite the integrand as

$$\left(\ell n \, P_U - \ell n \, P_D \right) dP_U = d\left(P_U \, \ell n \, P_U + P_D \, \ell n \, P_D \right) = - dS \, / \, Nk$$

so that
$$E = T \int_{Initial}^{Final} dS = T \, \Delta S$$

which is the basic result we are trying to establish, namely that the metastable state of all upspins can in principle be used to construct a battery that could deliver upto

$$T \, \Delta S = NkT \, \ell n \, 2$$

of energy to an external load.

17.2. Fuel Value Comes From Knowledge

A key point that might bother a perceptive reader is the following. We said that the state with all spins up has a higher free energy than that for a random collection of spins: $F > F_0$, and that this excess free energy can in principle be extracted with a suitable device.

But what is it that makes the random collection different from the ordered collection. As Feynman put it, we all realize that it is unusual to see a car with a license plate # 9999. But it is just as remarkable to see a car with any specific predetermined number say 1439. Similarly if we really knew the spins to be in a very  specific but seemingly random configuration like the one sketched here, its entropy would be zero, just like the all up configuration. The possibility of extracting energy comes not from the all up nature of the initial state, ***but from knowing exactly what state it is in.***

But how would we construct our conceptual battery to extract the energy from a random but known configuration? Consider a simple configuration that is not very random: The top half is up and the bottom half is down. Ordinarily this would not give us any open circuit voltage, the top half cancels the bottom half. But we could connect it as shown in Fig.17.4 reversing the contacts for the left and right halves and extract energy.

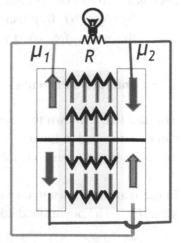

Fig.17.4. A suitably designed device can extract energy from any known configuration of spins.

Following the same principle we could construct a device to extract energy from a more random collection too. The key point is to know the exact configuration so that we can design the contacts accordingly.

Of course these devices would be much harder to build than the one we started with for the all-up configuration. But these devices are just intended to be conceptual constructs intended to illustrate a point. The point is that information consists of a system being in an out of equilibrium state and our knowing exactly which state it is in. This information can *in principle* be used to create a battery and traded for energy.

In the field of information theory, Shannon introduced the word entropy as a measure of the information content of a signal composed of a string of symbols *i* that appear with probability *{p_i}*

$$H = -\sum_i p_i \, \ell n \, p_i \qquad (17.11)$$

This expression looks like the thermodynamic entropy (see Eq.(16.17)) except for the Boltzmann constant and there are many arguments to this day about the connection between the two. One could argue that if we had a system with states *i* with equilibrium probabilities *{p_i}*, then *k*H* represents the entropy of an equilibrium system carrying no information. As soon as someone tells us which exact state it is in, the entropy becomes zero so that the entropy is lowered by *(Nk)*H* increasing its free energy by *(NkT)*H*. In principle, at least we could construct a battery to extract this excess free energy *(NkT)*H*.

17.3. Landauer's Principle

The idea that a known metastable state can be used to construct a battery can be connected to Landauer's principle which talks about the minimum energy needed to erase one bit of information.

In our language, erasure consists of taking a system from an equilibrium state *(F_{eq})* to a known standard state *(F)*:

$$F_{eq} = -NkT \, \ell n \, 2 \quad \rightarrow \quad Erasure \quad \rightarrow \quad F = 0$$

Is there a minimum energy needed to achieve this? We have just argued that once the spin is in the standard state we can construct a battery to extract $(F - F_{eq})$ from it. In a cyclic process we could spend E_{erase} to go from F_{eq} to F, and then construct a battery to extract $(F - F_{eq})$ from it, so that the total energy spent over the cycle equals

$$E_{erase} - (F - F_{eq})$$

which must be greater than zero, or we would have a perpetual source of energy. Hence

$$E_{erase} \geq F - F_{eq} \qquad (17.12)$$

which in this case yields Landauer's principle:

$$E_{erase} \geq NkT \, \ell n2$$

It seems to us, that erasure need not necessarily mean putting the spins in an all-up state. More generally it involves putting them in a known state, analogous to writing a complicated musical piece on a magnetic disk. Also, the minimum energy of erasure need not necessarily be dissipated. It often ends up getting dissipated only because it is impractical to build an info-battery to get it back.

Fifty years ago Landauer asked deep questions that were ahead of his time. Today with the progress in nanoelectronics, the questions are becoming more and more relevant, and some of the answers at least seem fairly clear. Quantum mechanics, however, adds new features some of which are yet to be sorted out and are being actively debated at this time.

17.4. Maxwell's Demon

Our info-battery could be related to Maxwell's famous demon (see for example, Lex (2005)) who was conjectured to beat the second law by letting hot molecules (depicted black) go from left to right and cold molecules (depicted gray) go from right to left so that after some time the

right hand side becomes hot and the left hand side becomes cold
(Fig.17.5).

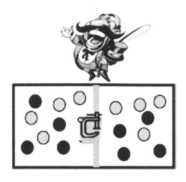
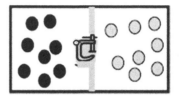

Fig.17.5. Maxwell's demon creates a temperature difference by letting hot molecules go
preferentially to the right.

To see the connection with our "info-battery" in Fig.17.3 we could draw
the following analogy:

$$Hot\ molecules \quad \leftrightarrow \quad up-spin\ electrons$$

$$Cold\ molecules \quad \leftrightarrow \quad down-spin\ electrons$$

$$Demon \quad \leftrightarrow \quad Collection\ of\ Localized\ Spins\ each\ with\ two\ states$$

$$Left,\ Right\ of\ Box \quad \leftrightarrow \quad Source,\ Drain\ Contacts$$

Our battery is run by a set of all up localized spins that flip electrons up
and send them to the source, while replacing the down-spin from the
drain. The demon sends hot molecules to the left and cold molecules to
the right, which is not exactly the same process, but similar.

The key point, however, is that the demon is making use of information
rather than energy to create a temperature difference just as our info-
battery uses the low entropy sate of the localized spins to create a
potential difference. Like our localized spins, the demon too must
gradually transition into a high entropy state that will stop it from

discriminating between hot and cold molecules. Or as Feynman (1963) put it in one of his Lectures,

> " .. if we build a finite-sized demon, the demon himself gets so warm, he cannot see very well after a while."

Like our info-battery (Fig.17.3), eventually the demon stops functioning when the entropy reaches its equilibrium value and all initial information has been lost.

We started in Lecture 1 by noting how transport processes combine two very different types of processes, one that is force-driven and another that is entropy-driven. In these last two Lectures, my objective has been to give readers a feeling for the concept of an *"entropic force"* that drives many everyday phenomena.

The fully polarized state with $S=0$ spontaneously goes to the unpolarized state with $S = Nk\,\ell n\,2$, but to make it go the other way we need to connect a battery and do work on it.

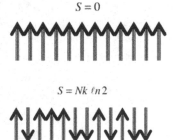

This directed flow physically arises from the fact that the fully polarized state represents a single state while the unpolarized state represents numerous (2^N) possibilities. It is this sheer number that drives the impurities spontaneously from the low entropy to the high entropy state and not the other way. Many real life phenomena are driven by such entropic forces which are very different from ordinary forces that take a system from a single state to another single state.

What makes transport so complicated is the intertwining of these two types of phenomena in the channel region. We have seen how the elastic resistor allows us to separate the two, with entropic processes confined to the contacts, and the channel described by semi-classical mechanics. It is now time to look at the quantum version of the problem with the channel described by quantum mechanics.

III. Contact-ing Schrödinger

Lecture 18

The Model

18.1. Schrödinger Equation
18.2. Electron-Electron Interactions
18.3. Differential to Matrix Equation
18.4. Choosing Matrix Parameters

Over a century ago Boltzmann taught us how to combine Newtonian mechanics with entropy-driven processes

$$\text{Classical Dynamics} + \text{⚡} = \text{BTE}$$

and the resulting Boltzmann transport equation (BTE) is widely accepted as the cornerstone of semiclassical transport theory. Most of the results we have discussed so far can be (and generally are) obtained from the Boltzmann equation, but the concept of an elastic resistor makes them more transparent by spatially separating force-driven processes in the channel from the entropy-driven processes in the contacts.

In this part of these lecture notes I would like to discuss the quantum version of this problem, using the non-equilibrium Green's function (NEGF) method to combine quantum mechanics described by the Schrödinger equation with "contacts"

$$\text{Quantum Dynamics} + \text{⚡} = \text{NEGF}$$

much as Boltzmann taught us how to combine classical dynamics with "contacts".

267

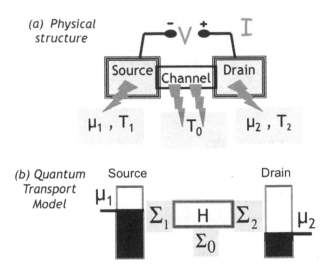

Fig.18.1. (a) Generic device structure that we have been discussing. (b) General quantum transport model with elastic channel described by a Hamiltonian *[H]* and its connection to each 'contact" described by a corresponding self-energy $[\Sigma]$.

The NEGF method originated from the classic works in the 1960's that used the methods of many-body perturbation theory to describe the distributed entropy-driven processes along the channel. Like most of the work on transport theory (semiclassical

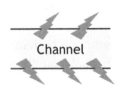

or quantum) prior to the 1990's, it was a *"contact-less"* approach focused on the interactions occurring throughout the channel, in keeping with the general view that the physics of resistance lay essentially in these distributed entropy generating processes.

As with semiclassical transport, our bottom-up perspective starts at the other end with the elastic resistor with entropy-driven processes confined to the contacts. This makes the theory less about interactions and more about "connecting contacts to the Schrödinger equation", or more simply, about *contact-ing Schrödinger*.

But let me put off talking about the NEGF model till the next Lecture, and use subsequent lectures to illustrate its application to interesting problems in quantum transport. As indicated in Fig.18.1b the NEGF method requires two types of inputs: the Hamiltonian, *[H]* describing the dynamics of an elastic channel, and the self-energy $[\Sigma]$ describing the connection to the contacts, using the word "contacts" in a broad figurative sense to denote all kinds of entropy-driven processes. Some of these contacts are physical like the ones labeled "1" and "2" in Fig.18.1b, while some are conceptual like the one labeled "0" representing entropy changing processes distributed throughout the channel.

In this Lecture let me just try to provide a super-brief but self-contained introduction to how one writes down the Hamiltonian *[H]*. The $[\Sigma]$ can be obtained by imposing the appropriate boundary conditions and will be described in later Lectures when we look at specific examples applying the NEGF method.

We will try to describe the procedure for writing down *[H]* so that it is accessible even to those who have not had the benefit of a traditional multi-semester introduction to quantum mechanics. Moreover, our emphasis here is on something that may be helpful even for those who have this formal background. Let me explain.

Most people think of the Schrödinger equation as a differential equation which is the form we see in most textbooks. However, practical calculations are usually based on a discretized version that represents the differential equation as a matrix equation involving the Hamiltonian matrix *[H]* of size NxN, N being the number of "basis functions" used to represent the structure.

This matrix *[H]* can be obtained from first principles, but a widely used approach is to represent it in terms of a few parameters which are chosen to match key experiments. Such semi-empirical approaches are often used because of their convenience and because they can often explain a wide range of experiments beyond the key ones that are used as input, suggesting that they capture a lot of essential physics.

In order to follow the rest of the Lectures it is important for the readers to get a feeling for how one writes down this matrix *[H]* given an accepted energy-momentum *E(p)* relation (Lecture 5) for the material that is believed to describe the dynamics of conduction electrons with energies around the electrochemical potential.

But I should stress that the NEGF framework we will talk about in subsequent lectures goes far beyond any specific model that we may choose to use for *[H]*. The same equations could be (and have been) used to describe say conduction through molecular conductors using first principles Hamiltonians.

18.1. Schrödinger Equation

We started these Lectures by noting that the key input needed to understand current flow is the density of states, *D(E)* telling us the number of states available for an electron to access on its way from the source to the drain.

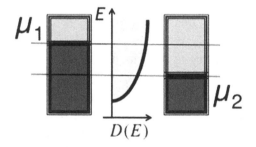

Theoretical models for *D(E)* all start from the Schrödinger equation which tells us the available energy levels. However, we managed to obtain expressions for *D(E)* in Chapter 5 without any serious brush with quantum mechanics by (1) starting from a given energy-momentum relation *E(p)*, (2) relating the momentum to the wavelength through the de Broglie relation (*p=h/wavelength*) and then (3) requiring an integer number of half wavelengths to fit into the conductor, the same way acoustic waves fit on a guitar string.

This heuristic principle is mathematically implemented by writing a wave equation which is obtained from a desired energy-momentum relation by making the replacements

$$E \rightarrow ih\frac{\partial}{\partial t}, \quad \vec{p} \rightarrow -ih\vec{\nabla} \qquad (18.1)$$

where the latter stands for

$$p_x \rightarrow -ih\frac{\partial}{\partial x}, \quad p_y \rightarrow -ih\frac{\partial}{\partial y}, p_z \rightarrow -ih\frac{\partial}{\partial z},$$

Using this principle, the classical energy-momentum relation

$$E_{classical}(\vec{p}) = \frac{p_x^2 + p_y^2 + p_z^2}{2m} \qquad (18.2a)$$

leads to the wave equation

$$ih\frac{\partial}{\partial t}\tilde{\psi} = -\frac{h^2}{2m}\left(\frac{\partial^2}{\partial x^2} + \frac{\partial^2}{\partial y^2} + \frac{\partial^2}{\partial z^2}\right)\tilde{\psi}(x,y,z,t) \qquad (18.2b)$$

whose solutions can be written in the form of exponentials of the form

$$\tilde{\psi}(x,y,z,t) = \psi_0\, e^{+ik_x x}\, e^{+ik_y y}\, e^{+ik_z z}\, e^{-iEt/h} \qquad (18.3)$$

where the energy E is related to the wavevector $\vec{k}$ by the dispersion relation

$$E(\vec{k}) = \frac{h^2(k_x^2 + k_y^2 + k_z^2)}{2m} \qquad (18.4)$$

Eq.(18.4) looks just like the classical energy-momentum relation (Eq.(18.2a)) of the corresponding particle with

$$\vec{p} = h\vec{k} \qquad (18.5)$$

which relates the particulate property $\vec{p}$ with the wavelike property $\vec{k}$. This can be seen to be equivalent to the de Broglie relation ($p=h/wavelength$) noting that the wavenumber k is related to the wavelength through

$$k = \frac{2\pi}{wavelength}$$

The principle embodied in Eq.(18.1) ensures that the resulting wave equation has a group velocity that is the same as the velocity of the corresponding particle

$$\underbrace{\frac{1}{\hbar}\vec{\nabla}_k E}_{\substack{Wave \\ group\ velocity}} = \underbrace{\vec{\nabla}_p E}_{\substack{Particle \\ velocity}} \quad \bullet$$

18.1.1. Spatially Varying Potential

The wave equation Eq.(18.2b) obtained from the energy-momentum relation describes free electrons. If there is a force described by a potential energy $U(\vec{r})$ so that the classical energy is given by

$$E_{classical}(\vec{r},\vec{p}) = \frac{p_x^2 + p_y^2 + p_z^2}{2m} + U(x,y,z) \qquad (18.6a)$$

then the corresponding wave equation has an extra term due to $U(\vec{r})$

$$i\hbar\frac{\partial}{\partial t}\tilde{\psi} = -\frac{\hbar^2}{2m}\nabla^2\tilde{\psi} + U(\vec{r})\tilde{\psi} \qquad (18.6b)$$

where $\vec{r} \equiv (x,y,z)$ and the Laplacian operator is defined as

$$\nabla^2 \equiv \frac{\partial^2}{\partial x^2} + \frac{\partial^2}{\partial y^2} + \frac{\partial^2}{\partial z^2}$$

Solutions to Eq.(18.6b) can be written in the form

$$\tilde{\psi}(\vec{r},t) = \psi(\vec{r})\,e^{-iEt/\hbar}$$

where $\psi(\vec{r})$ obeys the time-independent Schrödinger equation

$$E\psi(\vec{r}) = H_{op}\psi(\vec{r}) \qquad (18.7a)$$

where H_{op} is a differential operator obtained from the classical energy function in Eq.(18.6a), using the replacement mentioned earlier (Eq.(18.1)):

$$H_{op} = -\frac{\hbar^2}{2m}\nabla^2 + U(\vec{r}) \qquad (18.7b)$$

Quantum mechanics started in the early twentieth century with an effort to "understand" the energy levels of the hydrogen atom deduced from the experimentally observed spectrum of the light emitted from an incandescent source. For a hydrogen atom Schrödinger used the potential energy

$$U(\vec{r}) = -Z\frac{q^2}{4\pi\varepsilon_0 r}$$

where the atomic number $Z = 1$, due to a point nucleus with charge $+q$, and solved Eq.(18.7) analytically for the allowed energy values E_n (called the eigenvalues of the operator H_{op}) given by

$$E_n = -\frac{Z^2}{n^2}\frac{q^2}{8\pi\varepsilon_0 a_0} \qquad (18.8)$$

with
$$a_0 = \frac{4\pi\varepsilon_0\hbar^2}{mq^2}$$

and the corresponding solutions

$$\psi_{n\ell m}(\vec{r}) = R_{n\ell}(\vec{r})Y_\ell^m(\theta,\phi)$$

obeying the equation

$$E_n\psi_{n\ell m} = \left(-\frac{\hbar^2}{2m}\nabla^2 - \frac{Zq^2}{4\pi\varepsilon_0 r}\right)\psi_{n\ell m}(\vec{r})$$

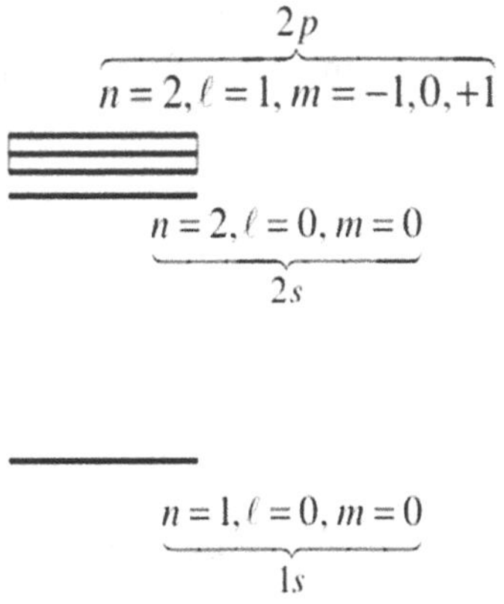

$$\overbrace{\qquad\qquad\qquad}^{2p}$$
$$n = 2, \ell = 1, m = -1, 0, +1$$

$$\underbrace{\qquad\qquad\qquad}_{2s}$$
$$n = 2, \ell = 0, m = 0$$

Fig.18.2.
Energy levels in atoms are catalogued with three
indices n, ℓ, m.

$$\underbrace{\qquad\qquad\qquad}_{1s}$$
$$n = 1, \ell = 0, m = 0$$

The energy eigenvalues in Eq.(18.8) were in extremely good agreement
with the known experimental results, leading to general acceptance of the
Schrödinger equation as **the** wave equation describing electrons, just as
acoustic waves, for example, on a guitar string are described by

$$\omega^2 u(z) = -\frac{\partial^2}{\partial z^2} u$$

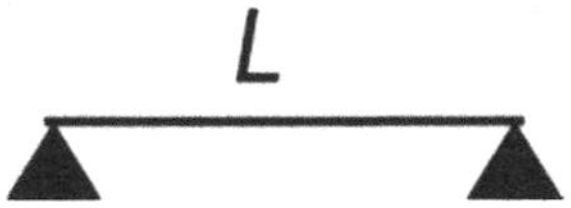

A key point of similarity to note is that when a guitar string is clamped
between two points, it is able to vibrate only at discrete frequencies
determined by the length L. Similarly electron waves when "clamped"
have discrete energies and most quantum mechanics texts start by
discussing the corresponding "particle in a box" problem.

Shorter the length L, higher the pitch of a guitar and hence the spacing
between the harmonics. Similarly smaller the box, greater the spacing
between the allowed energies of an electron. Indeed one could view the
hydrogen atom as an extremely small 3D box for the electrons giving rise
to the discrete energy levels shown in Fig.18.2. This is of course just a
qualitative picture. Quantitatively, we have to solve the time-independent
Schrödinger equation (Eq.(18.7)).

There is also a key dissimilarity between classical waves and electron waves. For acoustic waves we all know what the quantity $u(z)$ stands for: it is the displacement of the string at the point z, something that can be readily measured. By contrast, the equivalent quantity for electrons, $\psi(\vec{r})$ (called its wavefunction), is a complex quantity that cannot be measured directly and it took years for scientists to agree on its proper interpretation. The present understanding is that the real quantity $\psi\psi^*$ describes the probability of finding an electron in a unit volume around $\vec{r}$. This quantity, when summed over many electrons, can be interpreted as the average electron density.

18.2. Electron-electron interactions and the scf method

After the initial success of the Schrödinger equation in "explaining" the experimentally observed energy levels of the hydrogen atom, scientists applied it to increasingly more complicated atoms and by 1960 had achieved good agreement with experimentally measured results for all atoms in the periodic table (Herman and Skillman (1963)). It should be noted, however, that these calculations are far more complicated primarily because of the need to include the electron-electron (e-e) interactions in evaluating the potential energy (hydrogen has only one electron and hence no e-e interactions).

For example, Eq.(18.9) gives the lowest energy for a hydrogen atom as $E_1 = -13.6\ eV$ in excellent agreement with experiment. It takes a photon with at least that energy to knock the electron out of the atom ($E > 0$), that is to cause photoemission. Looking at Eq.(18.8) one might think that in Helium with $Z=2$, it would take a photon with energy $\sim 4*13.6\ eV = 54.5\ eV$ to knock an electron out. However, it takes photons with far less energy $\sim 30\ eV$ and the reason is that the electron is repelled by the other electron in Helium. However, if we were to try to knock the second electron out of Helium, it would indeed take photons with energy $\sim 54\ eV$, which is known as the second ionization potential. But usually what we want is the first ionization potential or a related quantity called the electron affinity. Let me explain.

Current flow involves adding an electron from the source to the channel and removing it into the drain. However, these two events could occur in either order.

Source Channel Drain

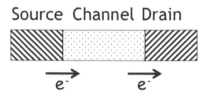

The electron could first be added and then removed so that the channel evolves as follows

> A. $N \rightarrow N+1 \rightarrow N$ electrons (Affinity levels)

But if the electron is first removed and then added, the channel would evolve as

> B. $N \rightarrow N-1 \rightarrow N$ electrons (Ionization levels)

In the first case, the added electron would feel the repulsive potential due to N electrons. Later when removing it, it would still feel the potential due to N electrons since no electron feels a potential due to itself. So the electron energy levels relevant to this process should be calculated from the Schrödinger equation using a repulsive potential due to N electrons. These are known as the *affinity levels*.

In the second case, the removed electron would feel the repulsive potential due to the other $N-1$ electrons. Later when adding an electron, it would also feel the potential due to $N-1$ electrons. So the electron energy levels relevant to this process should be calculated from the Schrödinger equation using a repulsive potential due to $N-1$ electrons. These are known as the *ionization levels*.

The difference between the two sets of levels is basically the difference in potential energy due to one electron, called the single electron charging energy U_0. For something as small as a Helium atom it is ~ 25

eV, so large that it is hard to miss. For large conductors it is often so small that it can be ignored, and it does not matter too much whether we use the potential due to *N* electrons or due to *N-1* electrons. For small conductors, under certain conditions the difference can be important giving rise to single-electron charging effects, which we will ignore for the moment and take up again later in Lecture 24.

Virtually all the progress that has been made in understanding "condensed matter," has been based on the self-consistent field (scf) method where we think of each electron as behaving quasi-independently feeling an average self-consistent potential $U(\vec{r})$ due to all the other electrons in addition to the nuclear potential. This potential depends on the electron density $n(\vec{r})$ which in turn is determined by the wavefunctions of the filled states. Given the electron density how one determines $U(\vec{r})$ is the subject of much discussion and research. The "zero order" approach is to calculate $U(\vec{r})$ from $n(\vec{r})$ based on the laws of electrostatics, but it is well-established that this so-called Hartree approximation will overestimate the repulsive potential and there are various approaches for estimating this reduction. The density functional theory (DFT) has been spectacularly successful in describing this correction for equilibrium problems and in its simplest form amounts to a reduction by an amount proportional to the cube root of the electron density

$$U(\vec{r}) = U_{Hartree} - \frac{q^2}{4\pi\varepsilon}\left(n(\vec{r})\right)^{1/3}$$

(18.9)

Many are now using similar corrections for non-equilibrium problems like current flow as well, though we believe there are important issues that remain to be resolved.

We should also note that there is a vast literature (both experiment and theory) on a regime of transport that cannot be easily described within an scf model. It is not just a matter of correctly evaluating the self-consistent potential. The very picture of quasi-independent electrons moving in a self-consistent field needs revisiting, as we will see in Lecture 23.

18.3. Differential to Matrix Equation

All numerical calculations typically proceed by turning the differential equation in Eq.(18.11) into a matrix equation of the form

$$E[S]\{\psi\} = [H]\{\psi\}$$

(18.10)

or equivalently

$$E \sum_m S_{nm} \psi_m = \sum_m H_{nm} \psi_m$$

by expanding the wavefunction in terms of a set of known functions $u_m(\mathbf{r})$ called the **basis functions:**

$$\psi(\vec{r}) = \sum_m \psi_m u_m(\vec{r})$$

(18.11a)

The elements of the two matrices *[S]* and *[H]* are given respectively by

$$S_{nm} = \int d\vec{r}\, u_n^*(\vec{r}) u_m(\vec{r})$$

$$H_{nm} = \int d\vec{r}\, u_n^*(\vec{r}) H_{op} u_m(\vec{r})$$

(18.11b)

These expressions are of course by no means obvious, but we will not go into it further since we will not really be making any use of them. Let me explain why.

18.3.1. Semi-empirical tight-binding (TB) models

There are a wide variety of techniques in use which differ in the specific basis functions they use to convert the differential equation into a matrix equation. But once the matrices *[S]* and *[H]* have been evaluated, the eigenvalues *E* of Eq.(18.10) (which are the allowed energy levels) are determined using powerful matrix techniques that are widely available.

In modeling nanoscale structures, it is common to use basis functions that are spatially localized rather than extended functions like sines or cosines. For example, if we were to model a hydrogen molecule, with two positive nuclei as shown (see Fig.18.3), we could use two basis functions, one localized around the left nucleus and one around the right nucleus. One could then work through the algebra to obtain *[H]* and *[S]* matrices of the form

$$H = \begin{bmatrix} \varepsilon & t \\ t & \varepsilon \end{bmatrix} \quad \text{and} \quad S = \begin{bmatrix} 1 & s \\ s & 1 \end{bmatrix} \quad (18.12)$$

where ε, t and s are three numbers.

The two eigenvalues from Eq.(18.10) can be written down analytically as

$$E_1 = \frac{\varepsilon - t}{1 - s} \quad \text{and} \quad E_2 = \frac{\varepsilon + t}{1 + s}$$

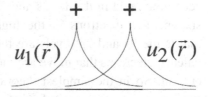

Fig.18.3. To model a Hydrogen molecule with two positive nuclei, one could use two basis functions, one localized around the left nucleus and one around the right nucleus.

What we just described above would be called a first-principles approach. Alternatively one could adopt a semi-empirical approach treating ε, t and s as three numbers to be adjusted to give the best fit to our "favorite" experiments. For example, if the energy levels $E_{1,2}$ are known from experiments, then we could try to choose numbers that match these. Indeed, it is common to assume that the [S] matrix is just an identity matrix ($s=0$), so that there are only two parameters ε and t which are then adjusted to match $E_{1,2}$. Basis functions with $s = 0$ are said to be "orthogonal."

18.3.2. *Size of matrix, N = n*b*

What is the size of the *[H]* matrix? Answer: (NxN), *N* being the total number of basis functions. How many basis functions? Answer: Depends on the approach one chooses. In the tight-binding (TB) approach, which we will use, the basis functions are the atomic wavefunctions for individual atoms, so that $N = n*b$, *n* being the number of atoms and b, the number of basis functions per atom. What is *b*? Let us look at specific examples.

4 electrons ▤
$2s,2p_{x,y,z}$

2 electrons ——
$1s$

Carbon, Z=6

————————————

Silicon, Z=14

Suppose we want to model current flow through **graphene** consisting of carbon atoms arranged in a two dimensional hexagonal sheet (see Fig.18.4). Carbon (*Z=6*) has six electrons which are accommodated in the 1s, 2s and 2p levels as shown. The electrons in the highest levels that is the 2s and 2p levels are the so called valence electrons that move around and carry current. So in the simplest theories, it is common to use the 2s and 2p levels on each atom as the basis functions, with *b=4*.

4 electrons ▤
$3s,3p_{x,y,z}$

8 electrons ▤
$2s,2p_{x,y,z}$

2 electrons ——
$1s$

The same is true of say silicon (*Z=14*), the most common semiconductor for electronic devices. Its fourteen electrons are accommodated as shown with the valence electrons in the 3s, 3p levels. Once again in the simplest models *b=4*, though some models include five 3d levels and/or the two 4s levels as part of the basis functions too.

One of the nice things about graphene is that the $2s,p_x,p_y$ orbitals are in the simplest approximation completely decoupled from the $2p_z$ orbitals, and for understanding current flow, one can get a reasonable description with just one $2p_z$ orbital for every carbon atom, so that *b=1*.

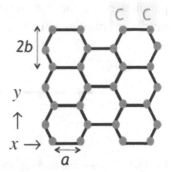

Fig.18.4. Graphene consists of a two-dimensional sheet of carbon atoms arranged in a two-dimensional hexagonal lattice.

In these simplest models, the matrix *[H]* is of size (nxn), n being the total number of carbon atoms. Its diagonal elements have some value ε, while the matrix element H_{nm} equals some value t if n and m happen to be nearest neighbors. If they are not nearest neighbors then one expects the value to be smaller since the functions ϕ_m and ϕ_n appearing in Eq.(18.14) do not overlap as much. In nearest neighbor tight-binding models it is common to set all such matrix elements to zero, so that we are finally left with just two parameters ε and t which are then adjusted to match known results.

18.4. Choosing Matrix Parametrs

One common way to select the parameters is to fit the known energy dispersion relation *E(k)*, also called the energy-momentum relation *E(p)* (Note that $\vec{p} = \hbar \vec{k}$) as discussed in Lecture 5. These relations have been arrived at through years of work combining careful experimental measurements with sophisticated first-principles calculations. If we can get our semi-empirical model to fit the accepted dispersion relation for a material, we have in effect matched the whole set of experiments that contributed to it.

18.4.1. One-Dimensional Conductor

Suppose we have a one-dimensional conductor that we would like to model with a nearest neighbor orthogonal tight-binding model with two parameters $\mathcal{E}$ and t representing the diagonal elements and the nearest neighbor coupling (Fig.18.5).

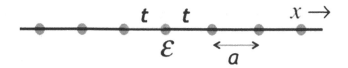

Fig.18.5. A one-dimensional array of atoms spaced by "a" modeled with a nearest neighbor orthogonal tight-binding model with two parameters $\mathcal{E}$ and t representing the diagonal elements and the nearest neighbor coupling.

How would we choose $\mathcal{E}$ and t so that we approximate a parabolic dispersion relation

$$E(k) = E_C + \frac{\hbar^2 k^2}{2m} \ ? \tag{18.13}$$

The answer is that our model represents a set of algebraic equations (see Eq.(18.10)) which for the orthogonal model reduces to

$$E \psi_n = \sum_m H_{nm} \psi_m \quad \Rightarrow \quad E = \sum_m H_{nm} \frac{\psi_m}{\psi_n}$$

If we assume a solution of the form

$$\psi_n = \exp(ik\,na)$$

we obtain the $E(k)$ relation corresponding to Eq.(18.10):

$$E(k) = \sum_m H_{nm} \exp(ik(m-n)a) \tag{18.14}$$

Can we always assume a solution of this form? No. In general Eq.(18.14) will give us different results for $E(k)$ depending on what value we choose for n when doing the summation and what we get for some particular choice of n is not very helpful. But if the structure is "translationally invariant" such that we get the same answer for all n then we get a unique $E(k)$ relation and Eq.(18.13) indeed represents an acceptable solution to our set of equations.

For our particular nearest neighbor model Eq.(18.14) yields straightforwardly

$$E(k) = \varepsilon + t\exp(+ika) + t\exp(-ika)$$

$$= \varepsilon + 2t\cos ka \qquad (18.15)$$

How would we make this match the desired parabolic relation in Eq.(18.13) ? Clearly one could not match them for all values of k, only for a limited range. For example, if we want them to match over a range of k-values around $k=0$, we can expand the cosine in a Taylor series around $ka = 0$ to write

$$\cos ka \approx 1 - \frac{(ka)^2}{2}$$

so that the best match is obtained by choosing

$$t = -\hbar^2 / 2ma^2$$

and

$$\varepsilon = E_C - 2t \qquad (18.16)$$

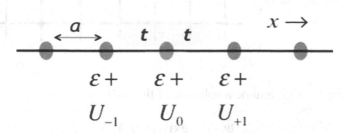

Fig.18.6. A spatially varying potential $U(x)$ along the channel is included by adding the local value of U to the diagonal element ε .

Finally I should mention that when modeling a device there could be a spatially varying potential $U(x)$ along the channel which is included by adding the local value of U to the diagonal element as indicated in Fig.18.6. We now no longer have the "translational invariance" needed for a solution of the form $exp(ikx)$ and the concept of a dispersion relation $E(k)$ is not valid. But a Hamiltonian of the form just described (Fig.18.6) can be used for numerical calculations and appear to be fairly accurate at least for potentials $U(x)$ that do not vary too rapidly on an atomic scale.

18.4.2. Two-Dimensional Conductor

A two-dimensional array of atoms (Fig.18.7) can be modeled similarly with a nearest neighbor orthogonal TB model, with the model parameters ε and t chosen to yield a dispersion relation approximating a standard parabolic effective mass relation:

$$E(k_x, k_y) = E_C + \frac{\hbar^2 (k_x^2 + k_y^2)}{2m} \qquad (18.17)$$

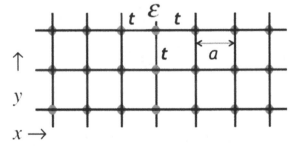

Fig.18.7. A two-dimensional nearest neighbor orthogonal TB model.

In this case we can assume a solution of the form

$$\psi_n = \exp(i\vec{k} \cdot \vec{r}_n)$$

where $\vec{k} = k_x \hat{x} + k_y \hat{y}$ and $\vec{r}_n$ denotes the location of atom n. Substituting into Eq.(18.10) we obtain the dispersion relation

$$E(\vec{k}) = \sum_m H_{nm} \exp\left(i\vec{k}.(\vec{r}_m - \vec{r}_n)\right)$$

(18.18a)

which for our nearest neighbor model yields

$$
\begin{aligned}
E(\vec{k}) =\ & \varepsilon + t\exp(+ik_x a) + t\exp(-ik_x a) \\
& + t\exp(+ik_y a) + t\exp(-ik_y a) \\
=\ & \varepsilon + 2t\cos(k_x a) + 2t\cos(k_y a)
\end{aligned}
$$

(18.18b)

Following the same arguments as in the 1-D case, we can make this match the parabolic relation in Eq.(18.17) by choosing

$$t = -\hbar^2/2ma^2$$

(18.19)

$$\varepsilon = E_C - 4t$$

18.4.3. TB parameters in B-field

It is shown in Appendix C that if we replace $\vec{p}$ with $\vec{p} + q\vec{A}$ in Eq.(18.6a)

$$E_{classical}(\vec{r}, \vec{p}) = \frac{(\vec{p} + q\vec{A}).(\vec{p} + q\vec{A})}{2m} + U(\vec{r})$$

yields the correct classical laws of motion of a particle of charge $-q$ in a vector potential $\vec{A}$. The corresponding wave equation is obtained using the replacement in Eq.(18.1): $\vec{p} \rightarrow -i\hbar\vec{\nabla}$.

To find the appropriate TB parameters for the Hamiltonian in a B-field we consider the homogeneous material with constant E_c and a constant vector potential. Consider first the 1D problem with

$$E(p_x) = E_c + \frac{(p_x + qA_x)(p_x + qA_x)}{2m}$$

so that the corresponding wave equation has a dispersion relation

$$E(k_x) = E_c + \frac{(\hbar k_x + qA_x)(\hbar k_x + qA_x)}{2m}$$

which can be approximated by a cosine function

$$E(k_x) = \varepsilon + 2t\cos\left(k_x a + \frac{qA_x a}{\hbar}\right)$$

with ε and t chosen according to Eq.(18.16). This means that we can model it with the 1-D lattice shown here, which differs from our original model in Fig.18.5 by the extra phase $qA_x a/\hbar$.

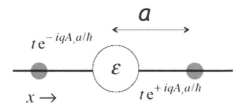

Similar arguments lead to a similar phase in the y-direction as well. This is included in the 2D tight-binding model by modifying the nearest neighbor coupling elements to include an appropriate phase in the nearest neighbor coupling elements as shown in Fig.18.8 with

$$\varphi_x = \frac{qA_x a}{\hbar} \quad , \quad \varphi_y = \frac{qA_y a}{\hbar}$$

Fig.18.8. The effect of a magnetic field in the z-direction is included in a tight-binding model by introducing a phase in the nearest neighbor coupling elements as discussed in the text.

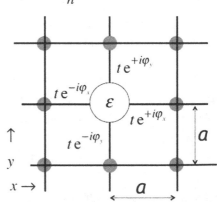

To include a B-field we have to let the vector potential vary spatially from one lattice point to the next such that

$$\vec{B} = \vec{\nabla} \times \vec{A}$$

For example, a B-field in the z-direction described in general by a vector potential $A_x(y)$ and/or $A_y(x)$ such that

$$B_z = \frac{\partial A_y}{\partial x} - \frac{\partial A_x}{\partial y}$$

For a given B-field the potential A is not unique, and it is usually convenient to choose a potential that does not vary along the direction of current flow.

18.4.4. Lattice with a "Basis"

We have seen how for any given TB model we can evaluate the $E(\vec{k})$ relation from Eq.(18.18) and then fit it to a desired function. However, Eq.(18.18) will not work if we have a "lattice with a basis". For example if we apply it to the graphene lattice shown in Fig.18.9, we will get different answers depending on whether we choose "n" to be the left carbon atom or the right carbon atom. The reason is that in a lattice like this these two carbon atoms are not in identical environments: One sees two bonds to the left and one bond to the right, while the other sees one bond to the left and two bonds to the right. We call this a lattice with a basis in the sense that two carbon atoms comprise a **unit cell:** if we view a pair of carbon atoms (marked A and B) as a single entity then the lattice looks translationally invariant with each entity in an identical environment.

We can then write the set of equations in Eq.(18.10) in the form

$$E\{\psi\}_n = \sum_m [H]_{nm} \{\psi\}_m$$

$$(18.20)$$

where $\{\psi_n\}$ is a (2x1) column vector whose components represent the two atoms comprising unit cell number n. Similarly $[H_{nm}]$ is a (2x2) matrix representing the coupling between the two components of unit cell n and unit cell m (see Fig.18.9).

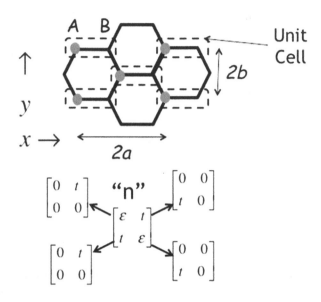

Fig.18.9: If we view two carbon atoms as a single entity then the lattice in Fig.18.4 looks translationally invariant with each entity in an identical environment. Viewing the two atoms in each unit cell as a single entity we can write the set of equations in the form shown in Eq.(18.20) with $[H]_{nm}$ given by (2x2) matrices as shown.

Now if we write the solution in the form

$$\{\psi\}_n = \{\psi\}_0 \exp(i\vec{k}.\vec{r}_n) \tag{18.21}$$

we obtain from Eq.(18.20)

$$E\{\psi\}_0 = \left[h(\vec{k})\right]\{\psi\}_0 \tag{18.22}$$

where

$$\left[h(\vec{k})\right] = \sum_m [H]_{nm} \exp\left(i\vec{k}.(\vec{r}_m - \vec{r}_n)\right) \tag{18.23}$$

Note that $[h(\vec{k})]$ obtained from Eq.(18.23) is a (2x2) matrix, which can be shown after some algebra to be

$$[h(\vec{k})] = \begin{bmatrix} \varepsilon & h_0^* \\ h_0 & \varepsilon \end{bmatrix}$$

(18.24)

where

$$h_0 \equiv t(1+2\cos(k_y b)\exp(+ik_x a))$$

(18.25)

Eq.(18.22) yields two eigenvalues for the energy E for each value of $\vec{k}$:

$$E(\vec{k}) = \varepsilon \pm \left| h_0(\vec{k}) \right|$$

(18.26)

Eq.(18.26) and (18.25) give a widely used dispersion relation for graphene. Once again to obtain a simple polynomial relation we need a Taylor series expansion around the k-value of interest. In this case the k-values of interest are those that make

$$h_0(\vec{k}) = 0,$$

so that

$$E(\vec{k}) = \varepsilon$$

This is because the equilibrium electrochemical potential is located at ε for a neutral sample for which exactly half of all the energy levels given by Eq.(18.26) are occupied.

It is straightforward to see that this requires

$$h_0(\vec{k}) = 0 \quad \rightarrow \quad k_x a = 0, k_y b = \pm 2\pi/3$$

(18.27)

Alternatively one could numerically make a grayscale plot of the magnitude of $h_0(\vec{k})$ as shown below and look for the dark spots where it is a minimum. Each of these spots is called a valley and one can do a Taylor expansion around the minimum to obtain an approximate dispersion relation valid for that valley. Note that two of the dark spots correspond to the points in Eq.(18.27), but there are other spots too and it requires some discussion to be convinced that these additional valleys do not need to be considered separately (see for example, Chapter 5, Datta (2005)).

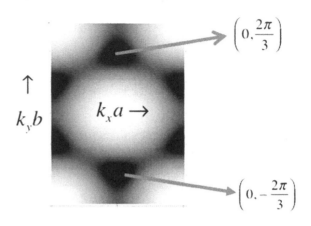

$$\left(0, \frac{2\pi}{3}\right)$$

$k_y b$ ↑

$k_x a \rightarrow$

$$\left(0, -\frac{2\pi}{3}\right)$$

A Taylor expansion around the points in Eq.(18.27) yields

$$h_0(\vec{k}) \approx \pm i t a \, (k_x \mp i \beta_y),$$

where

$$\beta_y \equiv k_y \mp 2\pi / 3b \qquad (18.28)$$

Using this approximate relation we obtain a simple dispersion relation:

$$E \;=\; \varepsilon \pm a t \sqrt{k_x^2 + k_y^2} \qquad (18.29)$$

which corresponds to the energy-momentum relation

$$E \;=\; v_0 p$$

that we stated in Lecture 5, if we set $\varepsilon = 0$. The two valleys correspond to the two values of $k_y b$ in Eq.(18.27).

In summary, although the differential form of the Schrödinger equation (Eq.(18.2)) is the well-known one that appears in most textbooks as well as on T-shirts, practical calculations are usually based on a discretized version that represents the Hamiltonian operator, H_{op} (Eq.(18.8)) as a matrix of size NxN, N being the number of basis functions used to represent the structure.

Given a set of basis functions, the matrix *[H]* can be obtained from first principles, but a widely used approach is to use the principles of bandstructure to represent the matrix in terms of a few parameters which are chosen to match key experiments. Such semi-empirical approaches are often used because of their convenience and can explain a wide range of experiments beyond the key ones used that are used as input, suggesting that they capture a lot of essential physics.

Our approach in these Lectures will be to
  (1) take accepted energy-momentum *E(p)* relations that are believed to describe the dynamics of conduction electrons with energies around the electrochemical potential μ_0 ,
  (2) extract appropriate parameters to use in tight-binding model by discretizing it.

Knowing the *[H]*, we can obtain the $[\Sigma_{1,2}]$ describing the connection to the physical contacts and possible approaches will be described when discussing specific examples in Lectures 20 through 22. A key difference between the *[H]* and $[\Sigma]$ matrices is that the former is Hermitian with real eigenvalues, while the latter is non-Hermitian with complex eigenvalues, whose significance we will discuss in the next Lecture.

As we mentioned at the outset, there are many approaches to writing *[H]* and $[\Sigma_{1,2}]$ of which we have only described the simplest versions. But regardless of how we chose to write these matrices, we can use the NEGF-based approach to be described in the next Lecture.

NEGF Method

19.1. One-level Resistor
19.2. Multi-level Resistors
19.3. Conductance Functions for Coherent Transport
19.4. Elastic Dephasing

In the last Lecture I tried to provide a super-brief but hopefully self-contained introduction to the Hamiltonian matrix *[H]* whose eigenvalues tell us the allowed energy levels in the channel. However, *[H]* describes an isolated channel and we cannot talk about the steady-state resistance of an isolated channel without bringing in the contacts and the battery connected across it. In this Lecture I will describe the NEGF-based transport model that can be used to model current flow, given *[H]* and the *[Σ]*'s (Fig.19.1).

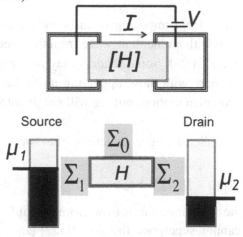

Fig.19.1. The NEGF-based quantum transport model described here allows us to model current flow given the Hamiltonian matrix *[H]* describing the channel, the self-energy matrices *[Σ]* describing the connection of the channel to the contacts, and *[Σ₀]* describing interactions within the channel.

As I mentioned in Lectures 1 and 18, the NEGF method originated from the seminal works of Martin and Schwinger (1959), Kadanoff and Baym (1962), Keldysh (1965) and others who used the methods of many-body perturbation theory (MBPT) to describe the distributed entropy-generating processes along the channel which were believed to constitute the essence of resistance. Since MBPT is an advanced topic requiring many semesters to master, the NEGF method is generally regarded as an esoteric tool for specialists.

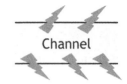

Channel

In our bottom-up approach we will start with elastic resistors for which energy exchange is confined to the contacts, and the problem of resistance can be treated within a one-electron picture by connecting contacts to the Schrödinger equation. Indeed our approach will be to start from the usual time-independent Schrödinger equation $E\{\psi\} = [H]\{\psi\}$ and add two terms to it representing the outflow and inflow from the contact

$$E\{\psi\} = [H]\{\psi\} + \underbrace{[\Sigma]\{\psi\}}_{OUTFLOW} + \underbrace{\{s\}}_{INFLOW}$$

These two terms arise from imposing open boundary conditions on the Schrödinger equation with an incident wave from the contact as shown in Chapters 8,9 of Datta (2005). Some readers may notice the similarity of the additional terms here with those appearing in the Langevin equation used to describe Brownian motion, but we will not go into it.

Using this modified Schrödinger equation, the wavefunction can be written as

$$\{\psi\} = [EI - H - \Sigma]^{-1}\{s\}$$

We will then argue that since the inflow from multiple sources *{s}* are incoherent, one cannot superpose the resulting $\{\psi\}$'s and it is more convenient to work in terms of quantities like (superscript '+' denotes conjugate transpose)

$$[G^n] \sim \{\psi\}\{\psi\}^+$$

$$[\Sigma^{in}] \sim \{s\}\{s\}^+$$

which can be superposed. Defining

$$G^R = [EI - H - \Sigma]^{-1} \tag{19.1}$$

$$and \quad G^A = [G^R]^+$$

we can write $\qquad \{\psi\} = [G^R]\{s\}$

so that $\qquad \underbrace{\{\psi\}\{\psi\}^+}_{G^n} = [G^R]\underbrace{\{s\}\{s\}^+}_{\Sigma^{in}}[G^A]$

giving us the second NEGF equation

$$G^n = G^R \Sigma^{in} G^A \tag{19.2}$$

Though we have changed the notation somewhat, writing Σ for Σ^R, and (see Chapter 8, Datta 1995)

$$G^n \quad for \quad -iG^< \qquad and \qquad \Sigma^{in} \quad for \quad -i\Sigma^<$$

Eqs.(19.1, 19.2) are essentially the same as Eqs.(75-77) in Keldysh (1965), which is one of the seminal founding papers on the NEGF method that obtained these equations using MBPT. Although for simplicity we have only discussed the time-independent version here, a similar derivation could be used for the time-dependent version too (See Appendix, Datta (2005)).

How could we obtain these results using elementary arguments, without invoking MBPT? Because we are dealing with an elastic resistor where all entropy-generating processes are confined to the contacts and can be handled in a relatively elementary manner. But should we call this NEGF?

It seems to us that NEGF has two aspects, namely
  A. Eqs.(19.1), (19.2) and
  B. calculating $[\Sigma]$, $[\Sigma^{in}]$ that appear in Eqs.(19.1), (19.2).

For historical reasons, these two aspects, A and B, are often intertwined in the literature, but they need not be. Indeed these two aspects are completely distinct in the Boltzmann formalism (Lecture 7). The Boltzmann transport equation (BTE)

$$\frac{\partial f}{\partial t} + \vec{v}.\vec{\nabla}f + \vec{F}.\vec{\nabla}_p f = S_{op}f \quad \text{(same as Eq.(7.5))}$$

is used to describe semiclassical transport in many different contexts, but the evaluation of the scattering operator S_{op} has evolved considerably since the days of Boltzmann and varies widely depending on the problem at hand.

Similarly it seems to me that the essence of NEGF is contained in Eqs.(19.1), (19.2) while the actual evaluation of the $[\Sigma]$'s may well evolve as we look at more and more different types of problems. The original MBPT–based approach may or may not be the best, and may need to be modified even for problems involving electron-electron interactions.

Above all we believe that by decoupling Eqs.(19.1) and (19.2) from the MBPT method originally used to derive them, we can make the NEGF method more transparent and accessible so that it can become a part of the standard training of physics and engineering students who need to apply it effectively to a wide variety of basic and applied problems that require connecting contacts to the Schrödinger equation.

I should also note briefly the relation between the NEGF method applied to elastic resistors with the scattering theory of transport or the transmission formalism widely used in mesoscopic physics. Firstly, the scattering theory works directly with the Schrödinger equation with open boundary conditions that effectively add the inflow and outflow terms we mentioned:

$$E\{\psi\} = [H]\{\psi\} + \underbrace{[\Sigma]\{\psi\}}_{OUTFLOW} + \underbrace{\{s\}}_{INFLOW}$$

However, as we noted earlier it is then important to add individual sources incoherently, something that the NEGF equation (Eq.(19.2)) takes care of automatically.

The second key difference is the handling of dephasing processes in the channel, something that has no classical equivalent. In quantum transport randomization of the phase of the wavefunction even without any momentum relaxation can have a major impact on the measured conductance. The scattering theory of transport usually neglects such dephasing processes and is restricted to *phase-coherent elastic resistors.*

Incoherence is commonly introduced in this approach using an insightful observation due to Büttiker that dephasing processes essentially remove electrons from the channel and re-inject them just like the voltage probes discussed in Section 12.2 and so one can include them phenomenologically by introducing conceptual contacts in the channel.

This method is widely used in mesoscopic physics, but it seems to introduce both phase and momentum relaxation and I am not aware of a convenient way to introduce pure phase relaxation if we wanted to. In the NEGF method it is straightforward to choose $[\Sigma_0]$ so as to include phase relaxation with or without momentum relaxation as we will see in the next Lecture. In addition, the NEGF method provides a rigorous framework for handling all kinds of interactions in the channel, both elastic and inelastic, using MBPT. Indeed that is what the original work from the 1960's was about.

Let me finish up this long introduction by briefly mentioning the two other key equations in NEGF besides Eqs.(19.1) and (19.2). As we will see, the quantity $[G^n]$ appearing in Eq.(19.2) represents a matrix version of the electron density (times 2π) from which other quantities of interest can be calculated. Another quantity of interest is the matrix version of the density of states (again times 2π) called the spectral function *[A]* given by

$$A = \quad G^R \Gamma G^A = \quad G^A \Gamma G^R$$
$$= \quad i[G^R - G^A] \tag{19.3a}$$

where G^R, G^A are defined in Eq.(19.1) and the $[\Gamma]$'s represent the **anti-Hermitian** parts of the corresponding $[\Sigma]$'s

$$\Gamma = i[\Sigma - \Sigma^+] \tag{19.3b}$$

which describe how easily the electrons in the channel communicate with the contacts.

There is a component of $[\Sigma]$, $[\Gamma]$, $[\Sigma^{in}]$ for each contact (physical or otherwise) and the quantities appearing in Eqs.(19.1-19.3) are the total obtained summing all components. The current at a specific contact m, however, involves only those components associated with contact m:

$$\tilde{I}_m = \quad \frac{q}{h} Trace[\Sigma_m^{in} A - \Gamma_m G^n] \tag{19.4}$$

Note that $\tilde{I}_m(E)$ represents the current per unit energy and has to be integrated over all energy to obtain the total current. In the following four Lectures we will look at a few examples designed to illustrate how Eqs.(19.1)-(19.4) are applied to obtain concrete results.

But for the rest of this Lecture let me try to justify these equations. We start with a one-level version for which all matrices are just numbers (Section 19.1), then look at the full multi-level version (Section 19.2), obtain an expression for the conductance function $G(E)$ for coherent transport (Section 19.3) and finally look at the different choices for the dephasing self-energy $[\Sigma_0]$ (Section 19.4).

19.1.One-level resistor

To get a feeling for the NEGF method, it is instructive to look at a particularly simple conductor having just one level and described by a 1x1 *[H]* matrix that is essentially a number: $[H] = \varepsilon$.

Starting directly from the Schrödinger equation we will see how we can introduce contacts into this problem. This will help set the stage for Section 19.3 when we consider arbitrary channels described by (NxN) matrices instead of the simple one-level channel described by (1x1) "matrices."

19.1.1. Semiclassical treatment

It is useful to first go through a semiclassical treatment as an intuitive guide to the quantum treatment. Physically we have a level connected to two contacts, with two different occupancy factors

$$f_1(\varepsilon) \quad and \quad f_2(\varepsilon)$$

Let us assume the occupation factor to be one for the source and zero for the drain, so that it is only the source that is continually trying to fill up the level while the drain is trying to empty it. We will calculate the resulting current and then multiply it by

$$f_1(\varepsilon) - f_2(\varepsilon)$$

to account for the fact that there is injection from both sides and the net current is the difference.

Fig.19.2.
Filling and emptying a level: Semiclassical picture

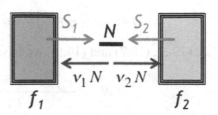

With $f_1=1$ in the source and $f_2=0$ in the drain, the average number N of electrons ($N<1$) should obey an equation of the form

$$\frac{d}{dt}N = -(v_1+v_2)N + S_1 + S_2 \tag{19.5}$$

where v_1 and v_2 represent the rates (per second) at which an electron escapes into the source and drain respectively, while S_1 is the rate at which electrons try to enter from the source. The steady state occupation is obtained by setting

$$\frac{d}{dt}N = 0 \rightarrow N = \frac{S_1+S_2}{v_1+v_2} \tag{19.6}$$

We can fix S_1, by noting that if the drain were to be disconnected, N should equal the Fermi function $f_1(\varepsilon)$ in contact 1, which we will assume one for this discussion. This means

$$\frac{S_1}{v_1} = f_1(\varepsilon) \quad and \quad \frac{S_2}{v_2} = f_2(\varepsilon)$$
$$\tag{19.7}$$

The current can be evaluated by writing Eq.(19.5) in the form

$$\frac{dN}{dt} = (S_1 - v_1 N) + (S_2 - v_2 N) \tag{19.8}$$

and noting that the first term on the right is the current from the source while the second is the current into the drain. Under steady state conditions, they are equal and either could be used to evaluate the current that flows in the circuit:

$$I = q(S_1 - v_1 N) = q(v_2 N - S_2) \tag{19.9}$$

From Eqs.(19.6), (19.7) and (19.9), we have

$$N = \frac{v_1 f_1(\varepsilon) + v_2 f_2(\varepsilon)}{v_1+v_2} \tag{19.10a}$$

$$I = q\frac{v_1 v_2}{v_1 + v_2}\left(f_1(\varepsilon) - f_2(\varepsilon)\right) \qquad (19.10b)$$

and

19.1.2. Quantum treatment

Let us now work out the same problem using a quantum formalism based on the Schrödinger equation. In the last Chapter we introduced the matrix version of the time-independent Schrödinger equation

$$E\{\psi\} = [H]\{\psi\}$$

which can be obtained from the more general time-dependent equation

$$i\hbar\frac{\partial}{\partial t}\{\tilde{\psi}(t)\} = [H]\{\tilde{\psi}(t)\} \qquad (19.11a)$$

by assuming

$$\{\tilde{\psi}(t)\} = \{\psi\}\, e^{-iEt/\hbar} \qquad (19.11b)$$

For problems involving steady-state current flow, the time-independent version is usually adequate, but sometimes it is useful to go back to the time-dependent version because it helps us interpret certain quantities like the self-energy functions as we will see shortly.

In the quantum formalism the squared magnitude of the electronic wavefunction $\tilde{\psi}(t)$ tells us the probability of finding an electron occupying the level and hence can be identified with the average number of electrons $N\ (< 1)$. For a single isolated level with $[H] = \varepsilon$, the time evolution of the wavefunction is described by

$$i\hbar\frac{d}{dt}\tilde{\psi} = \varepsilon\tilde{\psi}$$

which with a little algebra leads to

$$\frac{d}{dt}\left(\tilde{\psi}\tilde{\psi}^*\right) = 0$$

showing that for an isolated level, the number of electrons $\tilde{\psi}\tilde{\psi}^*$ does not change with time.

Our interest, however, is not in isolated systems, but in channels connected to two contacts. Unfortunately the standard quantum mechanics literature does not provide much guidance in the matter, but we can do something relatively simple using the rate equation in Eq.(19.4) as a guide.

We introduce **contacts into the Schrödinger equation** by modifying it to read

$$i\hbar\frac{d}{dt}\tilde{\psi} = \left(\varepsilon - i\frac{\gamma_1+\gamma_2}{2}\right)\tilde{\psi} \qquad (19.12a)$$

so that the resulting equation for

$$\frac{d}{dt}\tilde{\psi}\tilde{\psi}^* = -\left(\frac{\gamma_1+\gamma_2}{\hbar}\right)\tilde{\psi}\tilde{\psi}^* \qquad (19.12b)$$

looks just like Eq.(19.5) except for the source term S_1 which we will discuss shortly.

We can make Eq.(19.12b) match Eq.(19.5) if we choose

$$\gamma_1 = \hbar\nu_1 \qquad (19.13a)$$

$$\gamma_2 = \hbar\nu_2 \qquad (19.13b)$$

We can now go back to the **time-independent version** of Eq.(19.12a):

$$E\psi = \left(\varepsilon - i\frac{\gamma_1+\gamma_2}{2}\right)\psi \qquad (19.14)$$

obtained by assuming a single energy solution:

$$\tilde{\psi}(t) = \psi(E)\,e^{-iEt/\hbar}$$

Eq.(19.14) has an obvious solution $\psi=0$, telling us that at steady-state there are no electrons occupying the level which makes sense since we have not included the source term S_1. All electrons can do is to escape into the contacts, and so in the long run the level just empties to zero.

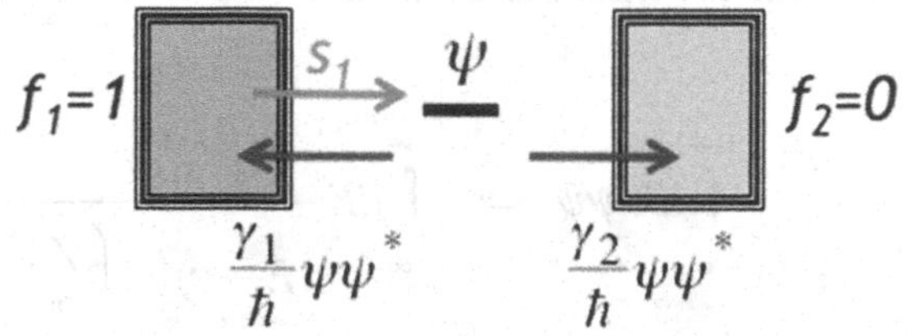

Fig.19.3.
Filling and emptying a level: Quantum picture

Introducing a source term into Eq.(19.11) and defining $\gamma \equiv \gamma_1 + \gamma_2$, we have

$$E\psi \;=\; \left(\varepsilon - i\frac{\gamma}{2}\right)\psi \; + s_1 \qquad (19.15)$$

Unlike the semiclassical case (Eq.(19.5)) we are introducing only one source rather than two. The reason is subtle and we will address it later at the end of this Section. From Eq.(19.15), we can relate the wavefunction to the source

$$\psi \;=\; \frac{s_1}{E-\varepsilon+i(\gamma/2)} \qquad (19.16)$$

Note that the wavefunction is a maximum when the electron energy E equals the energy ε of the level, as we might expect. But the important point about the quantum treatment is that the wavefunction is not significantly diminished as long as E differs from ε by an amount less than γ. This is an example of "broadening" or energy uncertainty that a semiclassical picture misses.

To obtain the **strength of the source** we require that the total number of electrons on integrating over all energies should equal our rate equation result from Eq.(19.5). that is,

$$\int_{-\infty}^{+\infty} dE \, \psi\psi^* = \frac{v_1}{v_1 + v_2} = \frac{\gamma_1}{\gamma_1 + \gamma_2} \tag{19.17}$$

where we have made use of Eq.(19.13). We now use Eqs.(19.16), (19.17) to evaluate the right hand side in terms of the source

$$\int_{-\infty}^{+\infty} dE \, \psi\psi^* = \int_{-\infty}^{+\infty} dE \, \frac{s_1 s_1^*}{(E-\varepsilon)^2 + \left(\dfrac{\gamma}{2}\right)^2} = \frac{2\pi s_1 s_1^*}{\gamma} \tag{19.18}$$

where we have made use of a standard integral

$$\int_{-\infty}^{+\infty} dE \, \frac{\gamma}{(E-\varepsilon)^2 + \left(\dfrac{\gamma}{2}\right)^2} = 2\pi \tag{19.19}$$

From Eqs.(19.17) and (19.18) we obtain, noting that ,

$$2\pi s_1 s_1^* = \gamma_1 \tag{19.20}$$

The strength of the source is thus proportional to the escape rate which seems reasonable: if the contact is well coupled to the channel and electrons can escape easily, they should also be able to come in easily.

Just as in the semiclassical case (Eq.(19.9)) we obtain the **current** by looking at the rate of change of N from Eq.(19.12b)

$$\frac{d}{dt} \tilde{\psi}\tilde{\psi}^* = (\textit{Inflow from 1}) - \frac{\gamma_1}{\hbar} \tilde{\psi}\tilde{\psi}^* - \frac{\gamma_2}{\hbar} \tilde{\psi}\tilde{\psi}^*$$

where we have added a term "Inflow from 1" as a reminder that Eq.(19.12a) does not include a source term. Both left and right hand sides of this equation are zero for the steady-state solutions we are considering. But just like the semiclassical case, we can identify the current as either the first two terms or the last term on the right:

$$\frac{I}{q} = (\textit{Inflow from 1}) \quad -\frac{\gamma_1}{\hbar}\tilde{\psi}\tilde{\psi}* \quad = \quad \frac{\gamma_2}{\hbar}\tilde{\psi}\tilde{\psi}*$$

Using the second form and integrating over energy we can write

$$I = q \int\limits_{-\infty}^{+\infty} dE \frac{\gamma_2}{\hbar}\psi\psi* \tag{19.21}$$

so that making use of Eqs.(19.16) and (19.20), we have

$$I = \frac{q}{\hbar}\frac{\gamma_1\gamma_2}{2\pi} \int\limits_{-\infty}^{+\infty} dE \frac{1}{(E-\varepsilon)^2 + (\gamma/2)^2} \tag{19.22}$$

which can be compared to the semiclassical result from Eq.(19.10) with $f_1=1$, $f_2=0$ (note: $\gamma = \gamma_1 + \gamma_2$)

$$I = \frac{q}{\hbar}\frac{\gamma_1\gamma_2}{\gamma_1+\gamma_2}$$

19.1.3. Quantum broadening

Note that Eq.(19.22) involves an integration over energy, as if the quantum treatment has turned the single sharp level into a continuous distribution of energies described by a density of states $D(E)$:

$$D = \frac{\gamma/2\pi}{(E-\varepsilon)^2 + (\gamma/2)^2} \tag{19.23}$$

Quantum mechanically the process of coupling inevitably spreads a single discrete level into a state that is distributed in energy, but integrated over all energy still equals one (see Eq.(19.15)). One could call it a consequence of the **uncertainty relation**

$$\gamma t \geq h$$

relating the length of time t the electron spends in a level to the uncertainty " γ " in its energy. The stronger the coupling, shorter the time and larger the broadening.

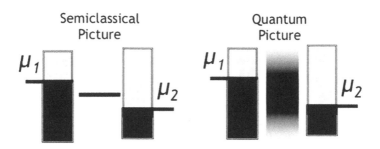

Is there any experimental evidence for this energy broadening (Eq.(19.23)) predicted by quantum theory? A hydrogen molecule has an energy level diagram like the one-level resistor we are discussing and experimentalists have measured the conductance of a hydrogen molecule with good contacts and it supports the quantum result (Smit et al. 2002). Let me elaborate a little.

Comparing Eq.(19.22) with Eq.(3.3) for elastic resistors we can write the conductance function for a one-level device including quantum broadening as

$$G(E) = \frac{q^2}{h} \frac{\gamma_1 \gamma_2}{(E - \varepsilon)^2 + \left(\dfrac{\gamma}{2}\right)^2}$$

If we assume (1) equal coupling to both contacts:

$$\gamma_1 = \gamma_2 = \frac{\gamma}{2}$$

and (2) a temperature low enough that the measured conductance equals $G(E=\mu_0)$, μ_0 being the equilibrium electrochemical potential, we have

$$G \approx G(E = \mu_0) = \frac{q^2}{h} \frac{(\gamma/2)^2}{(\mu_0 - \varepsilon)^2 + (\gamma/2)^2}$$

So the quantum theory of the one-level resistor says that the measured conductance should show a maximum value equal to the quantum of conductance q^2/h when μ_0 is located sufficiently close to ε. The experimentally measured conductance is equal to $2q^2/h$, the extra factor of 2 being due to spin degeneracy, since levels come in pairs and what we have is really a two-level rather than a one-level resistor.

19.1.4. Do Multiple Sources Interfere?

In our quantum treatment we considered a problem with electrons injected only from the source ($f_1 = 1$) with the drain empty ($f_2 = 0$) (Eq.(19.15)), unlike the semiclassical case where we started with both sources S_1 and S_2 (Eq.(19.5)).

This is not just a matter of convenience. If instead of Eq.(19.15) we start from

$$E\psi = \left(\varepsilon - i\frac{\gamma}{2}\right)\psi + s_1 + s_2$$

we obtain

$$\psi = \frac{s_1 + s_2}{E - \varepsilon + i\frac{\gamma}{2}}$$

so that

$$\psi\psi^* = \frac{1}{(E-\varepsilon)^2 + \left(\frac{\gamma}{2}\right)^2} (s_1 s_1{}^* + s_2 s_2{}^* + \underbrace{s_1 s_2{}^* + s_2 s_1{}^*}_{\text{Interference Terms}})$$

which has two extra interference terms that are never observed experimentally because the electrons injected from separate contacts have uncorrelated phases that change randomly in time and average to zero.

The first two terms on the other hand add up since they are positive numbers. It is like adding up the light from two light bulbs: we add their powers not their electric fields. Laser sources on the other hand can be coherent so that we actually add electric fields and the interference terms can be seen experimentally. Electron sources from superconducting contacts too can be coherent leading to Josephson currents that depend on interference. But that is a different matter.

Our point here is simply that normal contacts like the ones we are discussing are incoherent and it is necessary to take that into account in our models. The moral of the story is that we cannot just insert multiple sources into the Schrödinger equation. We should insert one source at a time, calculate bilinear quantities (things that depend on the product of wavefunctions) like electron density and current and add up the contributions from different sources. Next we will describe the non-equilibrium Green function (NEGF) method that allows us to implement this procedure in a systematic way and also to include incoherent processes.

19.2. Quantum transport through multiple levels

We have seen how we can treat quantum transport through a one-level resistor with a time-independent Schrödinger equation modified to include the connection to contacts and a source term:

$$E\psi = \left(\varepsilon - i\frac{\gamma}{2}\right)\psi + s$$

How do we extend this method to a more general channel described by an NxN Hamiltonian matrix *[H]* whose eigenvalues give the *N* energy levels?

For an *N*-level channel , the wavefunction $\{\psi\}$ and source term $\{s_1\}$ are Nx1 column vectors and the modified Schrödinger equation looks like

$$E\{\psi\} = [H + \Sigma_1 + \Sigma_2]\{\psi\} + \{s_1\} \tag{19.24}$$

where Σ_1 and Σ_2 are NxN non-Hermitian matrices whose anti-Hermitian components

$$\Gamma_1 = i[\Sigma_1 - \Sigma_1^+]$$

$$\Gamma_2 = i[\Sigma_2 - \Sigma_2^+]$$

play the roles of $\gamma_{1,2}$ in our one-level problem.

Fig.19.4. Transport model for multi-level conductor.

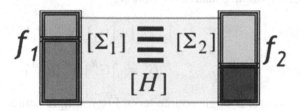

In Lecture 18 we discussed how for different structures we can write down the channel Hamiltonian *[H]* and in the next few Lectures I will present examples to show how the $[\Sigma]$ are obtained.

For the moment, let us focus on how the basic NEGF equations summarized earlier (Eqs.(19.1)-(19.4)) follow from our contact-ed Schrödinger equation, Eq.(19.24).

19.2.1. Obtaining Eqs.(19.1)

From Eq.(19.24) it is straightforward to write

$$\{\psi\} = [G^R]\{s_1\}$$

where G^R is given by Eq.(19.1) with

$$\Sigma = \Sigma_1 + \Sigma_2 \qquad (19.25)$$

19.2.2. Obtaining Eqs.(19.2)

The matrix electron density, G^n, defined as

$$G^n \rightarrow 2\pi \{\psi\}\{\psi\}^+ = 2\pi [G^R]\{s_1\}\{s_1\}^+ [G^A]$$

where the superscript "+" stands for conjugate transpose, and G^A stands for the conjugate transpose of G^R.

For the one-level problem $2\pi s_1 s_1^* = \gamma_1$ (see Eq.(19.20)): the corresponding matrix relation is

$$2\pi \{s_1\}\{s_1\}^+ = [\Gamma_1]$$

so that $$G^n = [G^R][\Gamma_1][G^A]$$

This is for a single source term. For multiple sources, the electron density matrices, unlike the wavefunctions, can all be added up with the appropriate Fermi function weighting to give Eq.(19.2),

$$G^n = [G^R][\Sigma^{in}][G^A] \qquad \text{(same as 19.2)}$$

with Σ^{in} representing an incoherent sum of all the independent sources:

$$[\Sigma^{in}] = [\Gamma_1] f_1(E) + [\Gamma_2] f_2(E) \qquad (19.26)$$

19.2.3. Obtaining Eq.(19.3)

Eq.(19.2) gives us the electron density matrix G^n, in terms of the Fermi functions f_1 and f_2 in the two contacts. But if both f_1 and f_2 are equal to one then all states are occupied, so that the matrix electron density becomes equal to the matrix density of states, called the spectral function matrix [A]. Setting $f_1 = 1$ and $f_2 = 1$, in Eq.(19.2) we have

$$[A] = [G^R][\Gamma][G^A] \qquad (19.27)$$

since $\Gamma = \Gamma_1 + \Gamma_2$. This gives us part of Eq.(19.3). The rest of Eq.(19.3) can be obtained from Eq.(19.1) using straightforward algebra as follows:

$$[G^R]^{-1} = EI - H - \Sigma \qquad (19.28a)$$

Taking conjugate transpose of both sides

$$\left[[G^R]^{-1}\right]^+ = \left[[G^R]^+\right]^{-1} = EI - H - \Sigma^+ \qquad (19.28b)$$

Subtracting Eq.(19.28b) from (19.28a) (note that G^A stands for $[G^R]^+$) and making use of Eq.(19.3b)

$$[G^R]^{-1} - [G^A]^{-1} = i[\Gamma] \qquad (19.28c)$$

Mutiplying with $[G^R]$ from the left and $[G^A]$ from the right we have

$$i\left[[G^R]-[G^A]\right] = G^R \Gamma G^A$$

thus giving us another piece of Eq.(19.3). The final piece is obtained by multiplying Eq.(19.28c) with $[G^A]$ from the left and $[G^R]$ from the right.

19.2.4. Obtaining Eq,(19.4): The Current Equation

Like the semiclassical treatment and the one-level quantum treatment, the current expression is obtained by considering the time variation of the number of electrons N. Starting from

$$i\hbar \frac{d}{dt} \{\psi\} = [H + \Sigma]\{\psi\} + \{s\}$$

and its conjugate transpose (noting that H is a Hermitian matrix)

$$-i\hbar \frac{d}{dt} \{\psi\}^+ = \{\psi\}^+ [H + \Sigma^+] + \{s\}^+$$

we can write

$$i\hbar\frac{d}{dt}\{\psi\}\{\psi\}^+ = \left(i\hbar\frac{d}{dt}\{\psi\}\right)\{\psi\}^+ + \{\psi\}\left(i\hbar\frac{d}{dt}\{\psi\}^+\right)$$

$$= \left([H+\Sigma]\{\psi\}+\{s\}\right)\{\psi\}^+ - \{\psi\}\left(\{\psi\}^+[H+\Sigma^+]+\{s\}^+\right)$$

$$= [(H+\Sigma)\psi\psi^+ - \psi\psi^+(H+\Sigma^+)] + [ss^+G^A - G^Rss^+]$$

where we have made use of the relations

$$\{\psi\} = [G^R]\{s\} \quad and \quad \{\psi\}^+ = \{s\}^+[G^A]$$

Since the trace of $[\psi\psi^+]$ represents the number of electrons, we could define its time derivative as a matrix current operator whose trace gives us the current. Noting further that

$$2\pi\{\psi\}\{\psi\}^+ = [G^n] \quad and \quad 2\pi\{s\}\{s\}^+ = [\Gamma]$$

we can write

$$I^{op} = \frac{[HG^n - G^nH] + [\Sigma G^n - G^n\Sigma^+] + [\Sigma^{in}G^A - G^R\Sigma^{in}]}{i2\pi\hbar} \tag{19.29}$$

We will talk more about the current operator in Lecture 22 when we talk about spins, but for the moment we just need its trace which tells us the time rate of change of the number of electrons N in the channel

$$\frac{dN}{dt} = \frac{-i}{h} Trace\left([\Sigma G^n - G^n\Sigma^+] + [\Sigma^{in}G^A - G^R\Sigma^{in}]\right)$$

noting that *Trace [AB] = Trace [BA]*. Making use of Eq.(19.3b)

$$\frac{dN}{dt} = \frac{1}{h} Trace\left[\Sigma^{in}A - \Gamma G^n\right]$$

Now comes a tricky argument. Both the left and the right hand sides of Eq.(19.29) are zero, since we are discussing steady state transport with no time variation. The reason we are spending all this time discussing something that is zero is that the terms on the left can be separated into two parts, one associated with contact 1 and one with contact 2. They tell us the currents at contacts 1 and 2 respectively and the fact that they add up to zero is simply a reassuring statement of Kirchhoff's law for steady-state currents in circuits.

With this in mind we can write for the current at contact m $(m=1,2)$

$$\tilde{I}_m = \frac{q}{h} Trace \left[\Sigma_m^{in} A - \Gamma_m G^n \right]$$

as stated earlier in Eq.(19.4). This leads us to the picture shown in Fig.19.5 where we have also shown the semiclassical result for comparison.

Classical

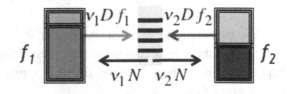

Fig.19.5:
Filling and emptying a channel: Classical and quantum treatment.

Quantum

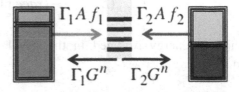

19.3. Conductance Functions for Coherent Transport

Finally we note that using Eqs.(19.2)-(19.3) we can write the current from Eq.(19.4) a little differently

$$\tilde{I}(E) = \frac{q}{h} Trace[\Gamma_1 G^R \Gamma_2 G^A] \left(f_1(E) - f_2(E) \right)$$

which is very useful for it suggests a quantum expression for the conductance function $G(E)$ that we introduced in Lecture 3 for all elastic resistors:

$$G(E) = \frac{q^2}{h} Trace \left[\Gamma_1 G^R \Gamma_2 G^A \right] \qquad (19.30)$$

More generally with multiterminal conductors we could introduce a self-energy function for each contact and show that

$$\tilde{I}_m = \frac{q}{h} \sum_r \bar{T}_{mn} \left(f_m(E) - f_n(E) \right) \qquad (19.31)$$

with $$\qquad \bar{T}_{mn} \equiv Trace \left[\Gamma_m G^R \Gamma_n G^A \right] \qquad (19.32)$$

For low bias we can use our usual Taylor series expansion from Eq.(2.8) to translate the Fermi functions into electrochemical potentials so that Eq.(19.31) looks just like the Büttiker equation (Eq.(12.3)) with the conductance function given

$$G_{m,n}(E) \equiv \frac{q^2}{h} Trace \left[\Gamma_m G^R \Gamma_n G^A \right] \qquad (19.33)$$

which is energy-averaged in the usual way for elastic resistors (see Eq.(3.1)).

$$G_{m,n} = \int\limits_{-\infty}^{+\infty} dE \left(-\frac{\partial f_0}{\partial E} \right) G_{m,n}(E)$$

19.4. Elastic Dephasing

So far we have focused on the physical contacts described by $[\Sigma_{1,2}]$ and the model as it stands describes coherent quantum transport where electrons travel coherently from source to drain in some static structure described by the Hamiltonian *[H]* without any interactions along the channel described by $[\Sigma_0]$ (Fig.19.1). In order to include $[\Sigma_0]$, however, no change is needed as far as Eqs.(19.1) through (19.4) is concerned. It is just that an additional term appears in the definition of Σ, Σ^{in}:

$$\Sigma = \Sigma_1 + \Sigma_2 + \Sigma_0$$

$$\Gamma = \Gamma_1 + \Gamma_2 + \Gamma_0$$

$$[\Sigma^{in}] = [\Gamma_1]f_1(E) + [\Gamma_2]f_2(E) + [\Sigma_0^{in}] \qquad (19.34)$$

What does $[\Sigma_0]$ represent physically? From the point of view of the electron a solid does not look like a static medium described by *[H]*, but like a rather turbulent medium with a random potential U_R that fluctuates on a picosecond time scale. Even at fairly low temperatures when phonons have been frozen out, an individual electron continues to see a fluctuating potential due to all the other electrons, whose average is modeled by the scf potential we discussed in Section 18.2. These fluctuations do not cause any overall loss of momentum from the system of electrons, since any loss from one electron is picked up by another. However, they do cause fluctuations in the phase leading to fluctuations in the current. What typical current measurements tell us is an average flow over nanoseconds if not microseconds or milliseconds. This averaging effect needs to be modeled if we wish to relate to experiments.

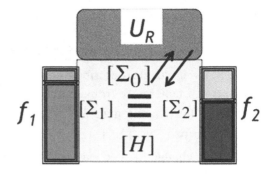

Fig.19.6.
Quantum transport model with
simple elastic dephasing.

As we mentioned earlier, the NEGF method was originally developed in the 1960's to deal with the problem of including inelastic processes into a quantum description of large conductors. For the moment, however, we will look at simple elastic dephasing processes leaving more general interactions for Lecture 23.

For such processes the self-energy functions are given by

$$[\Sigma_0] = D \times [G^R] \qquad (19.35a)$$

$$[\Sigma_0^{in}] = D \times [G^n] \qquad (19.35b)$$

where $\times$ denotes element by element multiplication. Making use of the relations in Eqs.(19.3), it is straightforward to show from Eq.(19.35a) that

$$[\Gamma_0] = D \times [A] \qquad (19.35c)$$

The elements of the matrix *[D]* represent the correlation between the random potential at location "i" and at location "j":

$$D_{ij} = \left\langle U_{Ri} U_{Rj} \right\rangle \qquad (19.36)$$

Two cases are of particular interest. The first is where the random potential is well-correlated throughout the channel having essentially the same value at all points "i'" so that the every element of the matrix *[D]* has essentially the same value D_0:

Model A: $$D_{ij} = D_0 \qquad\qquad (19.37)$$

The other case is where the random potential has zero correlation from one spatial point i to another j, so that

Model B: $D_{ij} = D_0$, $i = j$ and $= 0$, $i \ne j$ $\qquad (19.38)$

Real processes are usually somewhere between the two extremes represented by models A and B.

To see where Eqs.(19.35) come from we go back to our contact-ed Schrödinger equation

$$E\{\psi\} = [H + \Sigma_1 + \Sigma_2]\{\psi\} + \{s_1\}$$

and noting that a random potential U_R should lead to an additional term that could be viewed as an additional source term

$$E\{\psi\} = [H + \Sigma_1 + \Sigma_2]\{\psi\} + U_R\{\psi\} + \{s_1\}$$

with a corresponding inscattering term given by

$$\Sigma_0^{in} = 2\pi U_R U_R^*\{\psi\}\{\psi\}^+ = D_0 G^n$$

corresponding to Model A (Eq.(19.37)) and a little more careful argument leads to the more general result in Eq.(19.36). That gives us Eq.(19.35b). How about Eq.(19.35a) and (19.35c)?

The simplest way to justify Eq.(19.35c) is to note that together with Eq.(19.35b) (which we just obtained) it ensures that the current at terminal 0 from Eq.(19.4) equals zero:

$$I_0 = \frac{q}{h} \, Trace[\Sigma_0^{in} A - \Gamma_0 G^n]$$

$$= \frac{q}{h} \, Trace[G^n \Gamma_0 - \Gamma_0 G^n] = 0$$

This is a required condition since terminal 0 is not a physical contact where electrons can actually exit or enter from.

Indeed a very popular method due to Büttiker introduces incoherent processes by including a fictitious probe (often called a Büttiker probe) whose electrochemical potential is adjusted to ensure that it draws zero current. In NEGF language this amounts to assuming

$$\Sigma_0^{in} = \Gamma_0 \, f_P$$

with the number f_P is adjusted for zero current. This would be equivalent to the approach described here if the probe coupling Γ_0 were chosen proportional to the spectral function $[A]$ as required by Eq.(19.35c).

Note that our prescription in Eq.(19.35) requires a "self-consistent evaluation" since Σ, Σ^{in} depend on G^R and G^n which in turn depend on Σ, Σ^{in} respectively (see Eqs.(19.1), (19.2)).

Also, Model A (Eq.(19.37)) requires us to calculate the full Green's function which can be numerically challenging for large devices described by large matrices. Model B makes the computation numerically much more tractable because one only needs to calculate the diagonal elements of the Green's functions which can be done much faster using powerful algorithms.

In these Lectures, however, we focus on conceptual issues using "toy" problems for which numerical issues are not the "show stoppers." The important conceptual distinction between Models A and B is that the former destroys phase but not momentum, while the latter destroys momentum as well [Golizadeh-Mojarad et al. 2007].

The dephasing process can be viewed as extraction of the electron from a state described *by* $[G^n]$ and reinjecting it in a state described by $D \times G^n$. Model A is equivalent to multiplying $[G^n]$ by a constant so that the electron is reinjected in exactly the same state that it was extracted in, causing no loss of momentum, while Model B throws away the off-diagonal elements and upon reinjection the electron is as likely to go on way or another. Hopefully this will get clearer in the next Lecture when we look at a concrete example.

Another question that the reader might raise is whether instead of including elastic dephasing through a self-energy function $[\Sigma_0]$ we could include a potential U_R in the Hamiltonian itself and then average over a number of random realizations of U_R. The answer is that the two methods are not exactly equivalent though in some problems they could yield similar results. This too should be a little clearer in the next lecture when we look at a concrete example.

For completeness, let me note that in the most general case D_{ijkl} is a fourth order tensor and the version we are using (Eq.(19.35)) represents a special case for which D_{ijkl} is non-zero only if $i=k$, $j=l$ (see Appendix E).

Lecture 20

Can Two Offer Less Resistance than One?

20.1. Modeling 1D Conductors
20.2. Quantum Resistors in Series
20.3. Potential Drop Across Scatterer(s)

In the next three Lectures we will go through a few examples of increasing complexity which are interesting in their own right but have been chosen primarily as "do it yourself" problems that the reader can use to get familiar with the quantum transport model outlined in the last Lecture. The MATLAB codes are all included in Appendix F.

In this Lecture we will use 1D quantum transport models to study an interesting question regarding multiple scatterers or obstacles along a conductor. Are we justified in neglecting all interference effects among them and assuming that electrons diffuse like classical particles as we do in the semiclassical picture?

This was the question Anderson raised in his 1958 paper entitled "Absence of Diffusion in Certain Random Lattices" pointing out that diffusion could be slowed significantly and even suppressed completely due to quantum interference between scatterers. "Anderson localization" is a vast topic and we are only using some related issues here to show how the NEGF model provides a convenient conceptual framework for studying interesting physics.

For any problem we need to discuss how we write down the Hamiltonian *[H]* and the contact self-energy matrices $[\Sigma]$. Once we have these, the computational process is standard. The rest is about understanding and enjoying the physics.

321

20.1. Modeling 1D Conductors

For the one-dimensional examples discussed in this Lecture, we use the 1-D Hamiltonian from Fig.18.6, shown here in Fig.20.1. As we discussed earlier for a uniform wire the dispersion relation is given by

$$E(k) = \varepsilon + 2t \cos ka \qquad (20.1a)$$

which can approximate a parabolic dispersion

$$E = E_c + \frac{\hbar^2 k^2}{2m} \qquad (20.1b)$$

by choosing

$$E_c = \varepsilon + 2t \qquad (20.2a)$$

and $$-t \equiv t_0 \equiv \frac{\hbar^2}{2ma^2} \qquad (20.2b)$$

It is straightforward to write down the *[H]* matrix with ε on the diagonal and "t" on the upper and lower diagonals. What needs discussion are the *self-energy matrices.* The basic idea is to replace an infinite conductor described by the Hamiltonian *[H]* with a finite conductor described by $[H + \Sigma_1 + \Sigma_2]$ assuming **open boundary conditions** at the ends, which means that electron waves escaping from the surface do not give rise to any reflected waves, as a good contact should ensure.

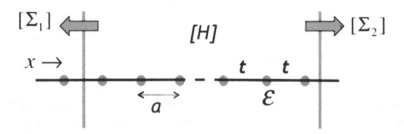

Fig.20.1. For the one-dimensional examples discussed in this Lecture, we use the 1-D Hamiltonian from Fig.18.6.

For a one-dimensional lattice the idea is easy to see. We start from the original equation for the extended system

$$E\psi_n = t\psi_{n-1} + \varepsilon\psi_n + t\psi_{n+1}$$

and then assume that the contact has no incoming wave, just an outgoing wave, so that we can write

$$\psi_{n+1} = \psi_n e^{ika}$$

which gives

$$E\psi_n = t\psi_{n-1} + (\varepsilon + t e^{ika})\psi_n$$

In other words the effect of the contact is simply to add $t*exp(+ika)$ to H_{nn} which amounts to adding the self-energy

$$\Sigma_1 = \begin{bmatrix} te^{ika} & 0 & 0 & \cdots \\ 0 & 0 & 0 & \\ 0 & 0 & 0 & \\ \cdots & & & \ddots \end{bmatrix} \quad (20.1a)$$

to the Hamiltonian. Note the only non-zero element is the $(1,1)$ element. Similarly at the other contact we obtain

$$\Sigma_2 = \begin{bmatrix} \ddots & & \cdots & \\ & 0 & 0 & 0 \\ \cdots & 0 & 0 & 0 \\ & 0 & 0 & te^{ika} \end{bmatrix}$$

(20.1b)

Note the only non-zero element is the (n,n) element.

In short, the self-energy function for each contact has a single non-zero element corresponding to the point that is connected to that contact.

20.1.1. 1D ballistic conductor

A good test case for any theory of coherent quantum transport is the conductance function for a length of uniform ballistic conductor: If we are doing things right, the conductance function $G(E)$ should equal the quantum of conductance q^2/h times an integer equal to the number of modes $M(E)$ which is one for 1D conductors (neglecting spin). This means that the transmission (see Eq.(19.30))

$$\bar{T}(E) = \ \textit{Trace}\left[\Gamma_1 G^R \Gamma_2 G^A\right] \tag{20.3}$$

should equal one over the energy range

$$0 \ < E - E_c \ < \ 4t_0$$

covered by the dispersion relation

$$E \ = \ \varepsilon + 2t\cos ka = \ E_c + 2t_0(1 - \cos ka) \tag{20.4}$$

but zero outside this range (see Fig.20.2 below with $U=0$). This is a relatively simple but good example to try to implement numerically when getting started. Obtaining a constant conductance across the entire band is usually a good indicator that the correct self-energy functions are being used and things have been properly set up.

Fig.20.2. Transmission through a single point scatterer in a 1D wire.

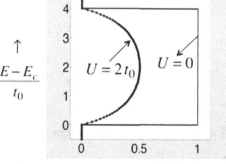

$$\frac{E - E_c}{t_0} \uparrow$$

$$Transmission, \bar{T}(E) \ \rightarrow$$

20.1.2. 1D conductor with one scatterer

Another good example is that of a conductor with just one scatterer whose effect is included in the Hamiltonian *[H]* by changing the diagonal element corresponding to that point to $\varepsilon + U$:

$$H = \begin{bmatrix} \ddots & & \cdots & & \\ & \varepsilon & t & 0 & \\ \vdots & t & \varepsilon+U & t & \\ & 0 & t & \varepsilon & \\ & & \cdots & & \ddots \end{bmatrix}$$

Fig.20.2 shows the numerical results for *U=0* (ballistic conductor) and for $U = 2t_0$. Actually there is a simple analytical expression for the transmission through a single point scatterer

$$\bar{T}(E) \;=\; \frac{(2t \sin ka)^2}{U^2 + (2t \sin ka)^2} \;=\; \frac{(\hbar v / a)^2}{U^2 + (\hbar v / a)^2} \tag{20.5}$$

that we can use to check our numerical results. This expression is obtained by treating the single point where the scatterer is located as the channel, so that all matrices in the NEGF method are (1x1) matrices, that is, just numbers:

$$\Sigma_1 = \begin{bmatrix} te^{ika} \end{bmatrix} \qquad H = \begin{bmatrix} \varepsilon + U \end{bmatrix} \qquad \Sigma_2 = \begin{bmatrix} te^{ika} \end{bmatrix}$$

$$\Gamma_1 = \begin{bmatrix} -2t \sin ka \end{bmatrix} \qquad\qquad\qquad \Gamma_2 = \begin{bmatrix} -2t \sin ka \end{bmatrix}$$

It is easy to see that the Green's function is given by

$$G^R(E) = \frac{1}{E-(\varepsilon+U)-2t\,e^{ika}} = \frac{1}{-U-i2t\sin ka}$$

making use of Eq.(20.2). Hence

$$\Gamma_1 G^R \Gamma_2 G^A = \frac{(2t\sin ka)^2}{U^2+(2t\sin ka)^2}$$

giving us the stated result in Eq.(20.3). The second form is obtained by noting from Eq.(20.2) that

$$\hbar v = \frac{dE}{dk} = -2at\sin ka \tag{20.6}$$

Once you are comfortable with the results in Fig.20.2 and are able to reproduce it, you should be ready to include various potentials into the Hamiltonian and reproduce the rest of the examples in this Lecture.

20.2. Quantum Resistors in Series

In Lecture 12 we argued that the resistance of a conductor with one scatterer with a transmission probability T can be divided into a scatterer resistance and an interface resistance (see Eqs.(12.1), (12.2))

$$R_1 = \frac{h}{q^2 M}\left(\underbrace{\frac{1-T}{T}}_{scatterer} + \underbrace{1}_{interface} \right)$$

What is the resistance if we have two scatterers each with transmission T?

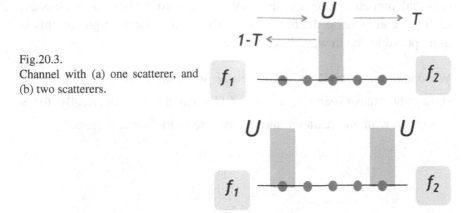

Fig.20.3.
Channel with (a) one scatterer, and
(b) two scatterers.

We would expect the scatterer contribution to double:

$$R_2 = \frac{h}{q^2 M}\left(2\underbrace{\frac{1-T}{T}}_{scatterer} + \underbrace{1}_{int\,erface}\right)$$

$$= \frac{h}{q^2 M}\frac{2-T}{T}$$

We can relate the two resistances by the relation:

$$R_2 = R_1(2-T)$$

If T is close to one we have the ballistic limit with $R_2 = R_1$: two sections in series have the same resistance as one of them, since all the resistance comes from the interfaces.

If $T << 1$, we have the Ohmic limit with $R_2 = 2R_1$: two sections have twice the resistance as one of them, since all the resistance comes from the channel.

But can R_2 ever be less than R_1? Not as long as electrons behave like classical particles. There is no way an extra roadblock on a classical highway can increase the traffic flow. But on a quantum highway this is quite possible due to wave interference.

We could use our 1D model to study problems of this type. Fig.20.4 shows the transmission functions $\bar{T}(E)$ calculated numerically for a conductor with one scatterer and a conductor with two scatterers.

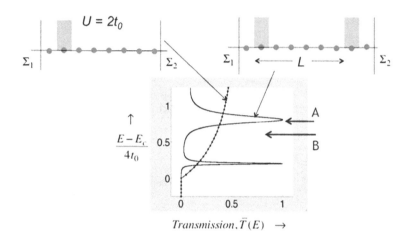

Transmission, $\bar{T}(E)$ $\rightarrow$

Fig.20.4. Normalized conductance for a wire with *M=1* with (a) one scatterer, and (b) two scatterers.

If the electrochemical potential happens to lie at an energy like the one marked "B", R_2 will be even larger than the Ohmic result R_1. But if the electrochemical potential lies at an energy like the one marked "A", R_2 is less than R_1.

At such energies, the presence of the second scatterer creates a reflection that cancels the reflection from the first one, because they are spaced a quarter wavelength apart. Such quarter wave sections are widely used to create anti-reflection coatings on optical lenses and are well-known in the world of waves, though they are unnatural in the world of particles.

Actually there is a class of devices called resonant tunneling diodes that deliberately engineer two strategically spaced barriers and make use of the resulting sharp peaks in conductance to achieve interesting current-voltage characteristics like the one sketched here where over a range of voltages, the slope dI/dV is negative ("negative differential

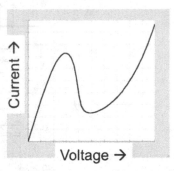

resistance, NDR"). We could use our elastic resistor model for the current from Eq.(3.3) and along with the conductance function from NEGF

$$G(E) \equiv \frac{q^2}{h}\overline{T}(E) = \frac{q^2}{h}Trace\left[\Gamma_1 G^R \Gamma_2 G^A\right]$$

to model devices like this, but it is important to include the effect of the applied electric field on the *[H]* as mentioned earlier (see Fig.18.6). In these Lectures we will focus more on low bias response for which this aspect can be ignored.

Consider for example a resistor with scatterers distributed randomly throughout the channel. If we were to use the quantum formalism to calculate the conductance function for a single-moded wire with random scatterers we would find that once the classical transmission $M\lambda/L$ drops below one, the quantum conductance is extremely low except for occasional peaks at specific energies (Fig.20.5). The result marked semiclassical is obtained by calculating T for a single scatterer and then increasing the scatterer contribution by a factor of six:

$$R_6 = \frac{h}{q^2 M}\left(\underbrace{6\,\frac{1-T}{T}}_{scatterer} + \underbrace{1}_{int\,erface}\right) = \frac{h}{q^2 M}\frac{6-5T}{T}$$

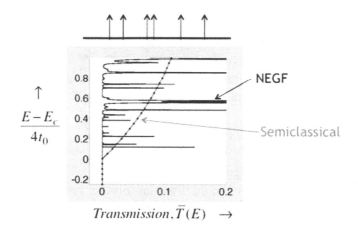

Fig.20.5. Normalized conductance for a wire with *M=1* with six scatterers

Comparing the classical and quantum results suggests that such conductors would generally show very high resistances well in excess of Ohm's law, with occasional wild fluctuations. In a multi-moded wire too quantum calculations show the same behavior once the classical transmission $M\lambda/L$ drops below one. Such conductors are often referred to as being in the regime of strong localization. Interestingly, even when $M\lambda/L$ is well in excess of one, the quantum conductance is a little (~ approximately one) less than the classical value and this is often referred to as the regime of weak localization.

However, localization effects like these are usually seen experimentally only at low temperatures. At room temperature there is seldom any evidence of deviation from Ohm's law. Consider for instance a copper wire with a cross-section of *10nm x 10nm* which should contain approximately 1000 atoms and hence should have *M~1000* (see discussion at end of Chapter 4). Assuming a mean free path of *40 nm* this suggests that a copper wire any longer than $M\lambda$ ~ 40 µm should exhibit strange non-Ohmic behavior, for which there is no experimental evidence. Why?

The answer is that localization effects arise from quantum interference and will be observed only if the entire conductor is **phase-coherent.** A copper wire 40 μm long is not phase coherent, certainly not at room temperature. Conceptually we can think of the real conductor as a series of individual coherent conductors, each of length equal to the phase coherence length L_P and whether we see localization effects will depend not on $M\lambda/L$, but on $M\lambda/L_P$.

The bottom line is that to describe real world experiments especially at room temperature it is often important to include a certain degree of dephasing processes as described at the end of the last Lecture. Unless we include an appropriate degree of dephasing our quantum models will show interference effects leading to resonant tunneling or strong localization which under certain conditions may represent real world experiments, but not always. Just because we are using quantum mechanics, the answer is not automatically more "correct."

This can be appreciated by looking at the potential variation along the channel using NEGF and comparing the results to our semiclassical discussion from Lecture 12.

20.3. Potential Drop across Scatterer(s)

In Lecture 12 we discussed the spatial variation of the occupation factor which translates to a variation of the electrochemical potential for low bias. A conductor with one scatterer in it (Fig.20.6), can be viewed (see Fig.12.5) as a normalized interface resistance of one in series with a normalized scatterer resistance of $(1-T)/T$, which can be written as

$$(Normalized) \; R_{scatterer} = \left(\frac{Ua}{\hbar v}\right)^2 \qquad (20.7)$$

using Eq.(20.5). The semiclassical potential profile in Fig.20.6 is then obtained by noting that since the current is the same everywhere, each section shows a potential drop proportional to its resistance.

The quantum profile is obtained using an NEGF model to calculate the effective occupation factor throughout the channel by looking at the ratio of the diagonal elements of G^n and A which are the quantum versions of the electron density and density of states respectively:

$$f(j) = \frac{G^n(j,j)}{A(j,j)} \qquad (20.8)$$

For low bias, this quantity translates linearly into a local electrochemical potential as noted in Lecture 2 (see Eq.(2.8)). If we choose $f = 0$ at one contact, $f=1$ at another contact corresponding to qV, then the f (j) obtained from Eq.(20.8) is simply translated into an electrochemical potential μ at that point:

$$\mu(j) = qV\, f(j) \qquad (20.9)$$

The occupation $f(j)$ shows oscillations due to quantum interference making it hard to see the potential drop across the scatterer (see solid black line marked NEGF).

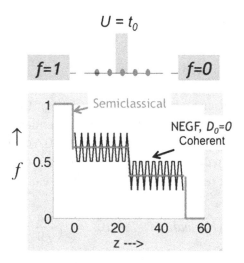

Fig.20.6. Potential drop across a scatterer calculated from the quantum formalism. (a) Physical structure, (b) Coherent NEGF calculation at $E = t_0$.

Experimentalists have measured profiles such as these using scanning probe microscopy (SPM) and typically at room temperature the quantum oscillations are not seen, because of the dephasing processes that are inevitably present at room temperature. This is another example of the need to include dephasing in order to model real world experiments especially at room temperature.

Indeed if we include pure phase relaxation processes (Eq.(19.37)) in the NEGF model we obtain a clean profile looking a lot like what we would expect from a semiclassical picture (see Fig.20.7a).

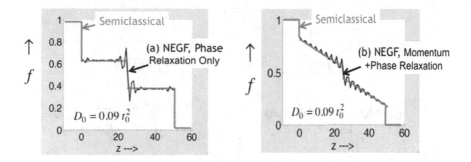

Fig.20.7. Potential drop for the structure in Fig.20.6 calculated from the NEGF method at $E=t_0$ with dephasing, (a) Phase-relaxation only, Eq.(19.37), (b) Phase and momentum relaxation, Eq.(19.38).

Interestingly, if we use a momentum relaxing model for Σ_0 (Eq.(19.38), the potential drops linearly across the structure (see Fig.20.7b), exactly what we would expect for a distributed classical resistor. The resistance per lattice site for this distributed resistor due to D_0 can be obtained by replacing U^2 with D_0 in Eq.(20.7):

$$(Normalized) \ R = \underbrace{\left(\frac{a}{\hbar v}\right)^2 D_0}_{\substack{Re \, sis \, tan \, ce \\ per \ lattice \ site}} \underbrace{\frac{L}{a}}_{\substack{\# \, of \ lattice \\ sites}}$$

Another interesting example is that of the two quantum resistors in series that we started with. We noted then that at energies corresponding to points A and B in Fig.20.2 we have constructive and destructive interference respectively. This shows up clearly in the potential profile for coherent transport with $D_0 = 0$ (see Fig.20.8). At $E=0.6t_0$ corresponding to destructive interference, the profile looks like what we might expect for a very large resistor showing a large drop in potential around it along with some sharp spikes superposed on it. At $E=0.81t_0$ corresponding to constructive interference, the profile looks like what we expect for a ballistic conductor with all the drop occurring at the two contacts and none across the scatterers.

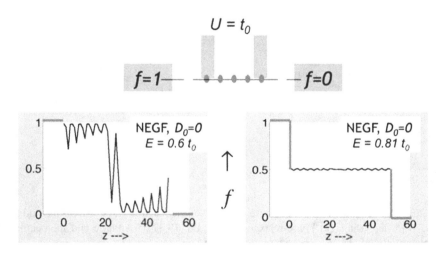

Fig.20.8. Potential drop across two scatterers in series calculated from the NEGF method without dephasing at two energies, $E = 0.81\ t_0$ and $E= 0.6\ t_0$ corresponding to points marked "A" and "B" respectively in Fig.20.4.

Clearly at $E=0.81t_0$ the answer to the title question of this Lecture is yes, two scatterers can offer less resistance than one. And this strange result is made possible by quantum interference. And once we introduce sufficient phase relaxation into the model using a non-zero D_0, the profile at both energies look much the same like any semiclassical resistor (Fig.20.9).

Before we move on, let me note that although it is straightforward to include dephasing into toy calculations like this, for large devices described by large matrices, it can be numerically challenging. This is because with coherent NEGF ($D_0 = 0$) or with the momentum relaxing model (Eq.(19.38)), it is often adequate to calculate just the diagonal elements of the Green's functions using efficient algorithms. But for pure phase relaxation (Eq.(19.37), it is necessary to calculate the full Green's function increasing both computational and memory burdens significantly.

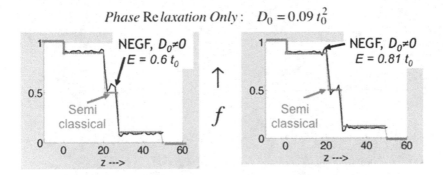

Phase Re *laxation Only* : $D_0 = 0.09 \, t_0^2$

Fig.20.9. Potential drop across two scatterers in series calculated from the NEGF method with pure phase rexation at two energies, $E = 0.81 \, t_0$ and $E = 0.6 \, t_0$ corresponding to points marked "A" and "B" respectively in Fig.20.4.

So a natural question to ask is whether instead of including dephasing through Σ_0 we could include the potential U_R in the Hamiltonian itself and then average our quantity of interest over a number of random realizations of U_R. Would these results be equivalent?

For short conductors like the one shown in Fig.20.4, this seems to be true, but for long conductors like the one in Fig.20.5 this may not be true. With a conductor in the regime of strong localization (Fig.20.5) it is hard to see how averaging the coherent quantum result over many configurations can lead to the semiclassical result.

NEGF with dephasing does not just average over many configurations, it also averages over different sections of the same configuration and that is why it is able to capture the semiclassical result which often describes real world experiments at room temperature quite well.

But could NEGF capture the localization effects observed at low temperatures through a proper choice of Σ_0 ? We believe so, but it would involve going beyond the simple dephasing models (technically known as the self-consistent Born approximation) for evaluating Σ_0 described in Section 19.4.

Lecture 21

Quantum of Conductance

21.1. 2D Conductor as 1D Conductors in Parallel
21.2. Contact self-energy for 2D Conductors
21.3. Quantum Hall Effect

As I mentioned our primary objective in Lectures 20-22 is to help the reader get familiar with the NEGF model through "do it yourself" examples of increasing complexity. The last Lecture used 1D examples. In this Lecture we look at 2D examples which illustrate one of the key results of mesoscopic physics, namely the observation of conductances that are an integer multiple of the conductance quantum q^2/h.

21.1. 2D Conductor as 1D Conductors in Parallel

Among the seminal experiments from the 1980's that gave birth to mesoscopic physics was the observation that the conductance of a ballistic 2D conductor went down in integer multiples of $2q^2/h$ as the width of the narrow region was decreased.

To understand this class of experiments we need a 2D model (Fig.21.1). As with 1D, two inputs are required: the Hamiltonian *[H]* and the contact self-energy matrices $[\Sigma]$. Once we have these, the rest is standard.

For *[H]*, we use the 2-D Hamiltonian from Fig.18.7 for conductors described by parabolic *E(k)* relations. As we discussed earlier for a uniform wire the dispersion relation is given by

$$E(k_x, k_y) = \varepsilon + 2t \cos k_x a + 2t \cos k_y a \qquad (21.1a)$$

which can approximate a parabolic dispersion

$$E = E_c + \frac{\hbar^2 k^2}{2m} \qquad (21.1b)$$

by choosing
$$E_c = \varepsilon + 4t \qquad (21.2a)$$

and
$$-t \equiv t_0 \equiv \frac{\hbar^2}{2ma^2} \qquad (21.2b)$$

(a) Schematic of physical structure

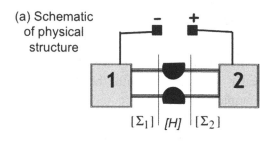

Fig.21.1.
(a) Schematic of structure for measuring the conductance of a short constriction created in a two dimensional conductor. (b) 2D model used for NEGF-based calculation and discussion. (c) Numerically computed transmission shows steps as a function of energy.

(b) 2D model used for NEGF-based calculation

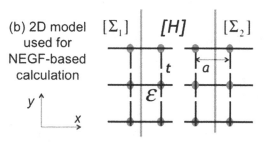

(c) Numerical Result

$$\frac{E - E_c}{t_0} \uparrow$$

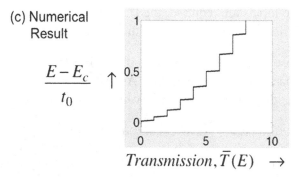

Transmission, $\bar{T}(E) \rightarrow$

Once again what needs discussion are the **self-energy matrices.**, [Σ], but before we get into it let us look at the transmission function

$$\bar{T}(E) = \quad Trace\left[\Gamma_1 G^R \Gamma_2 G^A\right] \quad \text{(same as Eq.(20.3))}$$

obtained directly from the numerical model (Fig.21.1), which shows steps at specific energies. How can we understand this?

The elementary explanation from Section 5.5 is that for a ballistic conductor the transmission function is just the number of modes $M(E)$ which equals the number of half de Broglie wavelengths that fits into the width W of the conductor (*Int(x)* denotes the highest integer less than x)

$$M = \quad Int\left(\frac{2W}{h/p}\right) = \quad Int\left(\frac{2W}{h}\sqrt{2m(E-E_c)}\right)$$

where we have used the parabolic relation $E - E_c = p^2/2m$. To compare with our numerical results we should use the cosine dispersion relation.

Experimentally what is measured at low temperatures is $M(E=\mu_0)$ and the steps are observed as the width is changed as first reported in van Wees et al. (1988) and Wharam et al. (1988). To compare with experimental plots, one could take a fixed energy $E=t_0$ and plot the transmission as a function of the number of points along the width to get something like this.

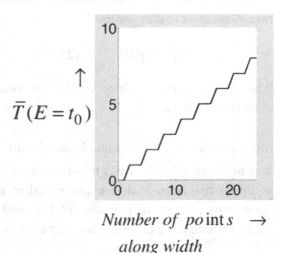

Number of *po*int *s* →
along width

Why does our numerical model show these steps? One way to see this is to note that our 2D model can be visualized as a linear 1D chain as shown in the adjoining figure where the individual elements α of the chain represent a column. For example if there are three sites to each column, we would have

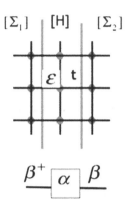

$$\alpha = \begin{bmatrix} \varepsilon & t & 0 \\ t & \varepsilon & t \\ 0 & t & \varepsilon \end{bmatrix} \qquad (21.3a)$$

while the coupling β from one column to the next is diagonal:

$$\beta = \begin{bmatrix} t & 0 & 0 \\ 0 & t & 0 \\ 0 & 0 & t \end{bmatrix} \qquad (21.3b)$$

Note that the matrix α describing each column has off-diagonal elements t, but we can eliminate these by performing a **basis transformation to diagonalize it:**

$$[\tilde{\alpha}] = [V]^{+}[\alpha][V] \qquad (21.3c)$$

where *[V]* is a matrix whose columns represent the eigenvectors of α.

Since the matrix β is essentially an identity matrix it is unaffected by the basis transformation, so that in this transformed basis we can visualize the 2D conductors as a set of independent 1D conductors, each of which has a different diagonal element

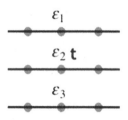

$$\varepsilon_1, \varepsilon_2, \varepsilon_3$$

equal to the eigenvalues of α. Each of these 1D conductors has a transmission of one in the energy range ($t_0 \equiv |t|$)

$$\varepsilon_n - 2t_0 \ < \ E < \ \varepsilon_n + 2t_0$$

as sketched below. Adding all the individual transmissions we obtain the transmission showing up-steps in the lower part and down-steps in the upper part.

Usually when modeling n-type conductors we use the lower part of the band as shown in Fig.21.1, and so we see only the up-steps occurring at

$$\varepsilon_n - 2t_0$$

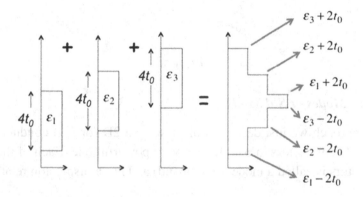

Now the ε_n's are the eigenvalues of α (see Eq.(21.3a)) which are given by

$$\varepsilon_n = \varepsilon - 2t_0 \cos k_n a, \quad with \quad k_n a = \frac{n\pi}{N+1} \qquad (21.4)$$

where N is the number of points along the width which determines the size of α. This result is not obvious, but can be shown analytically or checked easily using MATLAB.

Using Eq.(21.2) and (21.4) we can write the location of the steps as

$$\varepsilon_n - 2t_0 = E_c + 2t_0\left(1 - \cos\frac{n\pi}{N+1}\right)$$

which matches the numerical result obtained with $N=25$ very well as shown.

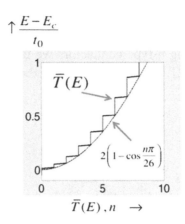

21.1.1. Modes or Subbands

The approach we just described of viewing a 2D (or 3D) conductor as a set of 1D conductors in parallel is a very powerful one. Each of these 1D conductors is called a mode (or subband) and has a dispersion relation

$$E_n(k_x) = \varepsilon_n - 2t_0 \cos k_x a$$

as shown below. These are often called the subband dispersion relations obtained from the general dispersion relation in Eq.(21.1a) by requiring k_y to take on quantized values given by

$$k_y a = \frac{n\pi}{N+1}$$

where each integer n gives rise to one subband as shown. If we draw a horizontal line at any energy E, then the number of dispersion relations it crosses is equal to twice the number of modes $M(E)$ at that energy, since each mode gives rise to two crossings, one for a state with positive velocity, and one for a state with negative velocity.

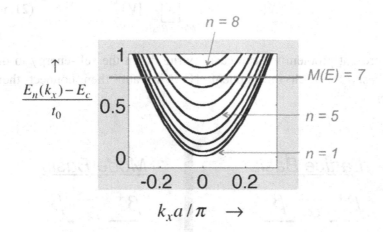

21.2. Contact Self-Energy for 2D Conductors

Let us now address the question we put off, namely how do we write the self-energy matrices for the contacts. Ideally the contact regions allow electrons to exit without any reflection and with this in mind, a simple way to evaluate $[\Sigma]$ is to assume the contacts to be just uniform extensions of the channel region and that is what we will do here.

21.2.1. Method of basis transformation

The viewpoint we just discussed in Section 21.1 allows us to picture a 2D conductor as a set of decoupled 1D conductors, by converting from the usual lattice basis to an abstract mode basis through a basis transformation :

$$\underbrace{[\tilde{X}]}_{Mode\ Basis} = [V]^{+} \underbrace{[X]}_{Lattice\ Basis} [V] \qquad (21.5a)$$

[X] being any matrix in the regular lattice basis. A unitary transformation like this can be reversed by transforming back:

$$\underbrace{[X]}_{\text{Lattice Basis}} = [V] \; \underbrace{[\tilde{X}]}_{\text{Mode Basis}} \; [V]^+ \qquad (21.5b)$$

In our present problem we can easily write down the self-energy in the mode basis for each independent 1D wire and then connect them together.

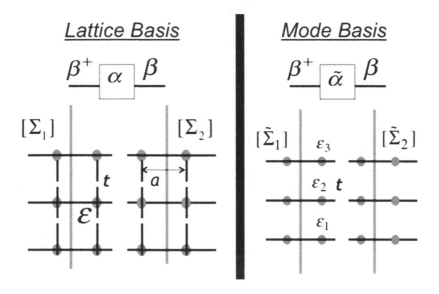

Fig.21.2. A 2D conductor can be pictured as a set of decoupled 1D conductors through a basis transformation.

For example if each wire consisted of just one site along x, then each wire would have a self-energy of te^{ika} , with the appropriate ka for that wire at a given energy E. For mode *n* we have

$$E = \varepsilon_n - 2t_0 \cos k_n a$$

so that overall we could write

$$[\tilde{\Sigma}_1] = \begin{bmatrix} te^{ik_1a} & 0 & 0 & \cdots \\ 0 & te^{ik_2a} & 0 & \\ 0 & 0 & te^{ik_3a} & \\ \cdots & & & \ddots \end{bmatrix}$$

and then transform it back to the lattice basis as indicated in Eq.(21.5b):

$$[\Sigma_1] = [V]\,[\tilde{\Sigma}_1]\,[V]^+$$

21.2.2. General Method

The method of basis transformation is based on a physical picture that is very powerful and appealing. However, I believe it cannot always be used at least not as straightforwardly, since in general it may not be possible to diagonalize both α and β simultaneously.

For the square lattice $\beta = t\,[I]$ (Eq.(21.3b)) making it "immune" to basis transformations. But in general this may not be so. The graphene lattice from Fig.18.9 pictured on the right is a good example. How do we write $[\Sigma]$ in such cases?

Any conductor with a uniform cross-section can be visualized as a linear 1-D chain of "atoms" each having an on-site matrix Hamiltonian $[\alpha]$ coupled to the next "atom" by a matrix $[\beta]$. Each of these matrices is of size (nxn), n being the number of basis functions describing each unit.

The *self-energy matrix* is zero except for the last (nxn) block at the surface

The non-zero block is given by

$$\beta g_2 \beta^+ \qquad\qquad (21.6a)$$

where g_2 is called the surface Green's function for contact 2, and is obtained by iteratively solving the equation:

$$\left[g_2\right]^{-1} = (E + i0^+)I - \alpha - \beta^+ g_2 \beta \qquad\qquad (21.6b)$$

for g2, where 0+ represents a positive infinitesimal. Eq.(21.6) is of course not meant to be obvious, but we have relegated the derivation to Appendix E. We will not go into the significance of the infinitesimal i0+ (see for example Datta (1995), Chapter 3 or Datta (2005), Chapter 8).

For the moment let me just note that for a 1-D conductor with

$$\alpha = \varepsilon \text{ and } \beta = t,$$

Eq.(21.2) reduces to an ordinary quadratic equation:

$$g_2(E + i0^+ - \varepsilon - t^2 g_2) = 1$$

whose solution gives two possible solutions $t e^{\pm ika}$ for the self-energy, and the one we want is that with the *negative imaginary part*, for which the corresponding broadening Γ is positive. More generally, we have a matrix quadratic equation (Eq.(21.6b)) and the infinitesimal $i0^+$ ensures that a numerical iterative solution converges on the solution for which Γ has all positive eigenvalues.

21.2.3. Graphene: Ballistic Conductance

As an example we have shown in Fig.21.3 the transmission $\bar{T}(E)$ calculated numerically for two common orientations of graphene, the so-called zigzag and armchair configurations with dimensions chosen so as to have roughly equal widths. Since these are ballistic conductors, the transmission is equal to the number of modes $M(E)$ and can be approximately described by the number of wavelengths that fit into the widths. The actual energy dependence is different from that obtained for the square lattice (see Eq.(21.3)) because of the linear $E(k)$ relation: $E = \hbar v_0 k = v_0 p$:

$$M = \mathrm{Int}\left(\frac{2W}{h/p}\right) = \mathrm{Int}\left(\frac{2W}{h}\frac{E}{v_0}\right) \qquad (21.7)$$

This applies equally to any orientation of graphene. Both the orientations shown have the same overall slope, but the details are quite different. For example, at $E = 0$, the armchair is non-conducting with M=0 while the zigzag is conducting with non-zero M.

For large dimensions the steps are close together in energy (compared to kT) and both appear to be semi-metallic. But for small dimensions the steps are much larger than kT. The zigzag now shows zero transmission $\bar{T}(E)$=0 at E=0 ("semiconducting") while the armchair shows non-zero conductance ("metallic"). These are clear observable differences that show up in experiments on samples of small width at low temperatures.

Another interesting observable difference is that between a flat graphene sheet and a cylindrical carbon nanotube (CNT). Mathematically, they are both described by the same Hamiltonian *[H]* but with different boundary conditions. Graphene like most normal conductors requires "hardwall boundary conditions" (HBC) where the lattice ends abruptly at the edges. CNT's on the other hand are among the few real conductors that require "periodic boundary conditions" (PBC) with no edges.

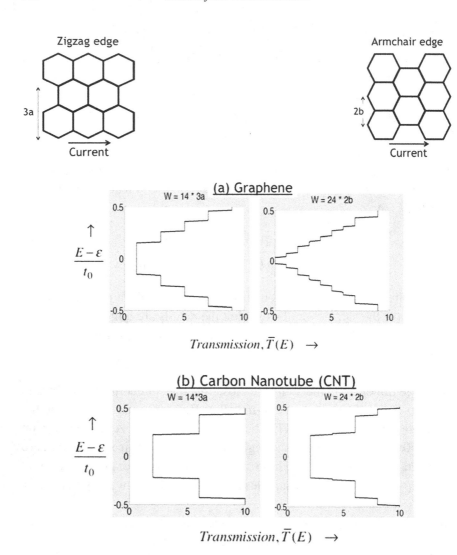

Fig.21.3. $\overline{T}(E)$ calculated from NEGF-based model for a ballistic (a) graphene sheet with armchair and zigzag edges as shown with roughly equal widths (*24*2b* ≈ *14*3a*), (b) carbon nanotube (CNT) obtained by rolling up the graphene sheet along the width.

The results for CNT are relatively easy to understand analytically, while those for graphene require a more extensive discussion. (see for example Brey and Fertig, 2006). As we mentioned in Lecture 5, PBC is mathematically simpler and that is why it is used so extensively for large conductors where it is known experimentally that the exact boundary conditions are not very relevant. But this of course is not true of small conductors and the difference is evident in Fig.21.2 for small conductors only a few nanometers in width. We will not go into this further. Our objective here is simply to show how easily our quantum transport formalism captures all the known physics.

The power of the numerical method lies in being able to calculate $M(E)$ automatically even before one has "understood" the results. However, one should use numerical calculations not as a substitute for understanding , but as an aid to understanding.

21.3.Quantum Hall Effect

The Hall effect (Lecture 13) provides another good example for a two-dimensional application of the quantum transport model. The basic structure involves a long conductor with side probes designed to measure the transverse Hall voltage developed in the presence of a magnetic field.

We use the same 2D Hamiltonian from Fig.21.1 but now including a magnetic field as explained in Section 18.4.3. As discussed in Lecture 13, the Hall resistance is given by the ratio of the Hall voltage to the current. In a theoretical model we could calculate the Hall voltage in one of two ways. We could attach a voltage probe to each side and use Büttiker's multiterminal method to find the potentials they float to.

Alternatively we could do what we explained in Section 20.3, namely calculate the fractional occupation of the states at any point j by looking at the ratio of the diagonal element of the electron density G^n and the density of states A and use the low bias Taylor expansion (Eq.(2.8)) to translate the occupation factor profile into a potential profile.

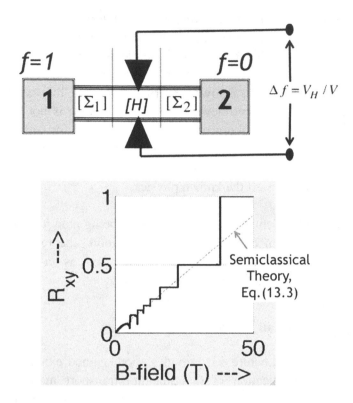

Fig.21.4
Normalized Hall resistance versus *B*-field for ballistic channel of width $W = 26a = 65\ nm$ calculated at an energy $E=t_0$ using a 2D model from Fig.21.1.

Fig.21.4 shows the calculated Hall resistance (normalized to the quantum of resistance h/q^2) as a function of the magnetic field. The striking result is of course the occurrence of plateaus at high fields known as the quantum Hall effect (von Klitzing et al. 1980). But first let us note the low field regime where the calculated result matches the Hall resistance expected from semiclassical theory

$$R_H \ = \ B/qn \qquad \text{(same as Eq.(13.4))}$$

The dashed line in Fig.21.4 is obtained from Eq.(13.4) assuming

$$\frac{N}{LW} = \frac{k^2}{4\pi} \qquad \text{(same as Eq.(5.12))}$$

and noting that the numerical calculation is carried out at $E = t$, corresponding to $ka = \pi/3$, with $a = 2.5\ nm$.

The semiclassical theory naturally misses the high field results which arise from the formation of Landau levels due to quantum effects. These are evident in the numerical plots of the local density of states at high B-field (20 T) shown in Fig.21.5.

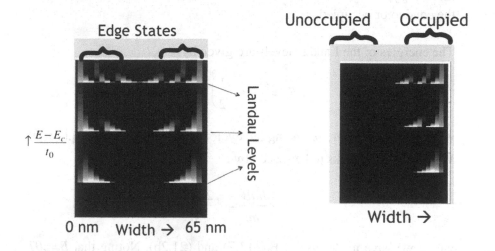

Fig.21.5. Grayscale plot of local density of states, obtained from the diagonal elements of $[A(E)]$ calculated at $B=20T$ from the NEGF method. Also shown on the right are the diagonal elelments of $[G^n(E)]$ calculated assuming $f_1 = 1, f_2 = 0$.

Usually the density of states varies relatively gently with position, but in the quantum Hall regime, there is a non -trivial modification of the local density of states which can be plotted from the NEGF method by looking at the diagonal elements of the spectral function $A(j,j;E)$. Fig.21.5 is a grayscale plot of $A(j,j;E)$ with energy E on the horizontal axis and the position j along the width on the vertical axis. The white streaks indicate a high density of states corresponding to the energy of Landau levels, which increase in energy along the edge forming what are called edge states.

As we mentioned in Lecture 13, the edge states can be pictured semiclassically in terms of "skipping orbits" that effectively isolate oppositely moving electrons from each other giving rise to a "divided highway" that provides an incredible level of ballisticity. This is evident if we plot the electron density from the diagonal elements of $[G^n]$ under non-equilibrium conditions assuming $f_1 = 1, f_2 = 0$. Only the edge states on one side of the sample are occupied. If we reverse the current flow assuming $f_1 = 0, f_2 = 1$, we would find the edge states on the other side of the conductor occupied.

The energies of the Landau levels are given by

$$E_n = \left(n + \frac{1}{2}\right)\hbar\omega_c \qquad (21.8)$$

where n is an integer, ω_c being the cyclotron frequency (see Eq.(13.7)). We expect the streaks to be spaced by

$$\hbar\omega_c = \frac{\hbar qB}{m} = \frac{2qBa^2}{\hbar}t_0$$

where we have made use of Eq.(13.7) and (21.2b). Noting that $B=20T$, $a=2.5$ nm, we expect a spacing of $\sim 0.37t_0$ between the streaks in approximate agreement with Fig.21.5.

Eq.(21.8) is a quantum result that comes out of the Schrödinger equation including the vector potential which is part of our numerical model. One can understand it heuristically by noting that semiclassically electrons describe circular orbits in a magnetic field, completing one orbit in a time (see Eq.(13.7))

$$t_c = \frac{2\pi}{\omega_c} = \frac{2\pi p}{qvB}$$

so that the circumference of one orbit of radius r_c is given by

$$2\pi r_c = vt_c = \frac{2\pi p}{qB}$$

If we now impose the quantum requirement that the circumference equal an integer number of de Broglie wavelengths h/p, we have

$$\frac{2\pi p}{qB} = integer * \frac{h}{p}$$

Semiclassically an electron can have any energy $E = p^2/2m$. But the need to fit an integer number of wavelengths leads to the condition that

$$p^2 = integer * hqB$$

suggesting that the allowed energies should be given by

$$E = integer * \frac{hqB}{2m} = integer * \frac{\hbar\omega_c}{2}$$

which is not exactly the correct answer (Eq.(21.8)), but close enough for a heuristic argument.

The resulting current equals

$$\frac{q^2}{h} V \left(Number\ of\ Edge\ States \right)$$

while the Hall voltage simply equals the applied voltage since one edge of the sample is in equilibrium with the source and other with the drain.

This leads to a quantized Hall resistance given by

$$\frac{h}{q^2} \frac{1}{Number\ of\ Edge\ States}$$

giving rise to the plateaus of *1/4, 1/3, 1/2, 1* seen in Fig.21.4, as the magnetic field raises the Landau levels, changing the number of edge states at an energy $E=t_0$ from *4* to *3* to *2* to *1*.

I should mention that the theoretical model does not include the two spins and so gives a resistance that is twice as large as the experimentally observed values which look more like

$$\frac{h}{2q^2} \frac{1}{Number\ of\ Edge\ States}$$

because edge states usually come in pairs, except at high *B*-fields.

Also, we have not talked at all about the fractional quantum Hall effect observed in pure samples at larger *B*-fields with Hall resistances that look like

$$\frac{h}{q^2} \frac{1}{a\ fraction}$$

This is a vast and rich area of research on its own beyond the scope of the simple NEGF model discussed here. As it stands it captures only the integer Hall effect though innovative extensions could take it beyond this regime.

Lecture 22

Rotating an electron

Back in Lecture 14, we discussed how magnets can be used to create spin potentials inside a non-magnetic channel that extend even outside the path of the current and can be measured using another magnetic voltage probe with a polarization $\vec{P}$ (Fig.22.1). This led naturally to the concept of a charge potential μ and a spin potential μ_s such that

$$\mu_{probe} = \mu + \frac{P\mu_s}{2}$$

At the end of the Lecture we stated the more general result

$$\mu_{probe} = \mu + \frac{\vec{P}.\vec{\mu}_s}{2} \qquad (22.1)$$

that can be used even when the magnet polarization $\vec{P}$ and the spin potential $\vec{\mu}_s$ in the channel are not collinear.

I am not sure if anyone has actually done the experiment shown in Fig.22.1, namely inject spins with a fixed magnet and measure the voltage with a magnet whose direction is rotated. What has been done, however, is to keep both magnets fixed and rotate the electron spin inside

355

the channel using an external magnetic field and more recently, an electric field.

All of these effects are interesting in their own right, but here I will use them primarily to illustrate how spin effects are included in the NEGF method described in Lecture 19. We will end by showing how the NEGF method leads to the result stated in Eq.(22.1). As we will see the quantum method requires a (2x2) *matrix potential* which can also be described in terms of a charge potential μ supplemented by a vector spin potential $\vec{\mu}_s$ as indicated in Eq.(22.1).

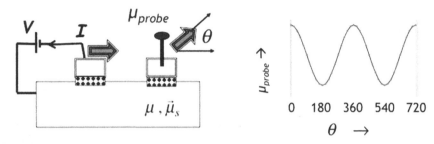

Fig.22.1. A magnetic probe can be used to measure the spin potential in the channel, even outside the current path. As the probe is rotated, the voltage probe should show an oscillatory signal.

As we mentioned in Lecture 14, electron spin is a lot like photon polarization, except for one key difference, namely that orthogonal directions are not represented by say z and x that are 90 degrees apart. Rather they are represented by up and down that are 180 degrees apart, which is why the maxima and minima of the oscillations in Fig.22.1 are separated by 180 degrees.

And that is why we need something other than vectors to represent electron spin, namely spinors. A vector $\hat{n}$ is described by three real components, namely the components along x, y and z, but spinors are described by two complex components, which are its components along up and down:

$$\begin{Bmatrix} n_x \\ n_y \\ n_z \end{Bmatrix} \quad , \quad \begin{Bmatrix} \psi_{up} \\ \psi_{dn} \end{Bmatrix}$$

$$\underbrace{\phantom{\begin{Bmatrix} n_x \\ n_y \\ n_z \end{Bmatrix}}}_{Vector} \qquad \underbrace{\phantom{\begin{Bmatrix} \psi_{up} \\ \psi_{dn} \end{Bmatrix}}}_{Spinor}$$

Nevertheless we visualize the spinor as an object pointing in some direction just like a vector. How do we reconcile the visual picture with the 2-component complex representation?

A spinor pointing along a direction described by a unit vector

$$\hat{n} \equiv \begin{Bmatrix} \sin\theta\cos\phi \\ \sin\theta\sin\phi \\ \cos\theta \end{Bmatrix} \quad (22.2)$$

has components given by

$$\begin{Bmatrix} \cos\dfrac{\theta}{2}e^{-i\phi/2} \equiv c \\[2mm] \sin\dfrac{\theta}{2}e^{+i\phi/2} \equiv s \end{Bmatrix} \quad (22.3)$$

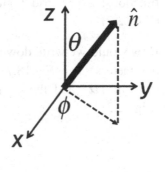

This is of course not obvious and later in the Lecture I will try to explain why Eqs.(22.2) and (22.3) represent isomorphic (more correctly "homomorphic") ways to represent an abstract rotatable object pointing in some direction. For the moment let us accept Eq.(22.3) for the components of a spinor and work out some of its consequences.

Although these subtleties of visualization and interpretation take some time to get used to, formally it is quite straightforward to incorporate spin into the quantum transport formalism from Lecture 19. The basic equations from Eq.(19.1) through (19.4) remain the same, but all the matrices like

$$[H], [\Sigma], [G^n], [A]$$

become twice as big (Fig.22.2).

(a) Without spin

(b) With spin

Fig.22.2.. Inclusion of spin in NEGF doubles the number of "grid points" or basis functions.

Ordinarily these matrices are of size (N x N), if N is the number of grid points (or more formally the number of basis functions) used to describe the channel. Inclusion of spin basically doubles the number of basis functions: every grid point turns into two points, an up and a down (Fig.22.2).

How would we write down *[H]* including spin? We can visualize the TB parameters (See Fig.18.7) exactly as before except that each on-site element $[\alpha]$ and the coupling elements $[\beta]$ are each (2x2) matrices (Fig.22.3).

Fig.22.3. 2D Lattice with each element given by a 2x2 matrix to reflect spin-related properties.

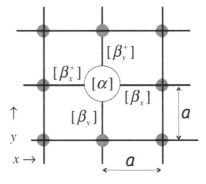

In the simplest case, we can imagine a "spin-innocent" channel that treats both spin components identically. Such a channel can be modeled by choosing the TB parameters as follows:

$$\alpha = \quad \varepsilon [I]$$

$$\beta_x = \quad t [I] \quad , \quad \beta_y = \quad t [I] \tag{22.4}$$

where *[I]* is the (2x2) identity matrix. We effectively have two identical decoupled Hamiltonians that includes no new physics.

Similarly we can write the self-energy $[\Sigma]$ for ordinary contacts that treat both spin components identically simply by taking our usual values and multiplying by *[I]*. This would again be in the category of a trivial extension that introduces no new physics. The results should be the same as what we would get if we worked with one spin only and multiplied by two at the end.

All spin-related phenomena like the ones we discussed in Lecture 14 arise either from non-trivial contacts described by $[\Sigma]$ with spin-related properties or from channels described by *[H]* with spin-related properties or both.

Let us now try to get a feeling for spin transport problems by applying the NEGF method to a series of examples, starting with a simple one-level version of the spin valve we started Lecture 14 with. From a *computational* point of view the only question is how to write down *[H]*, $[\Sigma]$. Once we have these, the rest is standard. One can then proceed to *understand and enjoy* the physics.

22.1. One-level Spin Valve

As we discussed in Lecture 14, a spin valve (Fig.22.4) shows different conductances G_P and G_{AP} depending on whether the magnetic contacts have parallel (P) or anti-parallel (AP) magnetizations. Using a simple model we showed in Lecture 14 that the magnetoresistance (MR) can be expressed as

$$MR \equiv \frac{G_P}{G_{AP}} - 1 = \frac{P^2}{1 - P^2}$$

where the polarization P was defined in terms of the interface resistances. In that context we noted that the standard expression for the

MR for magnetic tunnel junctions (MTJ's) has an extra factor of two (Eq.(14.4))

$$MR \equiv \frac{G_P}{G_{AP}} - 1 = \frac{2P^2}{1-P^2} \qquad (22.5)$$

which could be understood if we postulated that the overall resistance was proportional to the product of the interface resistances and not their sum.

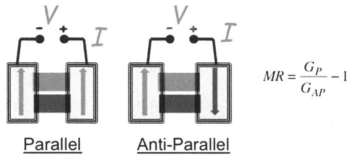

$$MR = \frac{G_P}{G_{AP}} - 1$$

Parallel   **Anti-Parallel**

Fig.22.4. The spin-valve has different conductances G_P and G_{AP} depending on whether the magnetic contacts have parallel (P) or anti-parallel (AP) magnetization.

We could obtain this result (Eq.(22.5)) including the factor of two directly from our NEGF model if we apply it to a one-level resistor and assume that the equilibrium electrochemical potential μ_0 is located many kT's below the energy ε of the level as sketched.

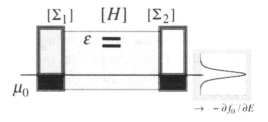

The 2x2 matrices *[H]*, *[Σ]* for this device are summarized below. Also shown for comparison are the corresponding 1x1 "matrices" (that is, just numbers) for the same device without spin. Note that the channel is assumed to treat both spins identically so that *[H]* is essentially an identity matrix, but the *[Σ]*'s have different values for the up and downspin components.

Using these matrices it is straightforward to obtain the Green's function

$$[G^R] = \begin{bmatrix} E - \varepsilon + \dfrac{i}{2}(\gamma_{1u} + \gamma_{2u}) & 0 \\ 0 & E - \varepsilon + \dfrac{i}{2}(\gamma_{1d} + \gamma_{2d}) \end{bmatrix}^{-1}$$

and hence the transmission

$$\bar{T} = Trace[\Gamma_1 G^R \Gamma_2 G^A]$$

$$= \frac{\gamma_{1u}\gamma_{2u}}{(E - \varepsilon)^2 + \left(\dfrac{\gamma_{1u} + \gamma_{2u}}{2}\right)^2} + \frac{\gamma_{1d}\gamma_{2d}}{(E - \varepsilon)^2 + \left(\dfrac{\gamma_{1d} + \gamma_{2d}}{2}\right)^2}$$

$$(22.6)$$

One-level spin-valve: Modifying the *[H]*, *[Σ]* for a spin-less one-level device to represent a one-level spin valve.

For the *parallel (P)* configuration we can assume both contacts to be identical so that we can write ($\alpha > \beta$)

$$\gamma_{1u} = \gamma_{2u} = \alpha \qquad (22.7a)$$

$$\gamma_{1d} = \gamma_{2d} \equiv \beta \qquad (22.7b)$$

while for the *anti-parallel (AP)* configuration the second contact has the roles of α and β reversed with respect to the former:

$$\gamma_{1u} = \gamma_{2d} = \alpha \qquad (22.7c)$$

$$\gamma_{1d} = \gamma_{2u} \equiv \beta \qquad (22.7d)$$

Inserting Eqs.(22.7a,b) into (22.5) we have the P- transmission

$$\bar{T}_P = \frac{\alpha^2}{(E-\varepsilon)^2 + \alpha^2} + \frac{\beta^2}{(E-\varepsilon)^2 + \beta^2}$$

while using Eqs.(22.7c,d) in Eq.(22.5) gives the AP- transmission

$$\bar{T}_{AP} = \frac{2\alpha\beta}{(E-\varepsilon)^2 + \left(\frac{\alpha+\beta}{2}\right)^2}$$

The measured conductance depends on the average transmission over a range of energies of a few kT around μ_0. Assuming that

$$\varepsilon - \mu_0 \gg kT, \alpha, \beta$$

we can write

$$G_P \sim \bar{T}_P(E = \mu_0) = \frac{\alpha^2}{(\mu_0 - \varepsilon)^2 + \alpha^2} + \frac{\beta^2}{(\mu_0 - \varepsilon)^2 + \beta^2} \approx \frac{\alpha^2 + \beta^2}{(\mu_0 - \varepsilon)^2}$$

and $$\qquad G_{AP} \sim \bar{T}_{AP}(E = \mu_0) \approx \frac{2\alpha\beta}{(\mu_0 - \varepsilon)^2}$$

This gives us

$$MR \equiv \frac{G_P}{G_{Ap}} - 1 = \frac{\alpha^2 + \beta^2}{2\alpha\beta} - 1 = \frac{2P^2}{1 - P^2}$$

as stated earlier in Eq.(22.5) with the polarization defined as

$$P \equiv \frac{\alpha - \beta}{\alpha + \beta} \tag{22.8}$$

Actually we could also obtain the result (Eq.(14.3a)) obtained from the resistor model in Lecture 14, if we assume that μ_0 is located right around the level ε, with $kT \gg \alpha, \beta$. But we leave that as an exercise. After all this is just a toy problem intended to get us started.

22.2. Rotating Magnetic Contacts

We argued in the last Section that for an anti-parallel spin valve, the second contact simply has the roles of α and β reversed relative to the first, so that we can write

$$\Gamma_1 = \begin{bmatrix} \alpha & 0 \\ 0 & \beta \end{bmatrix}, \quad \Gamma_2 = \begin{bmatrix} \beta & 0 \\ 0 & \alpha \end{bmatrix}$$

But how would we write the corresponding matrix for a contact if it were pointing along some arbitrary direction defined by a unit vector $\hat{n}$? The answer is

$$\Gamma = \frac{\alpha + \beta}{2} [I] + \frac{\alpha - \beta}{2} \begin{bmatrix} n_z & n_x - in_y \\ n_x + in_y & -n_z \end{bmatrix} \tag{22.9}$$

where n_x, n_y, n_z are the components of the unit vector $\hat{n}$ along x, y and z respectively. This result is of course not obvious and we will try to justify it shortly. But it is reassuring to note that the results for both the parallel and the anti-parallel contact come out as special cases of this general result (Eq.(22.9)):

$$\text{If} \quad n_z = +1, n_x = n_y = 0: \quad \Gamma = \begin{bmatrix} \alpha & 0 \\ 0 & \beta \end{bmatrix}$$

$$\text{If} \quad n_z = -1, n_x = n_y = 0: \quad \Gamma = \begin{bmatrix} \beta & 0 \\ 0 & \alpha \end{bmatrix}$$

One way to understand where Eq.(22.9) comes from is to note that the appropriate matrix describing a magnet pointing along $\hat{n}$ would be

$$\tilde{\Gamma} = \begin{bmatrix} \alpha & 0 \\ 0 & \beta \end{bmatrix} \tag{22.10}$$

if we were to take $+\hat{n}$ and $-\hat{n}$ as our reference directions instead of $+\hat{z}$ and $-\hat{z}$ as we normally do. How could we then transform the $\tilde{\Gamma}$ from Eq.(22.10) into the usual $\pm \hat{z}$ basis ?

Answer: Transform from the $\pm \hat{n}$ to the $\pm \hat{z}$ basis

$$
\begin{array}{c}
\begin{array}{cc} \hat{n} & -\hat{n} \end{array} \\
\begin{array}{c} \hat{z} \\ -\hat{z} \end{array}\!\!
\begin{bmatrix} c & -s^* \\ s & c^* \end{bmatrix}
\end{array}
\begin{array}{c}
\begin{array}{cc} \hat{n} & -\hat{n} \end{array} \\
\begin{array}{c} \hat{n} \\ -\hat{n} \end{array}\!\!
\begin{bmatrix} \alpha & 0 \\ 0 & \beta \end{bmatrix}
\end{array}
\begin{array}{c}
\begin{array}{cc} \hat{z} & -\hat{z} \end{array} \\
\begin{array}{c} \hat{n} \\ -\hat{n} \end{array}\!\!
\begin{bmatrix} c^* & s^* \\ -s & c \end{bmatrix}
\end{array}
\tag{22.11}
$$

$$\text{V} \qquad\qquad\qquad\qquad\qquad \text{V}^+$$

using the unitary transformation matrix *[V]* whose columns represent the components of a spinor pointing along $\pm\hat{n}$. The first column follows from the result we stated earlier in Eq.(22.3), while the second can be obtained from Eq.(22.3) if we set

$$\theta \rightarrow \pi - \theta, \quad \phi \rightarrow \pi + \phi$$

and remove a common phase factor from the two components.

Multiplying out the three matrices in Eq.(22.11) we have

$$\Gamma = \begin{bmatrix} c & -s^* \\ s & c^* \end{bmatrix}\begin{bmatrix} \alpha & 0 \\ 0 & \beta \end{bmatrix}\begin{bmatrix} c^* & s^* \\ -s & c \end{bmatrix} = \begin{bmatrix} c & -s^* \\ s & c^* \end{bmatrix}\begin{bmatrix} \alpha c^* & \alpha s^* \\ -\beta s & \beta c \end{bmatrix}$$

$$= \begin{bmatrix} \alpha cc * + \beta ss * & (\alpha - \beta) cs * \\ (\alpha - \beta) sc * & \alpha ss * + \beta cc * \end{bmatrix}$$

Making use of the definitions of c, s from Eq.(22.3) and some common trigonometric identities like

$$2\cos^2\frac{\theta}{2} = 1 + \cos\theta, \quad 2\sin^2\frac{\theta}{2} = 1 - \cos\theta,$$

$$\text{and} \quad 2\sin\frac{\theta}{2}\cos^2\frac{\theta}{2} = \sin\theta \tag{22.12}$$

we can rewrite this as

$$\Gamma = \frac{1}{2}\begin{bmatrix} (\alpha+\beta)+(\alpha-\beta)\cos\theta & (\alpha-\beta)\sin\theta\, e^{-i\phi} \\ (\alpha-\beta)\sin\theta\, e^{+i\phi} & (\alpha+\beta)-(\alpha-\beta)\cos\theta \end{bmatrix}$$

which leads to the result stated earlier in Eq.(22.9) if we make use of Eq.(22.2) for the x, y, z components of a unit vector.

Finally let me note that if we define the polarization as a vector whose magnitude is given by Eq. (22.8) and direction is given by $\hat{n}$:

$$\vec{P} \equiv P\hat{n} = \frac{\alpha-\beta}{\alpha+\beta}\,\hat{n} \tag{22.13}$$

then we could rewrite Eq.(22.9) as

$$\Gamma = \frac{\alpha+\beta}{2}\left([I] + \begin{bmatrix} P_z & P_x - iP_y \\ P_x + iP_y & -P_z \end{bmatrix}\right) \tag{22.14}$$

which can be rearranged as shown

$$\frac{[\Gamma]}{(\alpha+\beta)/2} = \underbrace{\begin{bmatrix} 1 & 0 \\ 0 & 1 \end{bmatrix}}_{I} + P_x\underbrace{\begin{bmatrix} 0 & 1 \\ 1 & 0 \end{bmatrix}}_{\sigma_x} + P_y\underbrace{\begin{bmatrix} 0 & -i \\ +i & 0 \end{bmatrix}}_{\sigma_y} + P_z\underbrace{\begin{bmatrix} 1 & 0 \\ 0 & -1 \end{bmatrix}}_{\sigma_z}$$

Any 2x2 matrix can be expressed in terms of the four matrices appearing here consist of the identity matrix *[I]* along with the three **Pauli spin matrices**

$$\sigma_x \equiv \begin{bmatrix} 0 & 1 \\ 1 & 0 \end{bmatrix}, \quad \sigma_y \equiv \begin{bmatrix} 0 & -i \\ +i & 0 \end{bmatrix} \text{ and } \sigma_z \equiv \begin{bmatrix} 1 & 0 \\ 0 & -1 \end{bmatrix} \tag{22.15}$$

which are widely used in the spin-related literature.

Making use of the Pauli spin matrices, we could write Eq.(22.14) compactly in the form

$$\Gamma = \frac{\alpha+\beta}{2}\left([I] + [\sigma_x]P_x + [\sigma_y]P_y + [\sigma_z]P_z \right)$$

$$= \frac{\alpha+\beta}{2}\left([I] + [\vec{\sigma}].\vec{P} \right) \tag{22.16}$$

This result applies to the self-energy matrices as well. For example, if

$$\tilde{\Sigma} = -\frac{i}{2}\begin{bmatrix} \alpha & 0 \\ 0 & \beta \end{bmatrix}$$

in the $\pm\hat{n}$ basis, then in the $\pm\hat{z}$ basis it is given by

$$\Sigma = -i\frac{\alpha+\beta}{4}[I] - i\frac{\alpha-\beta}{4}[\vec{\sigma}].\hat{n}$$

$$= -i\frac{\alpha+\beta}{4}\left([I] + [\vec{\sigma}].\vec{P} \right)$$

22.3. Spin Hamiltonians

Now that we have seen how to describe contacts with spin-dependent properties, let us talk briefly about channels with spin-dependent properties.

22.3.1. *Channel with Zeeman Splitting*

The commonest example is the Zeeman splitting that causes the energies of the up-spin state to go up by $\mu_{el}B$ and that of the down spin states to go down by $\mu_{el}B$, μ_{el} being the effective magnetic moment of the electron discussed in Section 14.3.

If the magnetic field points along $+\hat{n}$, then in the $\pm\hat{n}$ basis the corresponding Hamiltonian should look like

$$\mu_{el}\begin{bmatrix} +B & 0 \\ 0 & -B \end{bmatrix}$$

Following our discussion in the last Section we can write it in the $\pm\hat{z}$ basis as

$$H_B = \mu_{el}\,\vec{\sigma}.\vec{B}$$

The overall Hamiltonian is obtained by adding this to the spin-independent part multiplied by *[I]*. For parabolic dispersion this gives

$$H = \frac{\hbar^2}{2m}\left(k_x^2 + k_y^2\right)[I] + \mu_{el}\,\vec{\sigma}.\vec{B} \qquad (22.17)$$

while for a 2D square lattice we have (see Eq.(18.19))

$$H = \left(\varepsilon + 2t\cos k_x a + 2t\cos k_y a\right)[I] + \mu_{el}\,\vec{\sigma}.\vec{B} \qquad (22.18)$$

The corresponding parameters for the 2D lattice in Fig.22.3 (also shown here for convenience) are given simply by

$$\alpha = \varepsilon[I] + \mu_{el}\,\vec{\sigma}.\vec{B}$$

$$\beta_x = t[I]\,,\ \beta_y = t[I] \qquad (22.19)$$

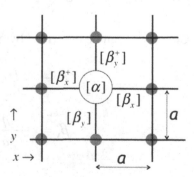

Only the on-site parameter α is changed relative to the spin independent channel (Eq.(22.4).

22.3.2. *Channel with Rashba Interaction*

A more complicated example is that of the Rashba spin-orbit coupling described by a Hamiltonian of the form

$$H_R = \eta \hat{z} \cdot (\vec{\sigma} \times \vec{k}) = \eta(\sigma_x k_y - \sigma_y k_x) \qquad (22.20)$$

whose effect has been observed in 2D surface conduction channels assumed to lie in the x-y plane. This is believed to be a relativistic effect whereby the extremely high atomic scale electric fields (that exist even at equilibrium) are perceived as an effective magnetic field by the electron and the resulting "Zeeman splitting" is described by H_R.

We will not go into the underlying physics of this effect any further here and simply address the question of how to include it in our 2D lattice model. With this in mind we approximate the linear terms with sine functions

$$H_R = \frac{\eta}{a}(\sigma_x \sin k_y a - \sigma_y \sin k_x a) \qquad (22.21)$$

which are written in terms of exponentials:

$$H_R = \frac{\eta}{2ia}\sigma_x(e^{+ik_y a} - e^{-ik_y a}) \; - \; \frac{\eta}{2ia}\sigma_y(e^{+ik_x a} - e^{-ik_x a})$$

Clearly H_R can be described by a Hamiltonian with

$$\beta_x = \frac{i\eta}{2a}\sigma_y, \quad \beta_x^+ = -\frac{i\eta}{2a}\sigma_y$$

$$\beta_y = -\frac{i\eta}{2a}\sigma_x, \quad \beta_y^+ = \frac{i\eta}{2a}\sigma_x$$

in order to ensure that if we write down the dispersion relation for the lattice we will indeed get back the original result in Eq.(22.17). Adding this to the usual spin-independent part from Eq.(22.4) along with any real magnetic field B we have the overall parameters:

$$\alpha = \varepsilon I + \mu_{el} \ \vec{\sigma}.\vec{B}$$

$$\beta_x = \ tI + \frac{i\eta}{2a}\sigma_y, \quad \beta_x^+ = \ tI - \frac{i\eta}{2a}\sigma_y$$

$$\beta_y = \ tI - \frac{i\eta}{2a}\sigma_x, \quad \beta_y^+ = \ tI + \frac{i\eta}{2a}\sigma_x \qquad (22.22)$$

22.4. Vectors and Spinors

One of the important subtleties that takes some time to get used to is that we represent spin with two complex components, but we visualize it as a rotatable object pointing in some direction, which we have learnt to represent with a vector having three real components. To see the connection between the spinor and the vector, it is instructive to consider the precession of a spin in a magnetic field from both points of view.

Consider the one-level device with $\varepsilon = 0$, and with a magnetic field in the z-direction so that the Schrödinger equation can be written as

$$\frac{d}{dt}\begin{Bmatrix}\psi_u \\ \psi_d\end{Bmatrix} \doteq \frac{\mu_{el}B_z}{i\hbar}\underbrace{\begin{bmatrix}1 & 0 \\ 0 & -1\end{bmatrix}}_{\sigma_z}\begin{Bmatrix}\psi_u \\ \psi_d\end{Bmatrix} \qquad (22.23)$$

These are two separate differential equations whose solution is easily written down:

$$\psi_u(t) = \ \psi_u(0) \, e^{-i\omega t/2}$$

$$\psi_d(t) = \ \psi_d(0) \, e^{+i\omega t/2}$$

where
$$\omega \equiv \frac{2\mu_{el}B_z}{\hbar}$$
(22.24)

So if the electron starts out at some angle (θ,ϕ) with a wavefunction

$$\left\{\begin{array}{l} \psi_u(0) = \cos\dfrac{\theta}{2}e^{-i\phi/2} \\[2mm] \psi_d(0) = \sin\dfrac{\theta}{2}e^{+i\phi/2} \end{array}\right\}$$

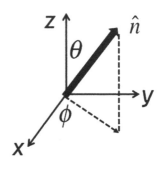

at t = 0, then at a later time it will have a
wavefunction given by

$$\left\{\begin{array}{l} \psi_u(t) = \cos\dfrac{\theta}{2}e^{-i\phi/2}e^{-i\omega t/2} \\[2mm] \psi_d(t) = \sin\dfrac{\theta}{2}e^{+i\phi/2}\,e^{+i\omega t/2} \end{array}\right\}$$

which means that the spin will be rotating around the z-axis such that the
angle θ remains fixed while the angle ϕ increases linearly with time:

$$\phi(t) = \phi(0) + \omega t$$
(22.25)

Making use of Eq.(22.2) for the x, y and z components of the vector $\hat{n}$
we can write

$$n_x = \sin\theta\cos\phi(t) \ , \ n_y = \sin\theta\sin\phi(t)$$

$$\text{and} \quad n_z = \cos\theta$$
(22.26)

From Eqs.(22.25) and (22.26) we can show that

$$\frac{dn_x}{dt} = -\omega n_y, \quad \frac{dn_y}{dt} = +\omega n_x$$

which can be written in matrix form

$$\frac{d}{dt}\begin{Bmatrix} n_x \\ n_y \\ n_z \end{Bmatrix} = \omega \underbrace{\begin{bmatrix} 0 & -1 & 0 \\ +1 & 0 & 0 \\ 0 & 0 & 0 \end{bmatrix}}_{R_z} \begin{Bmatrix} n_x \\ n_y \\ n_z \end{Bmatrix} \tag{22.27a}$$

For comparison we have rewritten the Schrödinger equation we started with (see Eq.(22.23)) in terms of the rotation frequency ω:

$$\frac{d}{dt}\begin{Bmatrix} \psi_u \\ \psi_d \end{Bmatrix} = \frac{\omega}{2i} \underbrace{\begin{bmatrix} 1 & 0 \\ 0 & -1 \end{bmatrix}}_{\sigma_z} \begin{Bmatrix} \psi_u \\ \psi_d \end{Bmatrix} \tag{22.27b}$$

If we wanted to describe the rotation of an electron due to a B-field pointing in the x-direction, it is easy to see how we would modify Eq.(22.27a): Simply interchange the coordinates, $x \to y, y \to z, z \to x$.

$$\frac{d}{dt}\begin{Bmatrix} n_x \\ n_y \\ n_z \end{Bmatrix} = \omega \underbrace{\begin{bmatrix} 0 & 0 & 0 \\ 0 & 0 & -1 \\ 0 & +1 & 0 \end{bmatrix}}_{R_x} \begin{Bmatrix} n_x \\ n_y \\ n_z \end{Bmatrix}$$

and we obtain R_x in place of R_z. But it is not as clear how to modify Eq.(22.27b). The correct answer is to replace σ_z with σ_x (Eq.(22.15))

$$\frac{d}{dt}\begin{Bmatrix} \psi_u \\ \psi_d \end{Bmatrix} = \frac{\omega}{2i} \underbrace{\begin{bmatrix} 0 & 1 \\ 1 & 0 \end{bmatrix}}_{\sigma_x} \begin{Bmatrix} \psi_u \\ \psi_d \end{Bmatrix}$$

but the reason is not as obvious.

Eqs.(22.27a) and (22.27b) both describe the same physics, namely the rotation of a spin about the z-axis due to an applied B-field in the z-direction, one in terms of three real components and the other in terms of two complex components.

But what do matrices like R in Eq.(22.27a) have in common with matrices like σ in Eq.(22.27b) that makes them "isomorphic" allowing them to describe the same physics? Answer: They obey the same "commutation relations". Let me explain.

It is easy to check that the matrices

$$R_x = \begin{bmatrix} 0 & 0 & 0 \\ 0 & 0 & -1 \\ 0 & +1 & 0 \end{bmatrix}, \quad R_y = \begin{bmatrix} 0 & 0 & +1 \\ 0 & 0 & 0 \\ -1 & 0 & 0 \end{bmatrix}$$

$$R_z = \begin{bmatrix} 0 & -1 & 0 \\ +1 & 0 & 0 \\ 0 & 0 & 0 \end{bmatrix}$$

obey the relations

$$R_x R_y - R_y R_x = R_z$$

$$R_y R_z - R_z R_y = R_x \qquad (22.28a)$$

$$R_z R_x - R_x R_z = R_y$$

The Pauli spin matrices obey a similar relationship with R replaced by $\sigma / 2i$:

$$\sigma_x \sigma_y - \sigma_y \sigma_x = 2i\sigma_z$$

$$\sigma_y \sigma_z - \sigma_z \sigma_y = 2i\sigma_x \qquad (22.28b)$$

$$\sigma_z \sigma_x - \sigma_x \sigma_z = 2i\sigma_y$$

The standard textbook introduction to spin starts from these commutation relations and argues that they are a property of the "rotation group". In order to find a mathematical representation with two components for a rotatable object, one must first write down three (2x2) matrices obeying these commutation properties which would allow us to rotate the spinor around each of the three axes respectively.

What are the components of a spinor that points along z? Since rotating it around the z-axis should leave it unchanged, it should be an eigenvector of σ_z that is,

$$\begin{Bmatrix}1\\0\end{Bmatrix} \quad or \quad \begin{Bmatrix}0\\1\end{Bmatrix}$$

which indeed represent an upspin and a downspin along z. Similarly if we want the components of a spinor pointing along x, then we should look at the eigenvectors of σ_x, that is,

$$\begin{Bmatrix}+1\\+1\end{Bmatrix}/\sqrt{2} \quad or \quad \begin{Bmatrix}+1\\-1\end{Bmatrix}/\sqrt{2}$$

which represent up and down spin along $+x$. If we consider a spinor pointing along an arbitrary direction described by a unit vector $\hat{n}$ (see Eq.(22.2)) and wish to know what its components are, we should look for the eigenvectors of

$$\vec{\sigma}.\hat{n} = \sigma_x \sin\theta\cos\phi + \sigma_y \sin\theta\sin\phi + \sigma_z \cos\theta$$

$$= \begin{bmatrix} \cos\theta & \sin\theta e^{-i\phi} \\ \sin\theta e^{+i\phi} & -\cos\theta \end{bmatrix} \tag{22.29a}$$

which can be written as (c and s defined in Eq.(22.3))

$$\begin{Bmatrix}c\\s\end{Bmatrix} \quad and \quad \begin{Bmatrix}-s*\\c*\end{Bmatrix} \tag{22.29b}$$

In short, the rigorous approach to finding the spinor representation is to first determine a set of three matrices with the correct commutation relations and then look at their eigenvectors. Instead in this Lecture I adopted a reverse approach stating the spinor components at the outset and then obtaining the matrices through basis transformations.

22.5. Spin Precession

We have already discussed how to write *[H]*, [Σ] including non-trivial spin-dependent effects and we could set up numerical models to calculate the electron density G^n, or the density of states *A*, or the current using the standard NEGF equations from Lecture 19. Consider for example, the non-local spin potential measurement we started this Lecture with (see Fig.22.5).

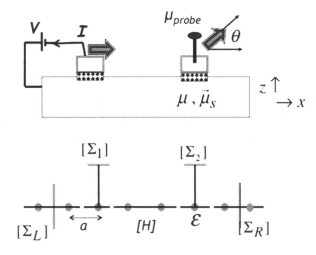

Fig.22.5. Spin potential measurement can be modeled with a 1D channel Hamiltonian having four contacts, two of which are magnetic described by Σ_1, Σ_2.

Fig.22.6 shows the result obtained from the numerical model which supports the basic result stated in Eq.(22.1). The measured voltage oscillates as a function of the angle of magnetization of the voltage probe. It has a constant part independent of the angle and an oscillatory component proportional to the polarization P_v of the voltage probe which can be understood in terms of Eq.(22.1) stated at the beginning of this Lecture.

As I mentioned earlier, I am not sure if the experiment shown in Fig.22.5 has been done, but what has been done is to keep both magnets fixed and rotate the electron spin inside the channel.

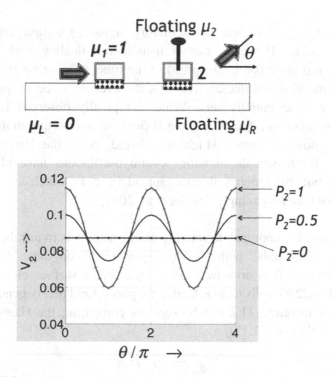

Fig.22.6. Voltage probe signal as the magnetization of the probe is rotated calculated from NEGF model.

How do we rotate the spin? One method that has been widely used is an external magnetic field B which causes the spin direction to precess around the magnetic field as we discussed in Section 22.4 with an angular frequency given by

$$\omega \equiv \frac{2\mu_{el}B_z}{\hbar} \qquad \text{(same as Eq.(22.24))}$$

This means that the spin voltage at the point where the probe is connected will rotate by an angle $\omega\tau$ where τ is the time it takes for the electron to travel from the point of injection to the voltage probe. Writing $\tau = L/v$, we have from Eq.(22.1) using Eq.(22.24)

$$\mu_{probe} = \mu + P_2\mu_s \cos\frac{2\mu_{el}L}{\hbar v}B_z \qquad (22.30)$$

One would expect to see an oscillatory signal as a function of the magnetic field. However, one is usually not dealing with ballistic transport, and there is a large spread in the time τ spent by an electron between injection and detection so that the average value of this signal over all τ is essentially zero. What is typically observed is not an oscillatory signal as a function of the B-field but a reduction in the signal from $P\mu_s$ down to zero, which is referred to as the Hanle signal. However, Hanle signals showing several oscillations have also been observed, but this requires that the spread in τ be much less than its mean value (see for example, Huang et al. 2007).

Another possible approach to rotating electron spins is to use the Rashba effect in materials with strong spin-orbit coupling. In many semiconductors, it is now well established that a surface electric field along z (Fig.22.6) leads to an effective magnetic field that depends on the electron momentum. This can be seen by comparing the Hamiltonians for the B-field (Eq.(22.17))

$$H_B = \mu_{el}\,\vec{\sigma}.\vec{B}$$

with that for the Rashba interaction (22.20) which can be rewritten as

$$H_R = \eta\hat{z}.(\vec{\sigma}\times\vec{k}) \rightarrow \eta\vec{\sigma}.(\hat{z}\times\vec{k})$$

suggesting that the effective B-field due to the Rashba interaction is given by

$$\mu_{el}\vec{B}_{eff} = \eta\hat{z}\times\vec{k}$$

so that from Eq.(22.30) we expect an oscillatory signal of the form

$$\mu_{probe} = \mu + P_2\mu_s \cos\frac{2\eta kL}{\hbar v}$$

with a period $\Delta\eta$ defined by

$$\frac{2kL}{\hbar v}\Delta\eta = 2\pi \quad \rightarrow \quad \Delta\eta = \frac{2\pi at_0}{kL}\sin ka$$

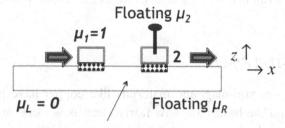

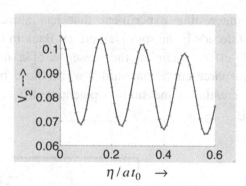

Fig.22.7. In materials with a large Rashba coefficient, a gate voltage should lead to an oscillatory output, if the source and drain magnets point along x, but not if they point along z.

This is in approximate agreement with the numerical result obtained from the NEGF method (Fig.22.7) using an energy E corresponding to

$ka = \pi / 3$, and a distance of about $L=40a$ between the injector and the detector.

In the structure shown in Fig.22.7 the electrons traveling along $+x$ should feel an effective B-field along y. Since the injected spins have a spin voltage $\vec{\mu}_s$ pointing along the source and drain magnets (x) it should be rotated. Note that the oscillation should not be observed if the source and drain magnets point along y rather than along x. There is some recent experimental evidence for this effect (Koo et al 2009, Wunderlich et al. 2010).

22.5.1. *Spin-Hall Effect*

Spin-diffusion in non-magnetic materials like copper have been studied extensively and the basics are now fairly clear as we saw in Lecture 14. What is relatively less understood is the nature of spin diffusion in materials with high spin-orbit coupling.

For example, an interesting experiment that has attracted a lot of attention in the last decade is the spin-Hall effect. Back in Lecture 13 we discussed the Hall effect where in the presence of a magnetic field, electrons from the source curve "upwards" while those from the drain curve downwards creating a measurable potential difference in the y-direction (Fig.22.8).

Fig.22.8. The spin-Hall effect can be understood in terms of an effective force that makes $+z$-spins and $-z$-spins veer in opposite directions, unlike the ordinary Hall effect that makes both spins veer in the same direction.

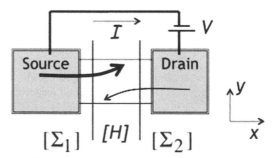

Since electrons feel an effective magnetic field in conductors with spin-orbit coupling, it seems natural to ask whether we could see a Hall effect without an external B-field simply due to this effective field.

The answer indeed is yes, but the phenomenon is subtle with $+z$-spins and $-z$-spins veering in opposite directions, unlike the ordinary Hall effect that makes both spins veer in the same direction (Eq.(13.1)). As a result, there is no buildup of charge and hence no ordinary Hall voltage that can be measured with ordinary contacts. But there is a spin voltage that can be measured with magnetic contacts. The first experiments detected this spin buildup using optical probes in large conductors (see for example Sih et al. 2006) but more recently magnetic contacts have been used to measure spin-Hall effects of this type in small conductors (see for example, Wunderlich et al. 2010).

Our NEGF-based transport model seems to capture this effect as shown in Fig.22.9 for a 2D conductor with spin-orbit coupling like the one shown in Fig.22.7 with $\eta = 1e - 11\, eV - m$, assuming an energy $E=0.05$ t_0. We use Eq.(22.31) to extract the z-component of the spin density from the G^n as discussed. The numerical results show a build-up of opposite spins on opposite sides of the conductor which reverses on reversing the current, in accordance with what we discussed above.

Fig.22.9. NEGF-Based transport model for spin-Hall effect. Numerical results show a pile-up of $+z$-spins and $-z$-spins on opposite sides of the sample which reverses on reversing the current.

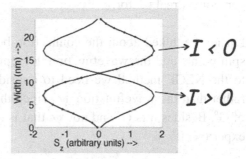

There is much activity currently in this general area of spin transport in materials with strong spin-orbit coupling like the "topological insulators" (see for example Xiu et al. 2011 and references therein) which exhibit a more striking version of the spin Hall effect.

The main point I want to make is that many striking non-intuitive results come automatically from the NEGF model, allowing one to explore new physics that we may not yet have understood clearly. However, I should mention that we have not discussed the calculation and interpretation of spin currents which are needed in order to couple the transport problem with the LLG equation describing the magnet dynamics (see Fig.14.10). We will not go into this and simply suggest a few references for interested readers.

22.6. From NEGF to Diffusion

Let me end this Lecture by talking a little about the connection between the diffusion equation approach from Lecture 14 with the NEGF-based approach of this Lecture. We have seen that the NEGF-based numerical examples in Fig.22.6, 7 are well described by Eq.(22.1) which follows heuristically from the discussions in Lecture 14. But the reason for this agreement is not obvious, especially since the formal method is based on two-component complex spinors, while $\vec{\mu}_s$ is an old-fashioned three-component real vector.

Earlier we talked about the connection between vector $\hat{n}$ along which a spin points and the wavefunction ψ representing it. To relate Eq.(22.1) to the NEGF method we need to consider quantities like $G^n \sim \psi\psi^+$ rather than the wavefunction ψ, since the NEGF is formulated in terms of G^n. Besides it is G^n and not ψ that is observable and can be related to experiment.

22.6.1. Matrix Electron Density

We have often referred to *[Gⁿ]* as the matrix electron density whose diagonal elements tell us the number of electrons at a point. With spin included, *[Gⁿ]* at a point is a (2x2) matrix and the elements of this matrix tell us the number of electrons N or the net number of spins $\vec{S}$.

To see this consider an electron pointing in some direction $\hat{n}$ represented by a spinor wavefunction of the form (see Eq.(22.3))

$$\psi = \left\{ \begin{array}{l} \cos\dfrac{\theta}{2}e^{-i\phi/2} \equiv c \\[2ex] \sin\dfrac{\theta}{2}e^{+i\phi/2} \equiv s \end{array} \right\}$$

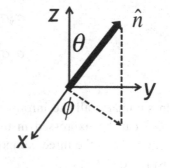

so that the corresponding (2x2) electron density G^n is given by

$$\psi\psi^+ = \left\{ \begin{array}{c} c \\ s \end{array} \right\} \{c* \quad s*\} = \begin{bmatrix} cc* & cs* \\ sc* & ss* \end{bmatrix}$$

Making use of Eq.(22.2) and Eq.(22.12) we have

$$\psi\psi^+ = \frac{1}{2}\begin{bmatrix} 1+n_z & n_x - in_y \\ n_x + in_y & 1-n_z \end{bmatrix} = \frac{1}{2}[I + \vec{\sigma}.\hat{n}]$$

For a collection of N electrons we can add up all their individual contributions to $\psi\psi^+$ to obtain the net *[Gⁿ]* given by

$$\frac{G^n}{2\pi} = \frac{1}{2}\begin{bmatrix} N+S_z & S_x - iS_y \\ S_x + iS_y & N-S_z \end{bmatrix}$$

$$= \frac{1}{2}\left(N[I] + \vec{\sigma}.\vec{S}\right)$$

Given a *[Gⁿ]* we can extract these quantities from the relations

$$N = \frac{1}{2\pi} Trace[G^n] \quad , \quad \vec{S} = \frac{1}{2\pi} Trace[\vec{\sigma} G^n] \qquad (22.31)$$

which follow from Eq.(22.30) if we make use of the fact that all three matrices (Eq.(22.15)) have zero trace, along with the following properties of the Pauli spin matrices that are easily verified.

$$\sigma_x^2 = \sigma_y^2 = \sigma_z^2 = I \qquad (22.32a)$$

$$\sigma_x \sigma_y = -\sigma_y \sigma_x = i\sigma_z \qquad (22.32b)$$

$$\sigma_y \sigma_z = -\sigma_z \sigma_y = i\sigma_x \qquad (22.32c)$$

$$\sigma_z \sigma_x = -\sigma_x \sigma_z = i\sigma_y \qquad (22.32d)$$

In summary, all the information contained in the 2x2 Hermitian matrix *[G_n]* can be expressed in terms of four real quantities consisting of a scalar N and the three components of a vector $\vec{S}$ which can be extracted using Eq.(22.31).

22.6.2. Measuring the Spin Potential

That brings us to the question of relating the result stated at the beginning of this Lecture

$$\mu_{probe} = \mu + \frac{\vec{P}.\vec{\mu}_s}{2}$$

(Same as Eq.(22.1))

to the NEGF method. The scalar version of this result was obtained in Lecture 14 (see Eq.(14.23)) using the semiclassical model for a probe shown and setting the probe current to zero. Now we can obtain the general vector version by starting from the NEGF model for a probe (Fig22.10) with the current given by (see Eq.(19.4))

$$I \sim Trace\,[\Gamma]\Big[f_{probe}[A]-[G^n]\Big]$$

so that for zero probe current we must have

$$f_{probe} = \frac{Trace\,[\Gamma][G^n]}{Trace\,[\Gamma][A]}$$

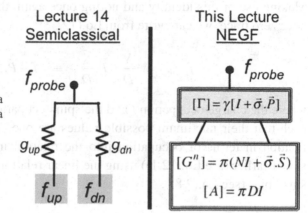

Fig.22.10. Model for a probe connected to a channel.

Making use of Eq.(22.30) for $[G^n]$, assuming that the density of states (D) is spin-independent

$$\frac{A}{2\pi} = \frac{D}{2}[I]$$

writing the probe coupling in the form (see Eq.(22.16))

$$\Gamma = \gamma\Big[I+\vec{\sigma}.\vec{P}\Big]$$

and noting that the Pauli matrices all have zero trace, we obtain

$$f_{probe} = Trace\,[I+\vec{\sigma}.\vec{P}]\left[\frac{N}{D}I+\vec{\sigma}.\frac{\vec{S}}{D}\right]$$

$$(22.33)$$

Once again there is an identity that can be used to simplify this expression: For any two vectors $\vec{P}$ and $\vec{B}$, it is straightforward (but takes some algebra) to verify that

$$[\vec{\sigma}.\vec{P}][\vec{\sigma}.\vec{B}] = (\vec{P}.\vec{B})[I] + i\vec{\sigma}.[\vec{P} \times \vec{B}] \text{ , so that}$$

$$[I + \vec{\sigma}.\vec{P}]\left[bI + \vec{\sigma}.\vec{B}\right] = (b + \vec{P}.\vec{B})[I] + \vec{\sigma}.[\vec{P} + \vec{B} + i\vec{P} \times \vec{B}] \quad (22.34)$$

Making use of this identity and noting once again that the Pauli matrices have zero trace, we can write from Eq.(12.33)

$$f_{probe} = \frac{N}{D} + \vec{P}.\frac{\vec{S}}{D} \equiv f + \vec{P}.\frac{\vec{f_s}}{2} \qquad (22.35)$$

where the charge occupation f and the spin occupation f_s are each defined such that their maximum possible values are one. We can translate this relation in terms of occupation into the relation in terms of potentials stated earlier (see Eq.(22.1)) using the linear relation between the two for small bias (see Eq.(2.8)).

22.6.3. *Four-Component Diffusion*

The key point we want to stress is that the quantum formalism naturally leads to a 2x2 complex matrix $[G^n]$ at each point, but it is straightforward to translate it into four physically transparent components, like N, $\vec{S}$ or μ, $\vec{\mu_s}$.

Back in Lecture 14, we saw how many spin transport phenomena can be understood in terms of the Valet-Fert equation along with a model for the spin-dependent interface resistances. However, this approach is limited to problems that involve spins in one direction (the z-direction) only.

Now we have the full NEGF model that can be used for spins pointing in any direction, but it is computationally more demanding and seems conceptually unrelated to the old approach. By translating the information in $[G^n]$ into N, $\vec{S}$ we can bridge this gap. For example, the spin-dependent interface conductances of Lecture 14 could be replaced by a 4x4 conductance matrix that relates the four components of the potential to the four components of the current:

$$\begin{Bmatrix} I \\ I_{sx} \\ I_{sy} \\ I_{sz} \end{Bmatrix} = \begin{bmatrix} 4 \times 4 \\ G \end{bmatrix} \begin{Bmatrix} \Delta\mu \\ \Delta\mu_{sx} \\ \Delta\mu_{sy} \\ \Delta\mu_{sz} \end{Bmatrix}$$

Similarly the two component Valet-Fert equation for μ, μ_s can be extended to a four component diffusion equation for

$$\mu, \quad \underbrace{\mu_{sx}, \mu_{sy}, \mu_{sz}}_{\bar{\mu}_s}$$

The question one could ask is whether these four-component formulations are equivalent to the NEGF method discussed here. The answer we believe is that they capture a subset of the effects contained in the NEGF and there are many problems where this subset may be adequate. Let me explain.

At the beginning of this Lecture I mentioned that including spin increases the size of the matrices by a factor of two since every point along z effectively becomes two points, an up and a down. So if there are three points in our channel, the matrix $[G^n]$ will be of size (6x6).

The four-component diffusion approach looks at the (2x2) diagonal blocks of this matrix and represents each block with four components. What it clearly misses is the information contained in the off-diagonal elements between two spatial elements which as we saw in Lecture 20 gives rise to quantum interference effects. But we may not be missing much, since as we discussed, dephasing processes often destroy these interference effects anyway.

Spin information is usually more robust. While phase relaxation times are often sub-picosecond, spin relaxation times are much longer, in nanoseconds. And so it is important to retain the information in the (2x2) diagonal blocks, even if we are throwing away the rest.

Formally we could do that starting from the NEGF method by defining a suitable D-matrix of the type discussed in Lecture 19 relating the inscattering to the electron density ($\times$ denotes element by element multiplication)

$$[\Sigma^{in}] = \quad D \times [G^n] \quad \text{(Same as Eq.(19.35b))}$$

The dephasing process can be viewed as extraction of the electron from a state described by $[G^n]$ and reinjecting it in a state described by $D \times G^n$. We introduced two models A and B with D defined by Eqs.(19.37) and (19.38) respectively. Model A was equivalent to multiplying $[G^n]$ by a constant so that the electron was reinjected in exactly the same state that it was extracted in, causing no loss of momentum, while Model B threw away the off-diagonal elements causing loos of momentum as we saw in the numerical example in Fig.20.7. We could define a **Model C** having a D-matrix that retains spin information while destroying momentum:

$$\frac{[D]}{D_0} =$$

	$1up$	$1dn$	$2up$	$2dn$	$3up$	$3dn$
$1up$	1	1	0	0	0	0
$1dn$	1	1	0	0	0	0
$2up$	0	0	1	1	0	0
$2dn$	0	0	1	1	0	0
$3up$	0	0	0	0	1	1
$3dn$	0	0	0	0	1	1

$$(22.36)$$

One could view this as Model B-like with respect to the lattice, but Model A-like with respect to spin. We could rewrite the NEGF equation

$$G^n = G^R \Sigma^{in} G^A$$

as
$$[G^n]_{i,i} = \sum_j [G^R]_{i,j} [\Sigma^{in}]_{j,j} [G^A]_{j,i}$$

$$= D_0 \sum_j [G^R]_{i,j} [G^n]_{j,j} [G^A]_{j,i} \qquad (22.37)$$

where the indices i, j refer to lattice points and we have made use of the fact that in our Model C, Σ^{in} is diagonal as far as the lattice is concerned.

We have seen earlier that at any point on the lattice the 2x2 matrix $[G^n]$ can be expressed in terms of four components, namely N and $\vec{S}$ so that with a little algebra we could rewrite Eq.(22.37) in the form

$$\left\{ \begin{matrix} N \\ S_x \\ S_y \\ S_z \end{matrix} \right\}_i = \sum_j \left[\begin{matrix} 4 \times 4 \\ "Hopping" \\ Matrix \end{matrix} \right]_{i,j} \left\{ \begin{matrix} N \\ S_x \\ S_y \\ S_z \end{matrix} \right\}_j \qquad (22.38)$$

where the (4x4) matrix could be viewed as describing the probability of the $(N, \vec{S})$ at a point "j" hopping to a point i in one time step. Indeed the (1x1) version of Eq.(22.38) resembles the standard description of Brownian motion on a lattice that leads to the drift-diffusion equation.

Spin diffusion equations based on alternative approaches like the Kubo formalism have been discussed in the past (see for example, Burkov et al. 2004). The main point I want to convey is that NEGF-based approaches can also be used to justify and benchmark spin diffusion models which could well capture the essential physics and provide insights that a purely numerical calculation misses. In the last lecture I will briefly explain how one might be able to adapt this approach to more complicated spin-like objects as well. But that is a topic for future research.

Lecture 23

Does NEGF Include "Everything?"

23.1. Coulomb Blockade
23.2. Fock Space Description
23.3. Entangled States

Back in Lecture 18 we used this picture to introduce our quantum transport model representing an elastic channel described by a Hamiltonian *[H]* and self-energies [Σ] describing the exchange of electrons with the physical contacts and energy with the surroundings which can be viewed as additional conceptual "contacts".

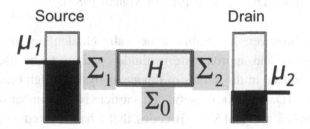

Given these inputs, the basic NEGF equations (see Eqs.(19.1)-(19.4)) tell us how to analyze any given structure. Since then we have been looking at various examples illustrating how one writes down *[H]* and [Σ] and uses the NEGF equations to extract concrete results and investigate the physics. One major simplification we have adopted is in our treatment of the interactions in the channel represented by Σ_0 which we have either ignored (coherent transport) or treated as an elastic dephasing process described by Eqs.(19.35).

This choice of self-energy functions leads to no exchange of energy with the surroundings, but it has an effect on transport due to the exchange of momentum and "phase". Basically we have been talking about elastic resistors like the ones we started these Lectures with, except that we are now including quantum mechanical effects. One could say that in the last few Lectures we have applied the general Non-Equilibrium Green's Function (NEGF) method to an elastic resistor, just as in Part one we applied the general Boltzmann Transport Equation (BTE) to an elastic resistor.

So how do we go beyond elastic resistors? For semiclassical transport, it is clear in principle how to include different types of interaction into the BTE for realistic devices and much progress has been made in this direction. Similarly for quantum transport, the NEGF tells us how to evaluate the self-energy Σ_0 for any given microscopic interaction. In these lectures we have talked only about elastic dephasing which is a small subset of the interactions considered in the classic work on NEGF (see for example, Danielewicz 1984 or Mahan 1987).

In practice, however, it remains numerically challenging to go beyond elastic resistors and approximate methods continue to be used widely. Readers interested in the details of device analysis at high bias may find an old article (Datta (2000)) useful. This article has a number of concrete results obtained using MATLAB codes that I had offered to share with anyone who asked me for it. Over the years many have requested these codes from me which makes me think they may be somewhat useful and we plan to have these available on our website for these notes.

I should mention that many devices are rather forgiving when it comes to modeling the physics of ineastic scattering correctly. Devices with energy levels that are equally connected to both contacts (Fig.9.5b.) do not really test the deep physics of inelastic transport and cannot distinguish between a good theory and a bad one. A good test for inelastic scattering models is the device shown in Fig.9.5a for which the entire terminal current is driven by inelastic processes. Only a fundamentally sound theory will predict results that comply with the requirements of the second law.

But practical issues apart, can the NEGF method model *"everything"*, at least in principle?

The formal NEGF method developed in the 1960's was based on many-body perturbation theory (MBPT) which provided clear prescriptions for evaluating the self-energy functions

$$\Sigma , \Sigma^{in}$$

for a given microscopic interaction up to any order in perturbation theory. It may seem that using MBPT we can in principle include everything. However, I believe this is not quite true since it is basically a perturbation theory which in a broad sense amounts to evaluating a quantity like $(1-x)^{-1}$ by summing a series like $1+x+x^2+x^3+ \ldots$, which works very well if x is much less than one. But if x happens to exceed one, it does not work and one needs non-perturbative methods, or perhaps a different perturbation parameter.

This is one of the reasons I prefer to decouple the NEGF equations (Eqs.(19.1) through (19.4)) from the MBPT-based methods used to evaluate the self-energy functions. The latter may well evolve and get supplemented as people find better approximations that capture the physics in specific situations.

With equilibrium problems, for example, density functional theory (DFT)-based techniques have proven to be very successful and are often used in quantum chemistry in place of MBPT. I believe one should be cautious about expecting the same success with non-equilibrium problems where a far greater spectrum of many body states are made accessible and can be manipulated through a judicious choice of contacts, but it is quite likely that people will find insightful approaches that capture the essential physics in specific problems.

Like the BTE for semiclassical transport, NEGF-based methods in their simplest form, seem to provide a good description of problems where electron-electron interactions can be treated within a *mean field theory* based on the widely used picture of quasi-independent electrons moving in a self-consistent potential U due to the other electrons (Section 18.2).

As we saw in Lecture 8, for low bias calculations one needs to consider only the equilibrium potential which is already included in the semi-empirical tight-binding (TB) parameters used to construct our Hamiltonian *[H]*. For real devices operating at high bias, the change in the potential due to any changes in the electron occupation in the channel are routinely included using the Poisson equation which is the simplest approximation to the very difficult problem of electron-electron interactions and there have been extensive discussions of how the self-consistent field (scf) can be corrected to obtain better agreement with experimental results.

However, there are examples where the self-consistent field approach itself seems to fail and some of the most intriguing properties arise from a failure of this simple picture. The purpose of this Lecture is to alert the reader that a straightforward application of NEGF may well miss these important experimentally observable effects. Future challenges and opportunities may well involve effects of this type, requiring insightful choices for Σ, Σ^{in} if we wish to use the NEGF method.

23.1. Coulomb Blockade

In the spirit of the bottom-up approach let us consider the simplest resistor that will show this effect, one that is only slightly more complicated than the one-level resistor we started with (Fig.3.1). We assume two levels, a spin up and a spin down, having the same energy ε, with the equilibrium chemical potential μ located right at ε, so that each level is half-filled since the Fermi function $f_0(E=\mu)$ equals *0.5*. Based on what we have discussed so far we would expect a high conductance since the electrochemical potential lies right in the middle of each broadened level as shown in the upper sketch in Fig.23.1.

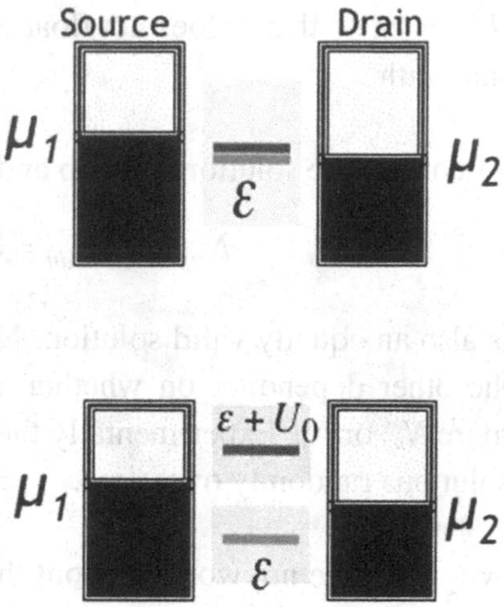

Fig.23.1.
The "bottom-up" view of Coulomb blockade: A two-level channel can show significantly lower density of states around $E=\mu$, and hence a higher resistance, if U_0 is large.

However, if the single electron charging energy U_0 is large then the picture could change to the lower one where one level floats up by U_0 due to the electron occupying the other level. Why doesn't the other level float up as well? Because no level feels any potential due to itself. This self-interaction correction is missed in the self-consistent field (SCF) model discussed in Lecture 8 where we wrote $U=U_0 N$. Instead we need an unrestricted SCF where each level i is not restricted to feeling the same potential. Instead it feels a potential U_i that depends on the change in the number of electrons occupying all levels except for i :

$$U_i = U_0 (N - N_i) \qquad (23.1)$$

If we were to use Eq.(23.1) instead of $U = U_0 N$ we would obtain a picture like the lower one in Fig.25.4, assuming that μ is adjusted to have approximately one electron inside the channel. We would find a self-consistent solution with

$$N_{dn} = 1, U_{up} = 0, N_{up} = 0, U_{dn} = 0$$

The down level will be occupied ($N_{dn} = 1$) and the resulting potential ($U_{up} = U_0$) will cause the up level to float up and be unoccupied ($N_{up} = 0$). Because it is unoccupied, the potential felt by the down level is zero ($U_{dn} = 0$), so that it does not float up, consistent with what we assumed to start with.

Of course, the solution with up and down interchanged

$$N_{up} = 1, U_{dn} = 0, N_{dn} = 0, U_{up} = 0$$

is also an equally valid solution. Numerically we will converge to one or the other depending on whether we start with an initial guess that has more N_{up} or N_d. Experimentally the system will fluctuate between the two solutions randomly over time.

Why have we not worried about this before? Because it is not observable unless the charging energy U_0 is well in excess of both kT and the broadening. U_0/q is the potential the channel would float to if one electron were added to it. For a large conductor this potential is microvolts or smaller and is unobservable even at the lowest of temperatures. After all, any feature in energy is spread out by kT which is ~ 25 meV at room temperature and ~ 200 µeV at ~ 1K. The single electron charging effect that we are talking about, becomes observable at least at low temperatures, once the conductor is small enough to make U_0 of the order of a meV. For molecular sized conductors, U_0 can be hundreds of meV making these effects observable even at room temperature.

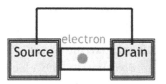

Fig.23.2. The single electron charging energy U_0 is the electrostatic energy associated with one extra electron in the channel.

However, there is a second factor that also limits the observability of this effect. We saw in Lecture 19 that in addition to the temperature broadening $\sim kT$, there is a second and more fundamental broadening, $\gamma \sim h/t$ related to the transfer time. Single electron charging effects will be observed only if the Coulomb gap U_0 exceeds this broadening: $U_0 \gg h/t$. For this reason we would not expect to see this effect even in the smallest conductors, as long as it has good contacts.

23.1.1 Current versus voltage

Let us now move on from the low bias conductance to the full current-voltage characteristics of the two-level resistor. For simplicity we will assume that the levels remain fixed with respect to the source and are unaffected by the drain voltage, so that we do not have to worry about the kind of issues related to simple electrostatics that we discussed in Lecture 8.

A simple treatment ignoring electron-electron interactions then gives the curve marked "non-interacting" in Fig.23.3. Once the electrochemical potential μ_2 crosses the levels at ε, the current steps up to its maximum value.

If we now include charging effects through a self-consistent potential $U=U_0$, the current step stretches out over a voltage range of $\sim U_0/q$, since the charging of the levels makes them float up and it takes more voltage to cross them completely.

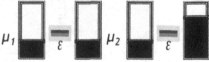

Fig.23.3.
Current-voltage characteristic of a two-level resistor with $U = 0$ and with $U= U_0 N$

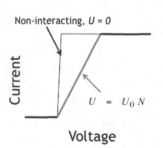

But if we include an SCF with self-interaction correction (Eq.(23.1)) we calculate a current-voltage characteristic with an intermediate plateau as shown in Fig.23.4 which can be understood in terms of the energy level diagrams shown. At first only the lower level conducts giving only half the maximum current and only when the voltage is large enough for μ_2 to cross $\varepsilon + U_0$ that we get the full current.

Such intermediate plateaus in the *I-V* characteristics have indeed been observed but the details are not quite right. The correct plateau current is believed to be *2/3* and not *1/2* of the total current of *2q/t*. This represents an effect that is difficult to capture within a one-electron picture, though it can be understood clearly if we adopt a different approach altogether, which we will now describe.

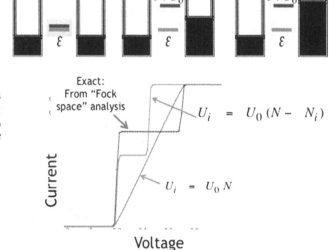

Fig.23.4. Current-voltage characteristic of a two-level resistor: Exact and with two different SCF potentials.

23.2. Fock Space Description

This approach is based on the Fock space picture introduced in Lecture 16. As we discussed earlier, in this new picture we do not think in terms of one-electron levels that get filled or emptied from the contacts. Instead we think in terms of the system being driven from one state to another.

Fig.23.5.
One-electron picture versus Fock space picture for a one-level channel.

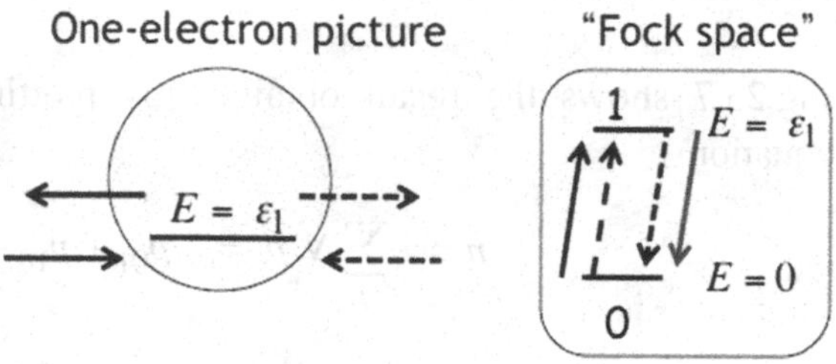

For example Fig.23.5 shows how we would view the one-level resistor in this Fock space picture where the system can be one of two states: *0* representing an empty state, and *1* representing a full state. Fig. 23.6 shows the two pictures for a two-level resistor. In general a *N*-level resistor will have 2^N Fock space states.

Fig.23.6.
One-electron picture versus Fock space picture for a two-level channel.

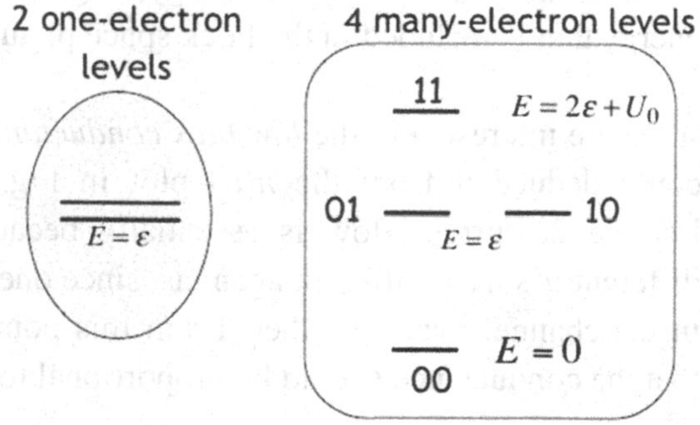

23.2.1. *Equilibrium in Fock space*

As we discussed in Lecture 16, there is a well-defined procedure for finding the probabilities of finding the system in a given eigenstate *i* at equilibrium.

$$p_i = \frac{1}{Z} e^{-(E_i - \mu N_i)/kT}$$

(Same as Eq.(16.14))

We could use this to calculate any equilibrium property. For example suppose we want to find the number of electrons, n occupying the two-level channel shown in Fig.23.6 if it is in equilibrium with an electrochemical potential μ.

Fig.23.7 shows the result obtained by plotting n versus μ from the equation

$$n = \sum_i N_i p_i = p_{01} + p_{10} + 2p_{11}$$

using the equilibrium probabilities from Eq.(16.14) cited above. Note how the electron number changes by one as μ crosses ε and then again when μ crosses $ε + U_0$ in keeping with the lower picture in Fig.23.1.

Note, however, that we did not assume the picture from Fig.23.1 with two one-electron states at different energies. We assumed two one-electron states with the same energy (Fig.23.6) but having an interaction energy that is included in the Fock space picture.

If we are interested in the *low bias conductance G* as a function of μ, we could deduce it from the n(μ) plot in Fig.23.7. As we discussed in Lecture 2, current flow is essentially because the two contacts with different μ's have different agendas, since one likes to see more electrons in the channel than the other. From this point of view one could argue that the conductance should be proportional to *dn/dμ* and show peaks at

$$\mu = \varepsilon \quad \text{and at} \quad \mu = \varepsilon + U_0$$

as shown. This is indeed what has been observed experimentally for the low bias conductance of small conductors in the single-electron charging regime where U_0 exceeds both the thermal energy kT and the energy broadening due to contacts.

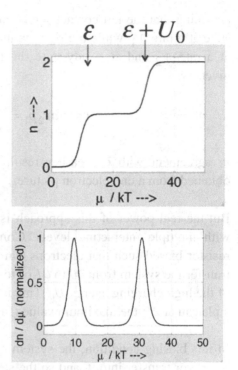

Fig.23.7. Equilibrium number of electrons, n in the two-level channel shown in Fig.23.6 as a function of μ, assuming $\varepsilon = 10kT$, $U_0 = 20kT$. The conductance can be argued to be proportional to the derivative *dn/dμ* showing peaks when μ equals ε and $\varepsilon + U_0$.

As we saw in Lecture 15, low bias conductance is an equilibrium property that can be deduced using the principles of equilibrium statistical mechanics. Current flow at higher voltages on the other hand requires the methods of non-equilibrium statistical mechanics. Let me explain briefly how one could understand the *2/3* plateau shown in Fig.23.4 by calculating the current at high bias in the Fock space picture.

23.2.2. Current in the Fock space picture

To calculate the current we write an equation for the probability that the system will be found in one of its available states, which must all add up to one. For example for the one level resistor we could write

$$\nu_1 p_0 = \nu_2 p_1 \quad \rightarrow \quad \frac{p_1}{p_0} = \frac{\nu_1}{\nu_2} \quad \rightarrow \quad p_1 = \frac{\nu_1}{\nu_1 + \nu_2}$$

assuming that the left contact sends the system from the *0* state to the *1* state at a rate V_1, while the right contact takes it in the reverse direction at a rate V_2 and at steady-state the two must balance. The current is given by

$$I = qV_2p_1 = q\frac{V_1V_2}{V_1+V_2} \tag{23.2}$$

in agreement with our earlier result in Lecture 19 (see Eq.(19.10b)) obtained from a one-electron picture.

But the real power of this approach is evident when we consider levels with multiple interacting levels. Consider for example the two-level resistor biased such that electrons can come in from the left contact and transfer the system from *00* to *01* or to *10*, but not to the *11* state because of the high charging energy U_0. This is the biasing condition that leads to a plateau at *2/3* the maximum value (Fig.23.4) that we mentioned earlier.

In this biasing condition, the system can only come out of the *11* state, but never transfer into it, and so the steady-state condition can be calculated simply by considering the kinetics of the three remaining states in Fock space, namely *00*, *01* and *10*:

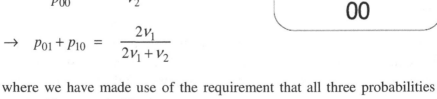

$$2V_1p_{00} = V_2(p_{01}+p_{10})$$

$$\rightarrow \frac{p_{01}+p_{10}}{p_{00}} = \frac{2V_1}{V_2}$$

$$\rightarrow p_{01}+p_{10} = \frac{2V_1}{2V_1+V_2}$$

where we have made use of the requirement that all three probabilities must add up to one. Hence

$$I \;=\; qv_2(p_{01} + p_{10}) \;=\; q\,\frac{2v_1v_2}{2v_1+v_2}$$

With $\quad v_1 = v_2 \;\rightarrow\; I = \dfrac{2}{3}qv_1$

which is *2/3* the maximum current as stated earlier.

It is important to note the very special nature of the solution we just obtained which makes it hard to picture within a one-electron picture. We showed that the system is equally likely to be in the states *00, 01* and the *10* states, but zero probability of being in the *11* state.

Fig.23.8
The intermediate plateau in the current corresponds to the channel being in a strongly correlated state.

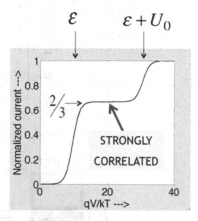

In other words, if we looked at the up-spin or the down-spin state (in the one-electron picture) we would find them occupied with *1/3* probability. If electrons were independent then we would expect the probability for both to be occupied to be the product = *1/9*.

Instead it is zero, showing that the electrons are correlated and cannot be described with a one-electron occupation factor f of the type we have been using throughout these lectures. Even with quantum transport we replaced the f's with a matrix G^n obtained by summing the $\psi\psi^+$ for individual electrons. This adds sophistication to our understanding of the one-electron state, but it still does not tell us anything about two-electron correlations.

23.3. Entangled states

What we just saw with one quantum dot is actually just the proverbial tip of the iceberg. Things get more interesting if we consider two or more quantum dots.

For example, with two coupled quantum dots we could write the one-electron Hamiltonian matrix as a 4x4 matrix using the up and down states in dots 1 and 2 as the basis functions as follows:

$$[H] =$$

$$
\begin{array}{c}
 \begin{array}{cccc} u_1 & u_2 & d_1 & d_2 \end{array} \\
\begin{array}{c} u_1 \\ u_2 \\ d_1 \\ d_2 \end{array}
\begin{bmatrix}
\varepsilon_1 & t & 0 & 0 \\
t & \varepsilon_2 & 0 & 0 \\
0 & 0 & \varepsilon_1 & t \\
0 & 0 & t & \varepsilon_2
\end{bmatrix}
\end{array}
$$

$$(23.3)$$

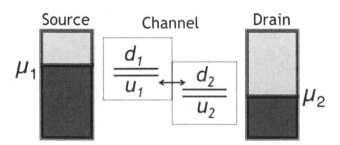

But what are the Fock space states ? With four one-electron states we expect a total of $2^4 = 16$ Fock space states, containing *0, 1, 2, 3* or *4* electrons. The number of *n*-electron states in Fock space is given by ${}^4C_n$: one with n=0, four with n=1, six with n=2, four with n=3 and one with n=4.

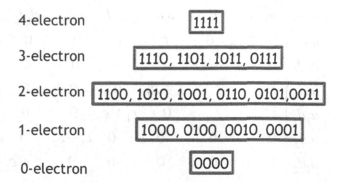

If there were no inter-dot coupling then these sixteen states would be the eigenstates and we could analyze their dynamics in Fock space just as we did for one dot. But in the presence of inter-dot coupling the true eigenstates are linear combinations of these states and these entangled states can lead to novel physics and make it much more interesting.

The 0-electron and 4-electron states are trivially composed of just one Fock space state, while the 1-electron state is essentially the same as the states in a one-electron picture. Indeed the 3-electron state also has a structure similar to the one-electron state and could be viewed as a 1-hole state.

The 2-electron states, however, have an interesting non-trivial structure. Consider the six 2-electron states which we label in terms of the two states that are occupied: u_1d_1, u_2d_2, u_1d_2, u_2d_1, u_1u_2, d_1d_2. Using these we can write the Fock space Hamiltonian *[HH]* as explained below.

The **diagonal** elements of *[HH]* are written straightforwardly by adding the one-electron energies plus an interaction energy U_0 if the two basis functions happen to be on the same dot making their Coulomb repulsion much stronger than what it is for two states on neighboring dots.

[HH] =

$$\begin{array}{c} \\ u_1d_1 \\ u_2d_2 \\ u_1d_2 \\ u_2d_1 \\ u_1u_2 \\ d_1d_2 \end{array}
\begin{array}{cccccc}
u_1d_1 & u_2d_2 & u_1d_2 & u_2d_1 & u_1u_2 & d_1d_2 \\
\begin{bmatrix} 2\varepsilon_1+U_0 & 0 & t & t & 0 & 0 \\ 0 & 2\varepsilon_2+U_0 & t & t & 0 & 0 \\ t & t & \varepsilon_1+\varepsilon_2 & 0 & 0 & 0 \\ t & t & 0 & \varepsilon_1+\varepsilon_2 & 0 & 0 \\ 0 & 0 & 0 & 0 & \varepsilon_1+\varepsilon_2 & 0 \\ 0 & 0 & 0 & 0 & 0 & \varepsilon_1+\varepsilon_2 \end{bmatrix}
\end{array}$$

(23.4)

The *off-diagonal* entries t are obtained by noting that this quantity couples the one electron states u_1 to u_2 and d_1 to d_2. With two electron states we have inserted t for non-diagonal elements that couples those states for which one state remains unchanged while the other changes from u_1 to u_2 or from d_1 to d_2.

The lowest eigenstate obtained from the two-electron Hamiltonian in Eq.(23.4) is with a wavefunction of the form (s_1, $s_2 < 1$)

$$S: \quad S\left(\{u_1d_2\}+\{u_2d_1\}\right)+s_1\{u_1d_1\}+s_2\{u_2d_2\}$$

(23.5)

is called the singlet state. Next comes a set of three states (called the triplets) that are higher in energy. These have the form

$$T1: \quad \frac{1}{\sqrt{2}}\left(\{u_1d_2\}-\{u_2d_1\}\right)$$

$$T2: \quad \{u_1u_2\}$$

$$T3: \quad \{d_1d_2\}$$

(23.6)

A system with two electrons is normally viewed as occupying two one-electron states. The states T2, T3 permit such a simple visualization. But the states S and T1 do not.

For example, each term in the state

$$T1: \quad \frac{1}{\sqrt{2}}\left(\{u_1 d_2\} - \{u_2 d_1\}\right)$$

permits a simple visualization: $\{u_1 d_2\}$ stands for an upspin electron in *1* and a downspin electron in *2* while $\{u_2 d_1\}$ represents an upspin in *2* and a downspin in *1*. But the real state is a superposition of these two "simple" or unentangled states and there is no way to define two one-electron states *a* and *b* such that the two-electron state could be viewed as *{ab}*. Such states are called entangled states which comprise the key entity in the emerging new field of quantum information and computing.

How would we compute the properties of such systems? The equilibrium properties are still described by the general law of equilibrium stated earlier

$$p_i = \frac{1}{Z} e^{-(E_i - \mu N_i)/kT}$$

(Same as Eq.(16.14))

and using the equilibrium properties to evaluate the average number of electrons.

$$n = \sum_i N_i p_i$$

The energies E_i are obtained by diagonalizing the Fock space Hamiltonian *[HH]* that we just discussed. Fig.23.9 shows the plot of n versus μ which looks like Fig.23.7, but the middle plateau now involves the entangled singlet state just discussed. There is also some additional structure that we will not get into. The main point we wanted to make is that the law of equilibrium statistical mechanics is quite general and can be used in this case.

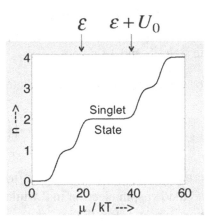

Fig.23.9. Equilibrium number of electrons, n in the two-level channel shown in Fig.23.6 as a function of μ, assuming $\varepsilon = 10kT$, $U_0 = 20kT$.

But the calculation of current at high bias is a non-equilibrium problem that is not as straightforward. Using the entangled states one could set up a rate equation as we did in the last Section and understand some of the interesting effects that have been observed experimentally including negative differential resistance (NDR), that is a decrease in current with increasing voltage (see for example Muralidharan et al. 2007). More generally one needs quantum rate equations to go beyond the simple rate equations we discussed and handle coherences (Braun et al. 2004, Braig and Brouwer 2005).

Can we model transport involving correlated and/or entangled states exactly if we use a Fock space picture instead of using NEGF and including interactions only approximately through self-energies? Sort of, but not quite.

There are two problems. The first is practical. A N-level problem in the one-electron picture escalates into a 2^N level problem n the Fock space picture. The second is conceptual.

We saw in Lecture 19 how the NEGF method allows us to include quantum broadening in the one-electron Schrödinger equation. To our knowledge there is no comparable accepted method for including broadening in the Fock space picture. So the rate equation approach from the last Section works fine for weakly coupled contacts where the resulting broadening is negligible, but the regime with broadening comparable to the charging energy stands out as a major challenge in transport theory. Even the system with two levels (Fig.23.7) shows interesting structure in n(μ) in this regime ("Kondo peak") that has occupied condensed matter physicists for many decades.

One could view Coulomb blockade as the bottom-up version of the Mott transition, a well-studied phenomenon in condensed matter physics. In a long chain of atoms, the levels ε and $\varepsilon + U_0$ (Fig.23.1) will each broaden into a band of width $\sim 2t_0$, $t_0 / \hbar$ being the rate at which electrons move from one atomic site to the next. These are known as the lower and upper Hubbard bands. If their separation U_0 exceeds the width $2t_0$ of each band we will have a Mott insulator where the electrochemical potential lies in the middle of the two bands with very low density of states and hence very low conductance. But if U_0 is small, then the two bands form a single half-filled band with a high density of states at $E = \mu_0$ and hence a high conductance.

Needless to say, the full theory of the Hubbard bands is far more complicated than this oversimplified description might imply and it is one of the topics that has occupied condensed matter theorists for over half a century. Since the late 1980's it has acquired an added significance with the discovery of a new class of superconductors operating at relatively high temperatures above *100K*, whose mechanism continues to be controversial and hotly debated.

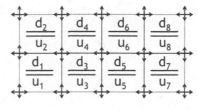

This problem remains one of the outstanding problems of condensed matter theory, but there seems to be general agreement that the essential physics involves a two-dimensional array of quantum dots with an inter-dot coupling that is comparable to the single dot charging energy.

Lecture 24

The Quantum and the Classical

Intel has a presentation, *From Sand to Circuits,* on their website
http://www.intel.com/about/companyinfo/museum/exhibits/sandtocircuits/index.htm
describing the amazing process that turns grains of sand into the chips
that have enabled the modern information age. As I explained at the
outset, these Lectures were about the physics that these "grains of sand"
and the associated technology have helped illuminate in the last 25 years,
the physics that helps validate the concept of an elastic resistor with a
clean separation of entropy-driven processes from the force-driven ones.

Interestingly much of this physics does not involve the quantum aspects
and can be understood within a semiclassical picture. Nearly all of what
we discussed in Parts 1 and 2 of these lectures follow from the
Boltzmann equation. Even though the modern nanotransistor that powers
today's laptops is only a few hundred atoms long, it remains in essence a
classical device controlled more by electrostatics than by quantum
subtleties. Indeed a recent paper reports classical behavior down to
atomic scale conductors (Weber et al. 2012).

Is this inevitable? Or could we harness our impressive progress in
nanofabrication and understanding to create a fundamentally different
class of quantum devices that will take us beyond today's charge-based
information processing paradigms. Let me end these lectures with a few
rambling thoughts (not answers!) on these lines.

24.1. Spin Coherence

The difference between quantum and classical is probably best exemplified by the spin of the electron. Consider for example, the experiment that we discussed in Lectures 14 and 22. An input magnet injects spins into the channel which produce a voltage on the output magnet given by

$$\mu_P = \mu + \frac{\vec{P}.\vec{\mu}_s}{2} \qquad\qquad (24.1)$$

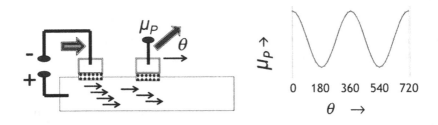

Fig.24.1. An input magnet injects spins into the channel which produce a voltage on the output magnet that depends on the cosine of the angle between the two magnets.

In Lecture 22 we saw how the NEGF model gives a complex matrix $[G^n]$ whose (2x2) components at each point along the diagonal can be expressed in terms of the physically transparent concepts of electron and spin density, N and $\vec{S}$:

$$
\begin{array}{c}
\begin{array}{cccccc} 1up & 1dn & 2up & 2dn & 3up & 3dn \end{array} \\
\begin{array}{c} 1up \\ 1dn \\ 2up \\ 2dn \\ 3up \\ 3dn \end{array}
\begin{bmatrix} N_1, \vec{S}_1 \\ & N_2, \vec{S}_2 \\ & & N_3, \vec{S}_3 \end{bmatrix}
\end{array}
$$

$\rightarrow z$

1   2   3

up

dn

$$\frac{[G^n]}{2\pi} = \begin{bmatrix} N + S_z & S_x - iS_y \\ S_x + iS_y & N - S_z \end{bmatrix} \qquad \text{(Same as Eq.(22.30))}$$

The NEGF equation for current then predicts the result in Eq.(24.1), with μ and $\bar{\mu}_s$ related to N and $\vec{S}$ respectively.

This simple example illustrates the relation between quantum and classical descriptions. If the transverse components of spin are negligible then we can describe the physics in terms of N and S_z only. We could interpret the non-zero components on the diagonal

$$(N + S_z) \quad as \ \ number \ of \ up \ electrons$$

$$(N - S_z) \quad as \ \ number \ of \ down \ electrons$$

(per unit energy) and then write semiclassical equations for the two types of electrons. That is essentially what we did in Lecture 14.

When does this work? One possibility is that the magnets are all collinear and there is no spin-orbit coupling so that we are restricted to angles θ that are multiples of 180 degrees (Fig.24.1). Another possibility is that various spin dephasing processes are strong enough to reduce transverse components of spin to negligible proportions. And if the z-components are reduced too, then we would not have to worry about spin at all.

What if we had collinear magnets but they point not along z, but along x? Now the $[G^n]$ matrix is not diagonal

$$\frac{[G^n]}{2\pi} \rightarrow \begin{bmatrix} N & S_x \\ S_x & N \end{bmatrix}$$

and it might appear that a semiclassical description is not possible. The trick is to choose the coordinates or more generally the "basis" correctly. What we should do is to select a basis in which up and down point along $+x$ and $-x$ respectively so that in this basis $[G^n]$ is diagonal

$$\frac{[G^n]}{2\pi} \rightarrow \begin{bmatrix} N+S_{up} & 0 \\ 0 & N+S_{dn} \end{bmatrix}$$

In a word, we should call the direction of the magnet z instead of x ! This sounds like a trivial observation, but can provide interesting insights that may not be quite so obvious. Let me explain.

24.2. Pseudo –spin

One of the nice things about the formalism of spin matrices (Lecture 22) is that it goes way beyond spins; it applies to any two-component complex quantity. For example in Lectures 18 and 19 we talked about the graphene lattice where the unit cell has an "A" atom (on the lattice sites marked with a red circle) and a "B" atom (on the unmarked lattice sites).

The wavefunction in a unit cell is described by a two component complex quantity:

$$\{\psi\} = \begin{Bmatrix} \psi_A \\ \psi_B \end{Bmatrix}$$

and we could look at the corresponding *[G^n]* and use our old relation from Eq.(22.30) to define a pseudo-spin

$$\frac{[G^n]}{2\pi} = \begin{bmatrix} \psi_A \psi_A{}^* & \psi_A \psi_B{}^* \\ \psi_B \psi_A{}^* & \psi_B \psi_B{}^* \end{bmatrix} \rightarrow \begin{bmatrix} N+S_z & S_x - iS_y \\ S_x + iS_y & N - S_z \end{bmatrix}$$

This has nothing to do with the real spin, just that they share the same mathematical framework. Once you have mastered the framework, there is no need to re-learn it, you can focus on the physics. In the literature on graphene, there are many references to pseudo spin and what direction it points in.

Let me point out a less familiar example of pseudospin involving an example we have already discussed. In Lecture 20, we discussed the potential variation across a single scatterer with transmission equal to T (Fig.20.7). Let us just look at the diagonal elements of *[G^n]* for the same problem. There are oscillations on the left of the barrier with a constatnt density on the right. The reason Fig.20.7 shows oscillations on the right as well is that we were looking at the occupation obtained from *G^n/A* and *A* has oscillations on the right. But let us not worry about that.

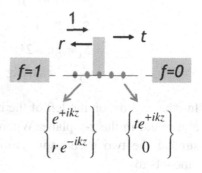

Let us see how we can use pseudospins to understand the spatial variation of the diagonal elements of *[G^n]*.

Let us view positive and negative going states as the up and down components of a pseudospin. The pseudospinor wavefunction on the left and right of the barrier have the form

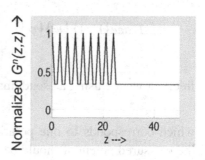

$$\{\psi\} \rightarrow \underset{\text{Left}}{\begin{Bmatrix} e^{+ikz} \\ r\,e^{-ikz} \end{Bmatrix}} , \quad \underset{\text{Right}}{\begin{Bmatrix} t\,e^{+ikz} \\ 0 \end{Bmatrix}}$$

$$\{\psi\}\{\psi\}^+ \rightarrow \underset{\text{Left}}{\begin{bmatrix} 1 & r^*e^{+i2kz} \\ r\,e^{-2ikz} & rr^* \end{bmatrix}} , \quad \underset{\text{Right}}{\begin{bmatrix} tt^* & 0 \\ 0 & 0 \end{bmatrix}}$$

$$\rightarrow \begin{bmatrix} N+S_z & S_x - iS_y \\ S_x + iS_y & N - S_z \end{bmatrix}$$

This suggests that the pseudospins to the left of the barrier are described by (assuming r, t are real)

Left	Right
$N = (1+r^2)/2$	$N = t^2/2$
$S_z = (1-r^2)/2$	$S_z = t^2/2$
$S_x = r\cos 2kz$	$S_x = 0$
$S_y = -r\sin 2kz$	$S_y = 0$

In other words, on the left of the barrier, the pseudospin is rotating round and round in the x-y plane. When we plot $G''(z,z)$, we are looking at the sum of the two pseudospin components and squaring the sum, which amounts to

$$Trace\{1 \ \ 1\}\{\psi\}\{\psi\}^+ \begin{Bmatrix} 1 \\ 1 \end{Bmatrix} \ = \ Trace \begin{pmatrix} 1 & 1 \\ 1 & 1 \end{pmatrix} \{\psi\}\{\psi\}^+$$

In effect we are using a pseudomagnet with $\Gamma = \begin{bmatrix} 1 & 1 \\ 1 & 1 \end{bmatrix}$

which corresponds to one polarized 100% along x. So from Eq.(24.1), the measured potential should be proportional to

	Left	Right
$N + \hat{x}.\vec{S} \ \rightarrow$	$\dfrac{1+r^2}{2} + r\cos 2kz$	$\dfrac{t^2}{2}$

which describes the numerical results quite well.

This is a relatively familiar problem where the concept of pseudospin probably does not add much to our undergraduate understanding of one-dimensional standing waves. The purpose was really to add our understanding of pseudospins!

24.3. Quantum Entropy

Now that we have seen how "spins" appear everywhere, let us talk briefly about the information content of a single spin which as we discussed in Lecture 17 is related to the thermodynamic entropy. We talked about the entropy of two examples of a collection of N spins obtained from the expression

A) $S = 0$

B) $S = Nk\ell n2$

$$\frac{S}{k} = -\sum_i p_i \, \ell n \, p_i \tag{24.2}$$

From a quantum mechanical point of view we could write the wavefunction of a single spin in collection A as

$$\psi = \begin{Bmatrix} 1 \\ 0 \end{Bmatrix} \rightarrow \psi\psi^+ = \begin{bmatrix} 1 & 0 \\ 0 & 0 \end{bmatrix}$$

and interpret the diagonal elements of $\psi\psi^+$ (*1* and *0*) as the $\{p_i\}$'s to use in Eq.(24.2). Writing $\psi\psi^+$ for a spin in collection B requires us to take a sum of two equally likely possibilities:

$$\psi\psi^+ = 0.5*\begin{bmatrix} 1 & 0 \\ 0 & 0 \end{bmatrix} + 0.5*\begin{bmatrix} 0 & 0 \\ 0 & 1 \end{bmatrix} = \begin{bmatrix} 0.5 & 0 \\ 0 & 0.5 \end{bmatrix}$$

Once again we can interpret the diagonal elements of $\psi\psi^+$ (both 0.5) as the $\{p_i\}$'s to use in Eq.(24.2) and get our semiclassical answers.

C) $S = 0$

What if we have collection C, which looks just like collection A, but the spins all pointing along x and not z. we then have

$$\psi = \frac{1}{\sqrt{2}}\begin{Bmatrix} 1 \\ 1 \end{Bmatrix} \rightarrow \psi\psi^+ = \begin{bmatrix} 0.5 & 0.5 \\ 0.5 & 0.5 \end{bmatrix}$$

If we just took the diagonal elements of $\psi\psi^+$ (both *0.5*) we obtain the same answer as we got for collection B which is obviously wrong. A collection with all spins pointing along x (C) should have the same entropy as a collection pointing along z (A) rather than a random collection (B).

The correct answer is obtained if we first diagonalize $\psi\psi^+$ and then use its diagonal elements (which are the eigenvalues) as the $\{p_i\}$'s in Eq.(24.2) . This is accomplished if we generalize Eq.(24.2) to write

$$\frac{S}{k} = -Trace[\rho \, \ell n \, \rho] \qquad (24.3)$$

where $\rho = \psi\psi^+$ is a 2x2 matrix (called the density matrix).

24.3.1. *How much information can one spin carry?*

Suppose we decide to use the spin of the electron, that is the direction of the input magnet in Fig.24.1 to convey information. It would seem that we could send large amounts of information, since there are now many possibilities. For example, suppose we choose a set of say *64* directions of the magnetization to convey information, it would seem that the entropy would be

$$S/k = \ell n \, 64$$

Note that we are using *64* figuratively to represent the number of magnetization directions we use, which could just as well be *10* or *100*.

We have seen in Lectures 14, 22 that a magnetic voltage probe making an angle θ with the injected spins measures a voltage proportional that depends on θ (Fig.24.1) and it would seem that we could measure the direction of spin simply by measuring the voltage. This would allow us to encode *64* possible values of θ thereby transmitting $\ell n \, 64$ rather than $\ell n \, 2$.

But how can this be correct? Didn't we argue earlier that for one spin $S/k = \ell n \, 2$ rather than $\ell n \, 64$? These two arguments can be reconciled by noting that in order to measure a voltage that depends on θ we need many many electrons so that we can take their average. An individual electron would either transmit or not transmit into the magnet with specific probabilities that depend on θ. Only by averaging over many electrons would we get the average values that we have discussed. This means that we could send $\ln 64$ worth of information, but only if we send many identically prepared electrons, so that the receiver can average over many measurements.

But couldn't we take one electron that we receive and create many electrons with the same wavefunction? After all, we can always copy a classical bit of information. There is a "no cloning theorem" that says we cannot copy the quantum state of an electron. The sender has to send us identically prepared electrons if we want to make many measurements and average.

These concepts are of course part of the emerging field of quantum information on which much has been written and will be written. My point here is simply that compared to charge, spin represents a qualitatively different state variable that could be the basis for fundamentally different approaches to information processing.

24.4. Does interaction increase the entropy?

Back in Lecture 17 we discussed how a perfect anti-parallel (AP) spin valve could function like an info-battery (Fig.17.3) that extracts energy from a collection of spins as it goes from the low entropy state A to the high entropy state B.

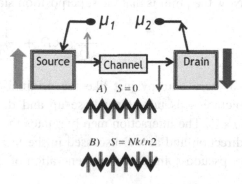

But exactly how does this increase in entropy occur? In Lecture 17 we described the interaction as a "chemical reaction"

$$u + D \quad \Leftrightarrow \quad U + d \qquad \text{(Same as Eq.(17.7))}$$

where u, d represent up and down channel electrons, while U, D represent up and down localized spins.

From a microscopic point of view the exchange interaction creates a superposition of wavefunctions as sketched below:

$$u \times D \quad \longrightarrow \quad \frac{1}{\sqrt{2}} u \times D + \frac{1}{\sqrt{2}} d \times U$$

We have shown equal superposition of the two possibilities for simplicity, but in general the coefficients could be any two complex numbers whose squared magnitude adds up to one.

Now the point is that the superposition state

$$\frac{1}{\sqrt{2}} u \times D + \frac{1}{\sqrt{2}} d \times U$$

has zero entropy just like the original state $u \times D$. Indeed we could picture a pseudo-spin whose up and down components are $u \times D$ and $d \times U$. The interaction merely rotates the pseudo-spin from the z to the x direction and as we discussed in the last Section, mere rotation of spins or pseudo-spins causes no generation of entropy.

So how does the increase in entropy occur? The itinerant electron eventually gets extracted from the channel. At that moment there is a "collapse of the wavefunction" either into a $u \times D$ or a $d \times U$ depending on whether the channel electron is extracted by the source as an up electron or by the drain as a down electron. The localized spin is left behind in a down or an up state with 50% probability each. This is when the entropy increases by $k\,\ell n2$.

24.5. Spins and Magnets

Spins have long been viewed as the quintessential quantum object and we have seen how we are learning to use solid state devices to control and manipulate not just single spins but even more sophisticated quantum objects like entangled spins. We have also seen that diverse quantum objects can all be viewed as two-component spins.

Actually this analogy need not be limited to two-component objects. As we mentioned earlier, the NEGF model gives a complex matrix $[G^n]$ whose (2x2) components at each point along the diagonal can be expressed in terms of the physically transparent concepts of electron and spin density, N and $\vec{S}$:

$$
\begin{array}{cccccc}
 & 1up & 1dn & 2up & 2dn & 3up & 3dn
\end{array}
$$

We could extend this viewpoint beyond the spin component. One could envision say a 10x10 block involving say multiple orbitals that retains its phase coherence for relatively long lengths of time and treat it as a kind of giant quasi spin-like object. Of course we could not describe it with just four numbers N, S_x, S_y, S_z. We would need *100* real components to represent all the information contained in a 10x10 Hermitian matrix and could use the NEGF equations to come up with a diffusion equation for this *100*-component object just as we discussed earlier for spin diffusion (Eq.(22.37)).

Could we use these giant quasi spins to encode and transmit information? Perhaps, but there is an important lesson the last twenty years have taught us. I am talking about the integration of what used to be two distinct fields, namely spintronics and magnetics.

Even ten years ago there was little overlap between these two fields. Spintronics was about the manipulation of individual spins motivated largely by basic low temperature physics, while magnetics was motivated largely by practical room temperature applications involving the manipulation of classical magnets whose magnetism is a result of enormous internal interactions that make all the elementary magnets (or spins) line up.

The first step in the integration of these two fields was the use of magnetic contacts to inject spins leading to the spin valve and eventually the magnetic tunnel junctions (MTJ's) whose change in resistance (R_P to R_{AP}, See Fig.14.1) is now routinely used to read information stored in magnets.

The second step was the demonstration of spin-torque which allows injected spins to turn magnets provided they are not more than a few atomic layers in thickness. This effect could be used to write information into nanomagnets. Memory devices utilizing spin for both reading and writing are now being actively developed.

The point I want to make is that for information processing it is not enough to have a spin, we also need a magnet. In standard charge-based architectures information is stored in capacitors and transmitted from capacitor to capacitor. Similarly we need a magnet to implement a spin capacitor and devices to transmit the information from magnet to magnet. Developments in the last decade have given us the basic ingredients. Whether we can build a information processing technology around it, remains to be seen.

It is tempting to go beyond simple spins and look at all kinds of exotic two-component pseudo-spins or multi-component quasi-spins that maintain phase coherence over useful lengths of time. But it seems to me that a key question one should ask is, *"do we have a quasi-magnet to generate and detect the quasi-spin?"*

These are the kinds of questions that will probably be explored and answered in the coming years. The purpose of these notes was to try to convey the insights from the last thirty years drawn from the remarkable experiments made possible by the advances in nanoelectronics. These experiments have helped establish the elastic resistor as a useful approximation to short conductors, and we have tried to show that it can also be used to understand long conductors by viewing them approximately as combinations of elastic resistors.

As I stated in the introductory lecture, I believe that the results we have discussed all follow directly from the formal theories of semiclassical or quantum transport. The real value of the bottom-up approach based on elastic resistors is in improving our physical understanding, and this will hopefully facilitate the insights needed to take us to the next level of understanding, discovery and innovation.

The last thirty years have taught us the significance of contacts, even the most mundane ones. We have also seen some of the interesting things that can be achieved with "smarter" contacts like the magnetic ones. Perhaps in the coming years we will learn to manipulate and control them creatively so that they become an active part of information processors.

References / Further Reading

It is the viewpoint presented here that we believe is unique, but not the topics we discuss. Each topic has its own associated literature that we cannot do justice to. What follows is a very incomplete list representing a small subset of the relevant literature.

General

These Lecture notes were used for an online course entitled "Fundamentals of Nanoelectronics" developed at Purdue, see *http://nanohub.org/u.* Some of these notes is based on

Datta S. (1995). *Electronic Transport in Mesoscopic Systems* (Cambridge University Press)

Datta S. (2005). *Quantum Transport: Atom to Transistor* (Cambridge University Press)

Datta S. (2010) Nanoelectronic Devices: A Unified View, Chapter 1, *The Oxford Handbook on Nanoscience and Technology: Frontiers and Advances,* eds. A. V. Narlikar and Y. Y. Fu, Oxford University Press (available on condmat, arXiv:0907.4122)

Part I: The New Ohm's Law

Lecture 1

Feynman R.P. (1963) *Lectures on Physics*, vol.II-2, Addison-Wesley.

One perspective on where things stand
Solomon P. and Theis T. (2010) It's Time to Reinvent the Transistor, *Science* **327**, 1600

Lecture 4

Berg H.C. (1993) *Random Walks in Biology*, Princeton University Press.

Lecture 5

Bolotin K.I. et al. (2008) Temperature-Dependent Transport in Suspended Graphene. *Phys. Rev. Lett.* **101**, 096802

A seminal paper in the field
Thouless, D. (1977). Maximum Metallic Resistance in Thin Wires, *Phys. Rev. Lett.* **39**, 1167

Lecture 7

Section 7.4 is based on
Salahuddin S. et al. (2005) Transport Effects on Signal propagation in Quantum Wires, *IEEE Trans. Electron Dev.* **52**,1734

Lecture 8

This discussion is based on a model due to Lundstrom that is widely used in the field. See
Rahman A. et al. (2003) Theory of Ballistic Transistors, *IEEE Trans. Electron Dev.* **50**, 1853 and references therein.

Part II: Old Topics in New Light

Lecture 10

Butcher P.N. (1990) Thermal and Electrical Transport Formalism for Electronic Microstructures with Many Terminals, *J. Phys. Condens. Matt.* **2**, 4869 and references therein.

On the thermoelectric effects in molecules:
Baheti K. et al. (2008) Probing the Chemistry of Molecular Heterojunctions Using Thermoelectricity, *Nano Letters* **8**, 715.
Paulsson M. and Datta S. (2003) Thermoelectric Effect in Molecular Electronics, *Phys. Rev. B* **67**, 241403(R)

Lecture 11

This discussion draws on my collaborative work with Mark Lundstron and Changwook Jeong. See

Jeong C. et al. (2011) Full dispersion versus Debye model evaluation of lattice thermal conductivity with a Landauer approach, *J.Appl.Phys.* **109**, 073718-8 and references therein.

A couple of other references on the subject:

Majumdar A. (1993) Microscale Heat Conduction in Dielectric Thin Films *Journal of Heat Transfer* **115,** 7

Mingo N. (2003) Calculation of Si nanowire thermal conductivity using complete phonon dispersion relations, *Phys. Rev. B* **68,** 113308

Lecture 12

This discussion is based on Chapters 2-3 of Datta (1995). A couple of papers by the primary contributors to the subject of discussion:

Buttiker M. (1988) Symmetry of Electrical Conduction, *IBM J. Res. Dev.* **32,** 317

Imry Y and Landauer R (1999) Conductance Viewed as Transmission, *Rev. Mod. Phys.* **71,** S306

Lecture 13

The pioneering paper that reported the first observation of the amazing quantization of the Hall resistance:

von Klitzing K. et al. (1980) New Method for High-Accuracy Determination of the Fine Structure Constant Based on Quantized Hall Resistance, *Phys.Rev.Lett.* **45,** 494.

For more on edge states in the quantum Hall regime the reader could look at Chapter 4 of Datta (1995) and references therein.

Lecture 14 (see references under Lecture 22 as well)

An article on spin injection by one of the inventors of the spin valve:

Fert A. et al. (2007) Semiconductors between Spin-Polarized Sources and Drains, *IEEE Trans. Electron Devices* 54, 921.

A review article
Schmidt G. (2005) Concepts for Spin Injection into Semiconductors – a Review *J.Phys.D: Appl. Phys.* **38,** R107

One of the widely cited papers on non-local spin voltages
Takahashi S. and Maekawa S. (2003) Spin Injection and Detection in Magnetic Nanostructures, Phys. Rev. B **67**, 052409.

One of the early papers on spin-torque based switching
Sun J.Z. (2000) Spin-current interaction with a monodomain magnetic body: A model study, *Phys. Rev. B*, **62**, 570, see references.

Lecture 15

Blanter Ya.M. and Büttiker M. (2000) Shot Noise in Mesoscopic Conductors, *Physics Reports*, **336**, 1

Doniach S. and Sondheimer E.H. (1974), *Green's Functions for Solid State Physicists*, Frontiers in Physics Lecture Note Series, Benjamin/Cummings

For an introduction to diagrammatic methods for conductivity calculation based on the Kubo formula, the reader could also look at Section 5.5 of Datta (1995).

For more on irreversible thermodynamics the reader could look at a book like
Yourgrau W., van der Merwe A., Raw G. (1982) *Treatise on Irreversible and Statistical Thermophysics*, Dover Publications

Lecture 16

Dill K. and Bromberg S. (2003) *Molecular Driving Forces, Statistical Thermodynamics in Chemistry and Biology*, Garland Science
Feynman R.P. (1965) *Statistical Mechanics*, Frontiers in Physics Lecture Note Series, Benjamin/Cummings

Lecture 17

For more on Maxwell's demon and related issues
Feynman R.P. (1963) *Lectures on Physics*, vol. I-46, Addison-Wesley.
Leff H.S. and Rex A.F. (2003), *Maxwell's Demon 2*, IOP Publishing

Part III: Contact-ing Schrodinger

Lecture 18

This discussion is based on Chapters 2-7 of Datta (2005).

Brey, L., and H. Fertig, (2006) Edge States and Quantized Hall Effect in
   Graphene, *Phys. Rev. B* **73**, 195408.
Herman F. and Skillman S. (1963) *Atomic Structure Calculations*,
   Prentice-Hall.

Lecture 19

This discussion is based on Chapter 8 of Datta (1995), and Chapters 8-10
   of Datta (2005).

An experiment showing approximate conductance quantization in a
   hydrogen molecule
Smit R.H.M. et al. (2002) Measurement of the Conductance of a
   Hydrogen Molecule, *Nature* **419**, 906.

The classic references on NEGF include
Kadanoff L.P. and and Baym G. (1962) *Quantum Statistical Mechanics*,
   Frontiers in Physics Lecture Note Series, Benjamin/Cummings
Keldysh L.V. (1965) Diagram Technique for non-equilibrium processes.
   Sov.Phys. JETP **20,** 1018
Martin P.C. and Schwinger J. (1959) Theory of many-particle systems I,
   Phys.Rev. **115,** 1342

In addition the reader could look at later references like
Danielewicz, P. (1984) Quantum Theory of Non-Equilibrium Processes,
   Ann.Phys., NY, **152**, 239.

Mahan G.D. (1987) Quantum Transport Equation for Electric and Magnetic Fields, *Phys. Rep. NY*, **145**, 251.

A standard text on NEGF
Haug H. and Jauho A.P. (1996) *Quantum Kinetics in Transport and Optics of Semiconductors*, Springer-Verlag.

The following reference is probably the first to apply NEGF to a device with contacts.
Caroli C. et al. (1972) A Direct Calculation of the Tunneling Current, *J.Phys. C: Solid State Physics,* 5, 21.

The NEGF method was related to the Landauer approach in
Datta, S. (1989) Steady-State Quantum Kinetic Equation, Phys. Rev., **B40**, 5830.
Meir Y. And Wingreen N. (1992) Landauer Formula for the Current through an Interacting Electron Region, *Phys. Rev. Lett.* **68**, 2512.

Extensive numerical results based on the method of Datta (1989) are presented in
McLennan M. et al. (1991) Voltage Drop in Mesoscopic Systems, *Phys. Rev. B*, **43**, 13846

The dephasing model described in Section 19.4 is based on
R.Golizadeh-Mojarad et al. (2007), Non-equilibrium Green's function based model for dephasing in quantum transport", *Phys. Rev.B* **75**, 081301 (2007).

A glimpse at the impressive progress in NEGF-based device modeling
Steiger S. et al. (2011) NEMO5: A Parallel Multiscale Nanoelectronics Modeling Tool, *IEEE Transactions on Nanotechnology*, **10**, 1464

Lecture 20

This discussion is based on Chapter 5 of Datta (1995) and Chapter 11 of Datta (2005).

A paper on localization by the person who pioneered the field along with many other seminal concepts.
Anderson P.W. et al. (1981) New Method for a Scaling Theory of Localization, *Phys. Rev.* B **23**, 4828

Resonant tunneling is discussed in more detail in Chapter 6 of Datta (1995) and Chapter 9 of Datta (2005).

Lecture 21

Two experiments reporting the discovery of quantized conductance in ballistic conductors.
van Wees, B.J et al. (1988) Quantized Conductance of Points Contacts in a Two-Dimensional Electron Gas, *Phys.Rev.Lett.* **60**, 848.
Wharam, D.A. et al. (1988) One-Dimensional Transport and the Quantisation of the Ballistic Resistance, *J.Phys.C.* **21**, L209.

Lecture 22 (see references under Lecture 14 as well)

Most quantum mechanics texts have a through discussion of spinors.
A non-standard reference that the reader may enjoy Misner C.W., Thorne K.S. and Wheeler J.A. (1970) *Gravitation,* Chapter 41, Freeman.

A recent textbook
Bandyopadhyay, S. and Cahay, M. (2008) *Introduction to Spintronics,* Taylor & Francis.

Spin circuits provide an attractive approximation to the full NEGF model that may be sufficiently accurate in many cases. See for example
Brataas A. et al. (2006) Non-Collinear Magnetoelectronics, *Phys. Rep.,* **427,** 157
Srinivasan S. et al. (2011) All-Spin Logic Device With Inbuilt Nonreciprocity, *IEEE Transactions On Magnetics,* **47,** 4026

A spin diffusion equation based on the Kubo formalism is described in
Burkov et al. (2004) Theory of spin-charge-coupled transport in a two-dimensional electron gas with Rashba spin-orbit interactions, *Phys. Rev. B* **70**, 155308

A paper stressing the distinction between transport spin currents and equilibrium spin currents.

Rashba E.I. (2003) Spin Currents in Thermodynamic Equilibrium: The Challenge of Discerning Transport Currents, *Phys.Rev.B* **68**, 241315R

A few interesting experiments showing the effects of spin coherence

Huang B. et al. (2007) Geometric Dephasing-limited Hanle Effect in Long Distance Lateral Silicon Spin Transport Devices, *Appl. Phys. Lett.* **93**, 162508

Koo H.C. et al. (2009) Control of Spin Precession in a Spin- Injected Field Effect Transistor, *Science* **325**, 1515

Sih V. et al. (2006) Generating Spin Currents in Semiconductors with the Spin Hall Effect, *Phys.Rev.Lett.* **97**, 2096605.

Wunderlich et al. (2010) Spin Hall Effect Transistor, *Science* **330**, 1801

A few papers using the NEGF method to model spin transport.

Nikolic B. et al. (2010) Spin Currents in Semiconductor Nanostructures: A Non-Equilibrium Green-Function Approach, Chapter 24, *The Oxford Handbook on Nanoscience and Technology: Frontiers and Advances,* eds. A. V. Narlikar and Y. Y. Fu, Oxford University Press (available on condmat, arXiv:0907.4122). See also Chapter 3 by R.Golizadeh-Mojarad et al. and Datta (2010).

Datta D. et al. (2012) Voltage Asymmetry of Spin Transfer Torques, *IEEE Transactions on Nanotechnology*, **11**, 261

Zainuddin A.N.M et al. (2011) Voltage-controlled Spin Precession, *Phys.Rev.B* **84**, 165306

An experimental paper showing evidence for spin-polarized surface states, including references to the literature on *topological insulators*.

Xiu F. et al. (2011) Manipulating Surface States in Topological Insulator Nanoribbons, *Nature Nanotechnology* **6**, 216 and references therein.

Lecture 23

Readers interested in device analysis at high bias may find this article useful. MATLAB codes available on our website.

Datta S. (2000) Nanoscale Device Modeling: The Green's Function Method, *Superlattices and Microstructures*, **28**, 253. See also Chapters 10, 11 of Datta (2005).

The rate equations described here are based on
Beenakker C.W.J. (1991) Theory of Coulomb Blockade Oscillations in the Conductance of a Quantum Dot, *Phys. Rev. B*, **44**, 1646

The reader may also find this reference helpful
Bonet E., Deshmukh M.M., and Ralph D.C. (2002) Solving Rate Equations for Electron Tunneling via Discrete Quantum States, *Phys. Rev. B* **65**, 045317

For an application of the rate equations to describe interesting current-voltage characteristics observed in double quantum dots
Muralidharan B. and Datta S. (2007) Generic model for current collapse in spin-blockaded transport *Phys. Rev.* **B76,** 035432

For a description of methods that go beyond the simple rate equation, the reader could look at
M Braun, J Koenig, and J Martinek (2004) Theory of Transport through Quantum-Dot Spin Valves in the Weak Coupling Regime, Phys Rev B, **70**, 195345
Braig S. and Brouwer P.W. (2005) Rate Equations for Coulomb Blockade with Ferromagnetic Leads, *Phys. Rev, B* **71**, 195324 (2005)

Lecture 24

A well-cited paper that discusses some issues we touched on
Zurek W. (2003) Decoherence, Einselection and the Quantum Origins of the Classical, *Rev. Mod. Phys.*, **75**, 715.

A recent paper reporting measurements on few-atom resistors:
Weber B. et al. (2012) Ohm's Law Survives to the Atomic Scale, *Science* **335,** 64

Appendix A

Fermi and Bose Function Derivatives

A.1. Fermi function:

$$f(x) \equiv \frac{1}{e^x + 1}, \qquad x \equiv \frac{E - \mu}{kT} \qquad (A.1)$$

$$\frac{\partial f}{\partial E} = \frac{df}{dx}\frac{\partial x}{\partial E} = \frac{df}{dx}\frac{1}{kT} \qquad (A.2a)$$

$$\frac{\partial f}{\partial \mu} = \frac{df}{dx}\frac{\partial x}{\partial \mu} = -\frac{df}{dx}\frac{1}{kT} \qquad (A.2b)$$

$$\frac{\partial f}{\partial T} = \frac{df}{dx}\frac{\partial x}{\partial T} = -\frac{df}{dx}\frac{E - \mu}{kT^2} \qquad (A.2c)$$

From Eqs.(A.2a, b, c),

$$\frac{\partial f}{\partial \mu} = -\frac{\partial f}{\partial E} \qquad (A.3a)$$

$$\frac{\partial f}{\partial T} = -\frac{E - \mu}{T}\frac{\partial f}{\partial E} \qquad (A.3b)$$

Eq.(2.8) in Lecture 2 is obtained from a Taylor series expansion of the Fermi function around the equilibrium point

$$f(E,\mu) \approx f(E,\mu_0) + \left(\frac{\partial f}{\partial \mu}\right)_{\mu=\mu_0} (\mu - \mu_0)$$

From Eq.(A.3a), $\qquad \left(\dfrac{\partial f}{\partial \mu}\right)_{\mu=\mu_0} = \left(-\dfrac{\partial f}{\partial E}\right)_{\mu=\mu_0}$

Letting $f(E)$ stand for $f(E,\mu)$, and $f_0(E)$ stand for $f(E, \mu_0)$, we can write

$$f(E) \approx f_0(E) + \left(-\frac{\partial f_0}{\partial E}\right)(\mu - \mu_0)$$

Rearranging, $\qquad f(E) \approx f_0(E) + \left(-\dfrac{\partial f_0}{\partial E}\right)(\mu - \mu_0)$

(same as Eq.(2.8))

A.2. Bose function:

$$n(x) \equiv \frac{1}{e^x - 1}, \qquad x \equiv \frac{\hbar\omega}{kT} \tag{A.4}$$

$$\frac{\partial n}{\partial T} = \frac{dn}{dx}\frac{\partial x}{\partial T} = -\frac{\hbar\omega}{kT^2}\frac{dn}{dx}$$

$$\hbar\omega \frac{\partial n}{\partial T} = -k x^2 \frac{dn}{dx}$$

$$= \frac{kx^2 e^x}{(e^x - 1)^2} \tag{A.5}$$

Appendix B

Angular averaging

B.1. *Two dimensions:*

$$\langle |v_z| \rangle = \frac{1}{\pi} \int_{-\pi/2}^{+\pi/2} d\theta\, v \cos\theta = v\frac{2}{\pi}$$

$$\langle v_z^2 \rangle = \frac{1}{\pi} \int_{-\pi/2}^{+\pi/2} d\theta\, v^2 \cos^2\theta$$

$$= \frac{v^2}{2\pi} \int_{-\pi/2}^{+\pi/2} d\theta\, (1+\cos 2\theta) = \frac{v^2}{2}$$

B.2. *Three dimensions:*

$$\langle |v_z| \rangle = \int_{0}^{+\pi/2} d\theta \sin\theta\, v \cos\theta = v\int_{0}^{1} dx\, x = \frac{v}{2}$$

$$\langle v_z^2 \rangle = \int_{0}^{+\pi/2} d\theta \sin\theta\, v^2 \cos^2\theta = v^2\int_{0}^{1} dx\, x^2 = \frac{v^2}{3}$$

Combining 2D and 3D,

$$\langle |v_z| \rangle = v \left\{ \underbrace{\frac{2}{\pi}}_{2D} , \underbrace{\frac{1}{2}}_{3D} \right\}$$

$$\langle v_z^2 \rangle = v^2 \left\{ \underbrace{\frac{1}{2}}_{2D} , \underbrace{\frac{1}{3}}_{3D} \right\}$$

Hamiltonian with E- and B- Fields

$$E(\vec{x}, \vec{p}) = \sum_{j} \frac{(\vec{p} - q\vec{A}) \cdot (\vec{p} - q\vec{A})}{2m} + U(\vec{x})$$

$$(B.1)$$

where $\qquad q\vec{F} = -\vec{\nabla} U \qquad$ Electric Field, $\vec{F}$

and $\qquad \vec{B} = \vec{\nabla} \times \vec{A} \qquad$ Magnetic Field, $\vec{B}$

Semiclassical laws of motion:

$$\vec{v} \equiv \frac{d\vec{x}}{dt} = \vec{\nabla}_p E$$

$$(B.2a)$$

$$\frac{d\vec{p}}{dt} = -\vec{\nabla} E$$

$$(B.2b)$$

where the gradient operators are defined as

$$\vec{\nabla} E \equiv \hat{x} \frac{\partial E}{\partial x} + \hat{y} \frac{\partial E}{\partial y} + \hat{z} \frac{\partial E}{\partial z}$$

$$\vec{\nabla}_p E \equiv \hat{x} \frac{\partial E}{\partial p_x} + \hat{y} \frac{\partial E}{\partial p_y} + \hat{z} \frac{\partial E}{\partial p_z}$$

From Eqs.(B.1) and (B.2a, b), we can show that

$$\vec{v} \;=\; (\vec{p} - q\vec{A})/m \tag{B.3a}$$

$$\frac{d(\vec{p} - q\vec{A})}{dt} \;=\; q\left(\vec{F} + \vec{v} \times \vec{B}\right) \tag{B.3b}$$

Proof:

$$E(\vec{x},\,\vec{p}) = \sum_j \frac{(p_j - qA_j(\vec{x}))^2}{2m} + U(\vec{x})$$

$$\rightarrow \quad v_i = \frac{\partial E}{\partial p_i} = \frac{p_i - qA_i(\vec{x})}{m}, \quad \rightarrow \quad \vec{v} = \frac{\vec{p} - q\vec{A}(\vec{x})}{m},$$

$$\frac{dp_i}{dt} = -\frac{\partial E}{\partial x_i} = -\frac{\partial U}{\partial x_i} + q\sum_j v_j \frac{\partial A_j}{\partial x_i}$$

$$\frac{d}{dt}(p_i - qA_i(\vec{x})) = -\frac{\partial U}{\partial x_i} + q\sum_j v_j \left(\frac{\partial A_j}{\partial x_i} - \frac{\partial A_i}{\partial x_j}\right)$$

$$= -\frac{\partial U}{\partial x_i} + q\sum_{j,n} v_j\, \varepsilon_{ijn} \left(\vec{\nabla} \times \vec{A}\right)_n$$

$$\rightarrow \quad \frac{d(\vec{p} - q\vec{A})}{dt} = q(\vec{F} + \vec{v} \ x \ \vec{B})$$

In Lecture 7 we used the 1D version of Eqs. (B.2a, b) to obtain the BTE (Eq.(7.5)). If we use full 3D version we obtain

$$\frac{\partial f}{\partial t} + \vec{v}.\vec{\nabla}f + \vec{F}.\vec{\nabla}_p f = \; S_{op}(f) \tag{B.4}$$

Appendix D

Transmission Line Parameters from BTE

In this Appendix, I will try to outline the steps involved in getting from

$$\frac{\partial \mu}{\partial t} + v_z \frac{\partial \mu}{\partial z} - \frac{\partial E}{\partial t} = -\frac{\mu(z,t) - \mu_0}{\tau} \qquad (7.19)$$

to

$$\frac{\partial (\mu/q)}{\partial z} = -(L_K + L_M)\frac{\partial I}{\partial t} - \frac{I}{\sigma A} \qquad (7.20a)$$

$$\frac{\partial (\mu/q)}{\partial t} = -\left(\frac{1}{C_Q} + \frac{1}{C_E}\right)\frac{\partial I}{\partial z} \qquad (7.20b)$$

First we separate Eq.(7.19) into two equations for μ^+ and μ^-,

$$\frac{\partial \mu^+}{\partial t} + v_z \frac{\partial \mu^+}{\partial z} - \frac{\partial E^+}{\partial t} = -\frac{\mu^+ - \mu_0}{\tau} \qquad (B.1a)$$

$$\frac{\partial \mu^-}{\partial t} - v_z \frac{\partial \mu^-}{\partial z} - \frac{\partial E^-}{\partial t} = -\frac{\mu^- - \mu_0}{\tau} \qquad (B.1b)$$

Next we add and subtract Eqs.(B.1a,b) and use the relations

$$I = (qM / h)(\mu^+ - \mu^-)$$

$$\mu = (\mu^+ + \mu^-)/2$$

to obtain

$$2\frac{\partial \mu}{\partial t} + \frac{v_z}{qM / h}\frac{\partial I}{\partial z} - \frac{\partial}{\partial t}\left(E^+ + E^-\right) = 0$$

$$\frac{1}{qM / h}\frac{\partial I}{\partial t} + 2v_z\frac{\partial \mu}{\partial z} - \frac{\partial}{\partial t}\left(E^+ - E^-\right) = -\frac{I}{(qM / h)\tau}$$

Rearranging

$$\frac{\partial(\mu/q)}{\partial t} = -\frac{1}{C_Q}\frac{\partial I}{\partial z} + \frac{1}{2q}\frac{\partial}{\partial t}\left(E^+ + E^-\right) \qquad \text{(B.2a)}$$

$$\frac{\partial(\mu/q)}{\partial z} = -L_K\frac{\partial I}{\partial t} + \frac{1}{2qv_z}\frac{\partial}{\partial t}\left(E^+ - E^-\right) - RI \qquad \text{(B.2b)}$$

where L_K, C_Q are the quantities defined in Eq.(7.21), and

Let us now consider the terms involving $E^{\pm}$. Assuming that the fields associated with a transverse electromagnetic (TEM) wave, they can be expressed in terms of $U(z,t)$ along with a vector potential $A_z(z,t)$ pointing along z, for which the energy is given by (Appendix C)

$$E = \frac{(p_z - qA_z(t))^2}{2m} + U(z,t)$$

Noting that
$$v_z = \frac{\partial E}{\partial p_z} = \frac{p_z - qA_z}{m}$$

we can write
$$\frac{\partial E}{\partial t} = v_z\left(-q\frac{\partial A_z}{\partial t}\right) + \frac{\partial U}{\partial t}$$

so that
$$\frac{\partial}{\partial t}\left(E^+ + E^-\right) = 2\frac{\partial U}{\partial t}$$

and
$$\frac{\partial}{\partial t}\left(E^+ - E^-\right) = 2v_z\left(-q\frac{\partial A_z}{\partial t}\right)$$

Eqs.(B.2a,b) can then be written as

$$\frac{\partial(\mu/q)}{\partial t} = -\frac{1}{C_Q}\frac{\partial I}{\partial z} + \frac{\partial(U/q)}{\partial t} \qquad (B.5a)$$

$$\frac{\partial(\mu/q)}{\partial z} = -L_K\frac{\partial I}{\partial t} - \frac{\partial A_z}{\partial t} - RI \qquad (B.5b)$$

which reduces to Eqs.(7.20a,b) noting that

$$A_z = L_M I, \, U/q = C_E Q$$

and making use of the continuity equation: $\partial Q/\partial t + \partial I/\partial z = 0$.

Appendix E

NEGF Equations

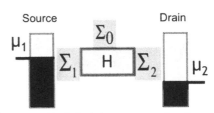

"Input":
1. H-matrix parameters chosen appropriately to match energy levels or dispersion relations.

2. Procedure for obtaining Σ_m for terminal 'm' is summarized below in Section E.1.

$$\Sigma \ = \ \Sigma_0 + \Sigma_1 + \Sigma_2 + \cdots$$

$$\Sigma^{in} \ = \ \Sigma_0^{in} + \Sigma_1^{in} + \Sigma_2^{in} + \cdots$$

$$\Gamma_j \ = \ i[\Sigma_j - \Sigma_j^+], \, j = 0,1,2,\cdots$$

1. Green's Function:

$$G^R \ = \ [EI - H - \Sigma]^{-1} \tag{19.1}$$

$$and \quad G^A = [G^R]^+$$

2. "Electron Density" times 2π:

$$G^n \ = \ G^R \ \Sigma^{in} \ G^A \tag{19.2}$$

3. "Density of states" times 2π:

$$A = \ G^R \Gamma G^A = \ G^A \Gamma G^R$$
$$= \ i[G^R - G^A] \tag{19.3a}$$

442

4a. *Current / energy at terminal "m"*

$$\tilde{I}_m = \frac{q}{h} Trace[\Sigma_m^{in} A - \Gamma_m G^n] \qquad (19.4)$$

4b. *Current / energy at terminal "m", to be used only if Σ_0 is zero*

$$I_m = \frac{q}{h} \sum_n \bar{T}_{mn} \left(f_m(E) - f_n(E) \right)$$

$$(19.31)$$

$$\bar{T}_{mn} \equiv Trace \left[\Gamma_m G^R \Gamma_n G^A \right] \qquad (19.32)$$

E.1.   Self-energy for contacts

(a) For 1-D problems, the self-energy function for each contact has a single non-zero element $t e^{ika}$ corresponding to the point that is connected to that contact (see Section 20.3).

(b) 2-D Hamiltonians for any conductor with a uniform cross-section can be visualized as a linear 1-D chain of "atoms" each having an on-site matrix Hamiltonian $[\alpha]$ coupled to the next "atom" by a matrix $[\beta]$.

Each of the matrices $[\alpha]$ and $[\beta]$ is of size (nxn), n being the number of basis functions describing each unit.

The *self-energy matrix* Σ_m for terminal m is zero except for the last

(nxn) block at the surface. This block is obtained from

$$\beta g \beta^+ \qquad \text{(Same as Eq.(21.6a))}$$

where the surface Green function g is calculated from a recursive relation:

$$[g]^{-1} = (E+i0^+)I - \alpha - \beta g \beta^+ \quad \text{(Same as Eq.(21.6b))}$$

In the rest of this Appendix, we will obtain Eqs.(21.6a, b).

To obtain these results, first we consider just the last point of the device and its connection to the infinite contact described by H_c:

$$\begin{bmatrix} \alpha & B \\ B^+ & H_c \end{bmatrix}$$

where $[B] \equiv [\beta \quad 0 \quad 0 \quad \cdots \quad \cdots \quad]$

The overall Green's function can be written as

$$\begin{bmatrix} A & -B \\ -B^+ & A_c \end{bmatrix}^{-1} = \begin{bmatrix} G^R & \cdots \\ \cdots & \cdots \end{bmatrix}$$

where
$$A \equiv (E+i0^+)I - \alpha \qquad \text{(E.1)}$$

$$A_c \equiv (E+i0^+)I_c - H_c \qquad \text{(E.2)}$$

With a little matrix algebra we can show that the top block of the Green's function, G^R is given by

$$G^R = [A - B A_c^{-1} B^+]^{-1} \qquad \text{(E.3)}$$

so that we can identify self-energy as

$$\Sigma = B A_c^{-1} B^+$$

Since B has only one non-zero element β, we can write

$$\Sigma = \beta g \beta^{+} \qquad \text{(same as Eq.(21.6a))}$$

where "g" represents the top block of $[A_c]^{-1}$, often

$$
\begin{bmatrix}
A & -\beta & 0 & 0 & \cdots \\
-\beta^{+} & A & -\beta & 0 & \cdots \\
0 & -\beta^{+} & A & -\beta & \cdots \\
& \cdots & \cdots & &
\end{bmatrix}^{-1}
\equiv
\begin{bmatrix}
g & \cdots & & \cdots \\
\cdots & \cdots & & \cdots \\
\cdots & \cdots & & \cdots \\
& & \cdots & \cdots
\end{bmatrix}
\tag{E.4}
$$

To obtain Eq.(21.6b), we apply Eq.(E.3) to the $(N \times N)$ matrix in Eq.(E.4) treating the first block A as the "device",
and the rest of the $(N-1) \times (N-1)$ as contact, to obtain

$$g_N = [A - \beta g_{N-1}\beta^{+}]^{-1} \tag{E.5}$$

where g_N represents the g on the right-hand side of Eq.(E.4) if the matrix on the left is of size $N \times N$. One could solve Eq.(E.5) recursively starting from g_1 to g_2 and so on till g_N is essentially the same as g_{N-1}. At that point we have the solution to Eq.(21.6b)

$$g = [A - \beta g \beta^{+}]^{-1} \qquad \text{(Same as Eq.(21.6b))}$$

E.2. Self-energy for Elastic Scatterers in Equilibrium:

$$[\Sigma]_{kl} = D_{kl,ij}[G]_{ij} \quad , \quad [\Sigma^{in}]_{kl} = D_{kl,ij}[G^n]_{ij} \tag{E.6}$$

where summation over repeated indices is implied and

$$D_{kl,ij} = \left\langle [U_s]_{ki}[U_s]^{*}_{lj} \right\rangle \tag{E.7a}$$

where $\langle \cdots \rangle$ denotes average value, and we are considering a general scattering potential with non-zero off-diagonal elements.

In Lecture 19 we assumed that only the diagonal elements are non-zero then the D can be simplified from a fourth-order tensor to a second-order tensor or in other words a matrix

$$D_{kl} \;=\; \left\langle [U_s]_{kk}\,[U_s]_{ll}^* \right\rangle \qquad \text{(E.7b)}$$

In terms of this matrix, Eq.(E.6) can be rewritten as

$$[\Sigma]_{kl} \;=\; [D]_{kl}\,[G]_{kl} \;,\; [\Sigma^{in}]_{kl} \;=\; [D]_{kl}\,[G^n]_{kl} \qquad \text{(E.8)}$$

which amounts to an element by element multiplication:

$$[\Sigma] \;=\; [D]\times[G],\; [\Sigma^{in}] \;=\; [D]\times[G^n] \text{ (see Eq.(19.35))}$$

E.3. Self-energy for Inelastic Scatterers
(Not discussed in these lectures, see Chapter 10, Datta(2005))

$$[\Sigma^{in}]_{kl} \;=\; D_{kl,\,ij}(\hbar\omega)\,[G^n(E-\hbar\omega)]_{ij} \qquad \text{(E.9a)}$$

$$[\Gamma]_{kl} \;=\; D_{kl,\,ij}(\hbar\omega)\,[G^n(E-\hbar\omega) + G^p(E+\hbar\omega)]_{ij} \qquad \text{(E.9b)}$$

where $G^p = A - G^n$, and summation over repeated indices is implied.

$$[\Sigma]_{kl} \;=\; \underbrace{[h]_{kl}}_{\substack{Hilbert \\ Transform \\ of\ \Gamma_{kl}}} -\; \frac{i}{2}[\Gamma]_{kl} \qquad \text{(E.9c)}$$

For scatterers in equilibrium with temperature T_s,

$$\frac{D_{kl,\,ij}(+\hbar\omega)}{D_{ji,\,lk}(-\hbar\omega)} \;=\; e^{-\hbar\omega/kT_s}$$

$$\text{(E.10)}$$

MATLAB Codes Used for Text Figures

These codes are included here mainly for their pedagogical value. It is planned to make soft copies available through our website

http://nanohub.org/groups/lnebook

F.1.   Lecture 8

```
% Fig.8.8. Saturation current
clear all

% Constants
hbar=1.06e-34;q=1.6e-19;

%Parameters
m=0.2*9.1e-31;g=2;ep=4*8.854e-12;
% mass, degeneracy factor, permittivity
kT=0.005;mu0=0.05;
% Thermal energy, equilibrium electrochemical potential
W=1e-6;L=1e-6;tox=2e-9;% dimensions

% Energy grid
dE=0.00001;E=[0:dE:2];NE=size(E,2);
D=(g*W*L*q*m/2/pi/hbar/hbar).*ones(1,NE);
% Density of states
M=(g*2*W/2/pi/hbar).*sqrt(2*m*q*E);% Number of modes

f0=1./(1+exp((E-mu0)./kT));
% Equilibrium Fermi function
```

447

```
N0=sum(dE*D.*f0);% Equilibrium electron number
U0=0*q*tox/ep/L/W;% Charging energy
alpha=0.05;% Drain induced barrier lowering

% I-V characteristics
ii=1;dV=0.01;for V=1e-3:dV:0.5
UL=-alpha*V;ii
 change=100;U=UL;

% Self-consistent loop
 while change>1e-6
   f1=1./(1+exp((E-mu0+U)./kT));
   f2=1./(1+exp((E-mu0+U+V)./kT));
   f=(f1+f2)./2;
  N=sum(dE*D.*f);% Electron number
  Unew=UL+U0*(N-N0);% Self-consistent potential
    change=abs(U-Unew);
    U=U+0.05*(Unew-U);
 end

    curr(ii)=(q*q*dE/2/pi/hbar)*sum(M.*(f1-f2));
    volt(ii)=V;Uscf(ii)=U;ns(ii)=N/W/L;ii=ii+1;
end
current=q*g*(4/3)*W*sqrt(2*m*q*mu0)*q*mu0/...
4/pi/pi/hbar/hbar;
max(curr)/current

hold on
h=plot(volt,curr,'r');
%h=plot(volt,curr,'r+');
%h=plot(volt,Uscf,'r');
%h=plot(volt,ns,'r');
set(h,'linewidth',[3.0])
set(gca,'Fontsize',[40])
xlabel(' Voltage (V_{D}) ---> ');
ylabel(' Current (A) ---> ');
```

grid on

% Fig.8.10. Electrostatic rectifier
clear all

```
% Constants
hbar=1.06e-34;q=1.6e-19;

%Parameters
m=0.2*9.1e-31;g=2;ep=4*8.854e-12;
% mass, degeneracy factor, permittivity
kT=0.025;mu0=0*0.05;
% Thermal energy, equilibrium electrochemicalpotential
W=1e-6;L=1e-6;tox=2e-9;% dimensions

% Energy grid
dE=0.0001;E=[0:dE:2];NE=size(E,2);
D=(g*W*L*q*m/2/pi/hbar/hbar).*ones(1,NE);
% Density of states
M=(g*2*W/2/pi/hbar).*sqrt(2*m*q*E);% Number of modes

f0=1./(1+exp((E-mu0)./kT));
% Equilibrium Fermi function
N0=sum(dE*D.*f0);% Equilibrium electron number
U0=0*q*tox/ep/L/W;% Charging energy
alpha=0;% Drain induced barrier lowering

% I-V characteristics
ii=1;dV=0.01;for V=-0.1:dV:0.1
UL=-alpha*V;ii
 change=100;U=UL;

% Self-consistent loop
 while change>1e-6
   f1=1./(1+exp((E-mu0+U)./kT));
   f2=1./(1+exp((E-mu0+U+V)./kT));
```

```
  f=(f1+f2)./2;
 N=sum(dE*D.*f);% Electron number
 Unew=UL+U0*(N-N0);% Self-consistent potential
   change=abs(U-Unew);
   U=U+0.05*(Unew-U);
end

   curr(ii)=(g*q*q*dE/2/pi/hbar)*sum(M.*(f1-f2));
   volt(ii)=V;Uscf(ii)=U;ns(ii)=N/W/L;ii=ii+1;
end

figure(1)
hold on
h=plot(volt,curr,'r');
%h=plot(volt,Uscf,'r');
%h=plot(volt,ns,'r');
set(h,'linewidth',[3.0])
set(gca,'Fontsize',[40])
xlabel(' Voltage (V_{D}) ---> ');
ylabel(' Current (A) ---> ');
grid on
```

F.2.   Lecture 20, NEGF1D

% Fig.20.2
```
clear all

t0=1;Np=11;X=[0:1:Np-1];
L=diag([1 zeros(1,Np-1)]);R=diag([zeros(1,Np-1) 1]);
zplus=i*1e-12;

H0=2*t0*diag(ones(1,Np))-t0*diag(ones(1,Np-1),1)...
-t0*diag(ones(1,Np-1),-1);
N1=3;N2=9;UB1=2*t0;UB2=0*2*t0;
```

```
H0(N1,N1)=H0(N1,N1)+UB1;H0(N2,N2)=H0(N2,N2)+UB2;H=H0;

VV=0;UV=linspace(0,-VV,Np);%Linear Potential

ii=1;dE=5e-2; for EE=[-.5:dE:4.5]*t0
   ck=(1-(EE-UV(1)+zplus)/(2*t0));ka=acos(ck);
s1=-t0*exp(i*ka);sig1=kron(L,s1);
     ck=(1-(EE-UV(Np)+zplus)/(2*t0));ka=acos(ck);
     s2=-t0*exp(i*ka);sig2=kron(R,s2);
   gam1=i*(sig1-sig1');gam2=i*(sig2-sig2');

   G=inv((EE*eye(Np))-H-diag(UV)-sig1-sig2);
   Tcoh(ii)=real(trace(gam1*G*gam2*G'));
   E(ii)=EE/t0;ii=ii+1;
end

hold on
h=plot(Tcoh,E,'k-o');
set(h,'linewidth',[1.2])
set(gca,'Fontsize',[36])
axis([-0.1 1.1 -.5 4.5]);
grid on

% Fig.20.4
clear all

t0=1;Np=11;X=[0:1:Np-1];
L=diag([1 zeros(1,Np-1)]);R=diag([zeros(1,Np-1) 1]);
zplus=i*1e-12;

H0=2*t0*diag(ones(1,Np))-t0*diag(ones(1,Np-1),1)...
-t0*diag(ones(1,Np-1),-1);
N1=3;N2=9;UB1=2*t0;UB2=2*t0;
H0(N1,N1)=H0(N1,N1)+UB1;H0(N2,N2)=H0(N2,N2)+UB2;H=H0;

VV=0;UV=linspace(0,-VV,Np); % Linear potential

ii=1;dE=5e-4; for EE=[-.25:dE:1.25]*t0
   ck=(1-(EE-UV(1)+zplus)/(2*t0));ka=acos(ck);
```

```
    theta(ii)=(real(ka)*(N2-N1+1)/pi);
    s1=-t0*exp(i*ka);sig1=kron(L,s1);
      ck=(1-(EE-UV(Np)+zplus)/(2*t0));ka=acos(ck);
      s2=-t0*exp(i*ka);sig2=kron(R,s2);
    gam1=i*(sig1-sig1');gam2=i*(sig2-sig2');

    G=inv((EE*eye(Np))-H-diag(UV)-sig1-sig2);
    Tcoh(ii)=real(trace(gam1*G*gam2*G'));
    E(ii)=EE/t0;ii=ii+1;
end

hold on
h=plot(Tcoh,E,'k');
set(h,'linewidth',[3.0])
set(gca,'Fontsize',[36])
axis([-0.1 1.1 -.25 1.25]);
grid on

% Fig.20.5
clear all
t0=1;
Np=100;Np1=11;Np2=23;Np3=7;Np4=31;
L=diag([1 zeros(1,Np-1)]);R=diag([zeros(1,Np-1) 1]);
zplus=i*1e-12;

H0=2*t0*diag(ones(1,Np))-t0*diag(ones(1,Np-1),1)...
-t0*diag(ones(1,Np-1),-1);
UB=2*t0;n=1;
H=H0+UB*diag([n zeros(1,Np1) 1 zeros(1,Np2) n zeros(1,Np3) n ...
    zeros(1,Np4) n zeros(1,Np-Np1-Np2-Np3-Np4-6) n]);

ii=1; for EE=[-.25:1e-3:1]*t0
% for EE=t0:-dE:t0
    ck=(1-(EE+zplus)/(2*t0));ka=acos(ck);
    s1=-t0*exp(i*ka);s2=-t0*exp(i*ka);
    sig1=kron(L,s1);sig2=kron(R,s2);
    gam1=i*(sig1-sig1');gam2=i*(sig2-sig2');

    G=inv((EE*eye(Np))-H-sig1-sig2);
```

```
A=real(diag(i*(G-G')));ii
   Gn=G*gam1*G';


Tcoh(ii)=real(trace(gam1*G*gam2*G'));TM(ii)=real(trace(gam2*Gn));
   E(ii)=EE/t0;ii=ii+1;
end

hold on
%h=plot(Tcoh./(6.-5*Tcoh),E,'k-o');
h=plot(Tcoh,E,'k');
set(h,'linewidth',[3.0])
set(gca,'Fontsize',[36])
axis([-.1 1.1 -.25 1])
grid on
```

% Figs.20.6,7
```
clear all

t0=1;Np=51;X=[0:1:Np-1];Nh=floor(Np/2);
L=diag([1 zeros(1,Np-1)]);R=diag([zeros(1,Np-1) 1]);
zplus=i*1e-12;D=9e-2*t0^2;

sigB=zeros(Np);siginB=zeros(Np);

H0=2*t0*diag(ones(1,Np))-t0*diag(ones(1,Np-1),1)...
-t0*diag(ones(1,Np-1),-1);
N1=Nh+1;UB1=1*t0;
H0(N1,N1)=H0(N1,N1)+UB1;H=H0;

EE=t0;
   ck=(1-(EE+zplus)/(2*t0));ka=acos(ck);
   v=2*t0*sin(ka);
% Semiclassical profile
   T=real(v^2/(UB1^2+v^2));R1=(1-T)/T;
   TT=real(v^2/(D+v^2));R2=1*(1-TT)/TT;
   RR=[0.5 R2*ones(1,Nh) R1 R2*ones(1,Nh) 0.5];
   RR=cumsum(RR);Vx=ones(1,Np+2)-(RR./RR(Np+2));
   Fclass=Vx([2:Np+1]);
```

%Based on resistance estimates

```
   s1=-t0*exp(i*ka);sig1=kron(L,s1);
     ck=(1-(EE+zplus)/(2*t0));ka=acos(ck);
     s2=-t0*exp(i*ka);sig2=kron(R,s2);
   gam1=i*(sig1-sig1');gam2=i*(sig2-sig2');

   G=inv((EE*eye(Np))-H-sig1-sig2);
   Tcoh=real(trace(gam1*G*gam2*G'));

   change=100;
   while change>1e-6
     G=inv((EE*eye(Np))-H-sig1-sig2-sigB);
     sigBnew=diag(diag(D*G));%sigBnew=D*G;
     change=sum(sum(abs(sigBnew-sigB)));
     sigB=sigB+0.25*(sigBnew-sigB);
   end
   A=real(diag(i*(G-G')));change=100;
   while change>1e-6
     Gn=G*(gam1+siginB)*G';
     siginBnew=diag(diag(D*Gn));%siginBnew=D*Gn;
     change=sum(sum(abs(siginBnew-siginB)));
     siginB=siginB+0.25*(siginBnew-siginB);
   end
   F=real(diag(Gn))./A;

hold on
h=plot(X,F,'k');
set(h,'linewidth',[3.0])
h=plot(X,Fclass,'r-o');
set(h,'linewidth',[1.2])
set(gca,'Fontsize',[36])
xlabel(' z ---> ')
grid on
axis([-10 60 0 1])
```

% Fig.20.8,9
clear all

```
t0=1;Np=51;X=[0:1:Np-1];Nh=floor(Np/2);
L=diag([1 zeros(1,Np-1)]);R=diag([zeros(1,Np-1) 1]);
zplus=i*1e-12;D=9e-20*t0^2;

sigB=zeros(Np);siginB=zeros(Np);

H0=2*t0*diag(ones(1,Np))-t0*diag(ones(1,Np-1),1)-...
t0*diag(ones(1,Np-1),-1);
N1=Nh-3;N2=Nh+3;UB1=2*t0;UB2=2*t0;
H0(N1,N1)=H0(N1,N1)+UB1;H0(N2,N2)=H0(N2,N2)+UB2;H=H0;

EE=0.6*t0;EE=0.81*t0;
  ck=(1-(EE+zplus)/(2*t0));ka=acos(ck);
  v=2*t0*sin(ka);

  %Semiclassical profile
  T=real(v^2/(UB1^2+v^2));R1=(1-T)/T;
  TT=real(v^2/(D+v^2));R2=0*(1-TT)/TT;
  RR=[0.5 R2*ones(1,Nh-4) R1 zeros(1,6) R1 R2*ones(1,Nh-4) 0.5];
  RR=cumsum(RR);Vx=ones(1,Np+1)-(RR./RR(Np+1));
  Fclass=Vx([2:Np+1]);
  %Based on resistance estimates

  s1=-t0*exp(i*ka);sig1=kron(L,s1);
    ck=(1-(EE+zplus)/(2*t0));ka=acos(ck);
    s2=-t0*exp(i*ka);sig2=kron(R,s2);
  gam1=i*(sig1-sig1');gam2=i*(sig2-sig2');

  G=inv((EE*eye(Np))-H-sig1-sig2);
  Tcoh=real(trace(gam1*G*gam2*G'));

  change=100;
  while change>1e-6
    G=inv((EE*eye(Np))-H-sig1-sig2-sigB);
    sigBnew=diag(diag(D*G));sigBnew=D*G;
    change=sum(sum(abs(sigBnew-sigB)));
    sigB=sigB+0.25*(sigBnew-sigB);
  end
  A=real(diag(i*(G-G')));change=100;
```

```
  while change>1e-6
    Gn=G*(gam1+siginB)*G';
    siginBnew=diag(diag(D*Gn));siginBnew=D*Gn;
    change=sum(sum(abs(siginBnew-siginB)));
    siginB=siginB+0.25*(siginBnew-siginB);
  end
  F=real(diag(Gn))./A;


hold on
h=plot(X,F,'k');
set(h,'linewidth',[3.0])
%h=plot(X,Fclass,'r-o');
%set(h,'linewidth',[1.2])
set(gca,'Fontsize',[36])
xlabel(' z ---> ')
grid on
axis([-10 60 0 1])
```

F.3.   Lecture 21, NEGF2D

```
% Fig.21.1
clear all


%Constants (all MKS, except energy which is in eV)
hbar=1.06e-34;q=1.6e-19;qh=q/hbar;B=0;

%inputs
a=2.5e-9;t0=1;
NW=25;Np=1;L=zeros(Np);R=L;L(1,1)=1;R(Np,Np)=1;zplus=i*1e-12;

%Hamiltonian
al=4*t0;by=-t0;bx=-t0;
alpha=kron(eye(NW),al)+kron(diag(ones(1,NW-
1),+1),by)+kron(diag(ones(1,NW-1),-1),by');
alpha=alpha+diag([1:1:NW].*0);
alpha=alpha+diag([zeros(1,8) 0*ones(1,9) zeros(1,8)]);
```

```
beta=kron(diag(exp(i*qh*B*a*a*[1:1:NW])),bx);
H=kron(eye(Np),alpha);
if Np>1
H=H+kron(diag(ones(1,Np-1),+1),beta)+kron(diag(ones(1,Np-1),-
1),beta');end

ii=0;for EE=[-0.05:1e-2:1.05]*t0
ii=ii+1;ig0=(EE+zplus)*eye(NW)-alpha;
if ii==1
gs1=inv(ig0);gs2=inv(ig0);end
  change=1;
  while change >1e-6
    Gs=inv(ig0-beta'*gs1*beta);
    change=sum(sum(abs(Gs-gs1)))/(sum(sum(abs(gs1)+abs(Gs))));
    gs1=0.5*Gs+0.5*gs1;
  end
  sig1=beta'*gs1*beta;sig1=kron(L,sig1);gam1=i*(sig1-sig1');

  change=1;
  while change >1e-6
    Gs=inv(ig0-beta*gs2*beta');
    change=sum(sum(abs(Gs-gs2)))/(sum(sum(abs(gs2)+abs(Gs))));
    gs2=0.5*Gs+0.5*gs2;
  end
  sig2=beta*gs2*beta';sig2=kron(R,sig2);gam2=i*(sig2-sig2');

    G=inv((EE*eye(Np*NW))-H-sig1-sig2);
     DD=real(diag(i*(G-G'))).2/pi;
  Tcoh(ii)=real(trace(gam1*G*gam2*G'));E(ii)=EE/t0;ii
end

ii=1;for kk=pi*[-1:0.01:1]
H=alpha+beta*exp(i*kk)+beta'*exp(-i*kk);
[V,D]=eig(H);EK(:,ii)=sort(abs(diag(D)))./t0;K(ii)=kk/pi;ii=ii+1;
end

X=linspace(0,9,101);Ean= 2*(1-cos(pi*X./(NW+1)));
hold on
figure(1)
```

```
h=plot(Tcoh,E,'k');
set(h,'linewidth',[3.0])
%h=plot(X,Ean,'k--');
%set(h,'linewidth',[1.2])
set(gca,'Fontsize',[36])
axis([0 10 -.1 1])
% Fig.21.1a, Transmission versus width, at E=t0
clear all

%Constants (all MKS, except energy which is in eV)
hbar=1.06e-34;q=1.6e-19;qh=q/hbar;B=0;

%inputs
a=2.5e-9;t0=1;
NW=25;Np=1;L=zeros(Np);R=L;L(1,1)=1;R(Np,Np)=1;zplus=i*1e-12;

%Hamiltonian
al=4*t0;by=-t0;bx=-t0;
alpha1=kron(eye(NW),al)+kron(diag(ones(1,NW-
1),+1),by)+kron(diag(ones(1,NW-1),-1),by');

ii=0;EE=t0*1;for NN=[0:1:NW-1]
ii=ii+1;

alpha=alpha1+diag([zeros(1,NN) 100*ones(1,NW-NN)]);
beta=kron(diag(exp(i*qh*B*a*a*[1:1:NW])),bx);
H=kron(eye(Np),alpha);
if Np>1
H=H+kron(diag(ones(1,Np-1),+1),beta)+kron(diag(ones(1,Np-1),-
1),beta');end

ig0=(EE+zplus)*eye(NW)-alpha;
if ii==1
gs1=inv(ig0);gs2=inv(ig0);end
   change=1;
   while change >1e-6
     Gs=inv(ig0-beta'*gs1*beta);
     change=sum(sum(abs(Gs-gs1)))/(sum(sum(abs(gs1)+abs(Gs))));
     gs1=0.5*Gs+0.5*gs1;
```

```
end
sig1=beta'*gs1*beta;sig1=kron(L,sig1);gam1=i*(sig1-sig1');

change=1;
while change >1e-6
   Gs=inv(ig0-beta*gs2*beta');
   change=sum(sum(abs(Gs-gs2)))/(sum(sum(abs(gs2)+abs(Gs))));
   gs2=0.5*Gs+0.5*gs2;
end
sig2=beta*gs2*beta';sig2=kron(R,sig2);gam2=i*(sig2-sig2');

   G=inv((EE*eye(Np*NW))-H-sig1-sig2);
      DD=real(diag(i*(G-G'))))./2/pi;
 Tcoh(ii)=real(trace(gam1*G*gam2*G'));E(ii)=NN;
 X(ii)=(NN+1)*(acos(1-(EE/2/t0)))/pi;
   X1(ii)=(NN+1)*sqrt(EE/2/t0)/pi;ii

end

hold on
figure(1)
h=plot(E,Tcoh,'k');
set(h,'linewidth',[3.0])
%h=plot(E,X,'k');
%h=plot(E,X1,'k--');
%set(h,'linewidth',[1.2])
set(gca,'Fontsize',[36])
axis([0 NW -.1 10])
grid on
```

■■

```
%Fig.21.3, Graphene, CNT's
clear all

%Constants (all MKS, except energy which is in eV)
hbar=1.06e-34;q=1.6e-19;qh=q/hbar;a=1e-9;

%inputs
```
■■

```
t0=-2.5;D=1e-50;ctr=0;zplus=i*1e-3;
NL=1;L=zeros(NL);R=L;L(1,1)=1;R(NL,NL)=1;
config=1;%1 for armchair, 2 for zigzag edge
NW=floor(14*sqrt(3));% Armchair
%NW=14;% Zigzag

%Hamiltonian
al=t0*[0 1 0 0;1 0 1 0;0 1 0 1;0 0 1 0];
if config==1
bL=t0*[0 0 0 0;0 0 0 0;0 0 0 0;1 0 0 0];
bW=t0*[0 0 0 0;1 0 0 0;0 0 0 1;0 0 0 0];end
if config==2
bW=t0*[0 0 0 0;0 0 0 0;0 0 0 0;1 0 0 0];
bL=t0*[0 0 0 0;1 0 0 0;0 0 0 1;0 0 0 0];end

n=4;% al=4;bW=-1;bL=-1;n=1;
alpha=kron(eye(NW),al)+kron(diag(ones(1,NW-
1),+1),bW)+kron(diag(ones(1,NW-1),-1),bW');
alpha=alpha+kron(diag(ones(1,1),1-NW),bW)+kron(diag(ones(1,1),NW-
1),bW');% for CNT's

sigB=zeros(NW*NL*n);siginB=zeros(NW*NL*n);

ii=0;for EE=t0*[-0.5:+0.01:+0.5]
ii=ii+1;
ig0=(EE+zplus)*eye(NW*n)-alpha;
if ii==1
gs1=inv(ig0);gs2=inv(ig0);end

BB=0;beta=kron(diag(exp(i*qh*BB*a*a*[1:1:NW])),bL);
%beta=kron(eye(NW),bL);
H=kron(eye(NL),alpha);if NL>1
H=H+kron(diag(ones(1,NL-1),+1),beta)+...
kron(diag(ones(1,NL-1),-1),beta');end
```

▪▪▪

```
change=1;
while change>1e-4
    Gs=inv(ig0-beta'*gs1*beta);
    change=sum(sum(abs(Gs-gs1)))/(sum(sum(abs(gs1)+abs(Gs))));
    gs1=0.5*Gs+0.5*gs1;
end
sig1=beta'*gs1*beta;sig1=kron(L,sig1);gam1=i*(sig1-sig1');
change=1;
while change>1e-4
    Gs=inv(ig0-beta*gs2*beta');
    change=sum(sum(abs(Gs-gs2)))/(sum(sum(abs(gs2)+abs(Gs))));
    gs2=0.5*Gs+0.5*gs2;
end
sig2=beta*gs2*beta';sig2=kron(R,sig2);gam2=i*(sig2-sig2');

G=inv((EE*eye(NW*NL*n))-H-sig1-sig2);
T(ii)=real(trace(gam1*G*gam2*G'));E(ii)=EE/t0;
if EE==0
    T(ii)=T(ii-1);end,EE
end


hold on
h=plot(T,E,'k');
set(h,'linewidth',[3.0])
set(gca,'Fontsize',[36]);
axis([0 10 -0.5 +0.5])
title(' W = 24 * 2b ')
grid on
```

■■■

```
% Fig.21.4, Hall Effect
clear all
```

```
%Constants (all MKS, except energy which is in eV)
hbar=1.06e-34;q=1.6e-19;m=0.1*9.1e-31;qh=q/hbar;

%inputs
a=2.5e-9;t0=(hbar^2)/(2*m*(a^2)*q);
NW=25;Np=1;L=zeros(Np);R=L;L(1,1)=1;R(Np,Np)=1;zplus=i*1e-12;

%Hamiltonian
al=4*t0;by=-t0;bx=-t0;
alpha=kron(eye(NW),al)+kron(diag(ones(1,NW-...
1),+1),by)+kron(diag(ones(1,NW-1),-1),by');
alpha=alpha+diag([1:1:NW].*0);

EE=t0;ii=0;for B=0:0.1:50
%B=0;ii=0;for EE=[-0.05:0.01:1]*t0
   ii=ii+1;E(ii)=B;
   ig0=(EE+zplus)*eye(NW)-alpha;
   if ii==1
      gs1=inv(ig0);gs2=inv(ig0);end

beta=kron(diag(exp(i*qh*B*a*a*[1:1:NW])),bx);
H=kron(eye(Np),alpha);
if Np>1
H=H+kron(diag(ones(1,Np-1),+1),beta)+kron(diag(ones(1,Np-1),-
1),beta');end

   change=1;
   while change >5e-5
      Gs=inv(ig0-beta'*gs1*beta);
      change=sum(sum(abs(Gs-gs1)))/(sum(sum(abs(gs1)+abs(Gs))));
      gs1=0.5*Gs+0.5*gs1;
   end
   sig1=beta'*gs1*beta;sig1=kron(L,sig1);gam1=i*(sig1-sig1');

   change=1;
   while change >5e-5
```

```
        Gs=inv(ig0-beta*gs2*beta');
        change=sum(sum(abs(Gs-gs2)))/(sum(sum(abs(gs2)+abs(Gs))));
        gs2=0.5*Gs+0.5*gs2;
    end
    sig2=beta*gs2*beta';sig2=kron(R,sig2);gam2=i*(sig2-sig2');

        G=inv((EE*eye(Np*NW))-H-sig1-sig2);
        Gn=G*gam1*G';

    A=i*(G-G');V=real(diag(Gn./A));
    Tcoh=real(trace(gam1*G*gam2*G'));TM=real(trace(gam2*Gn));
    %Y(ii)=Tcoh;ii
    Y(ii)=(V(1)-V(NW))/Tcoh;ii
end

hold on
h=plot(E,Y,'k');
set(h,'linewidth',[3.0])
set(gca,'Fontsize',[36])
xlabel(' B-field (T) ---> ')
ylabel(' R_{xy} ---> ')
grid on
```

∎∎

% Fig.21.5, Edge States

```
clear all

%Constants (all MKS, except energy which is in eV)
hbar=1.06e-34;q=1.6e-19;m=0.1*9.1e-31;qh=q/hbar;B=20;

%inputs
a=2.5e-9;t0=(hbar^2)/(2*m*(a^2)*q);
NW=25;Np=1;L=zeros(Np);R=L;L(1,1)=1;R(Np,Np)=1;zplus=i*1e-12;

%Hamiltonian
al=4*t0;by=-t0;bx=-t0;
alpha=kron(eye(NW),al)+kron(diag(ones(1,NW-
```

```
1),+1),by)+kron(diag(ones(1,NW-1),-1),by');
alpha=alpha+diag([1:1:NW].*0);
beta=kron(diag(exp(i*qh*B*a*a*[1:1:NW])),bx);
H=kron(eye(Np),alpha);
if Np>1
H=H+kron(diag(ones(1,Np-1),+1),beta)+kron(diag(ones(1,Np-1),-...
1),beta');end

ii=0;for EE=[-0.05:0.008:1.05]*t0
ii=ii+1;ig0=(EE+zplus)*eye(NW)-alpha;
if ii==1
gs1=inv(ig0);gs2=inv(ig0);end
   change=1;
   while change >1e-4
     Gs=inv(ig0-beta'*gs1*beta);
     change=sum(sum(abs(Gs-gs1)))/(sum(sum(abs(gs1)+abs(Gs))));
     gs1=0.5*Gs+0.5*gs1;
   end
   sig1=beta'*gs1*beta;sig1=kron(L,sig1);gam1=i*(sig1-sig1');

   change=1;
   while change >1e-4
     Gs=inv(ig0-beta*gs2*beta');
     change=sum(sum(abs(Gs-gs2)))/(sum(sum(abs(gs2)+abs(Gs))));
     gs2=0.5*Gs+0.5*gs2;
   end
   sig2=beta*gs2*beta';sig2=kron(R,sig2);gam2=i*(sig2-sig2');

     G=inv((EE*eye(Np*NW))-H-sig1-sig2);
     DD(:,ii)=real(diag(i*(G-G'))))./2/pi;
     Gn=G*gam1*G';
     NN(:,ii)=real(diag(Gn))./2/pi;
end

XX=DD;
lo=.4*min(min(XX));hi=.4*max(max(XX));

figure(1)
hold on
```

```
imagesc(XX,[lo hi])
colormap(gray)
set(gca,'Fontsize',[36])
grid on
axis([0 140 0 25])
```

F.4. Lecture 22, NEGF-Spin

```
%Fig.22.6, Magnet Rotation
clear all
hbar=1.06e-34;q=1.6e-19;m=0.1*9.1e-31;a=2.5e-
9;t0=(hbar^2)/(2*m*(a^2)*q);
sx=[0 1;1 0];sy=[0 -i;i 0];sz=[1 0;0 -1];zplus=1i*1e-12;

Np=50;N1=10;N2=20;X=1*[0:1:Np-1];
  L=diag([1 zeros(1,Np-1)]);
    R=diag([zeros(1,Np-1) 1]);
  L1=0.1*diag([zeros(1,N1-1) 1 zeros(1,Np-N1)]);
  L2=0.1*diag([zeros(1,N2-1) 1 zeros(1,Np-N2)]);

ii=0;for theta=[0:0.1:4]*pi
  P1=0.7*[0 0 1];
  P2=1*[sin(theta) 0 cos(theta)];ii=ii+1;

H0=diag(ones(1,Np));
  HR=diag(ones(1,Np-1),1);HL=diag(ones(1,Np-1),-1);

H=2*t0*kron(H0,eye(2))-t0*kron(HL,eye(2))-t0*kron(HR,eye(2));

EE=t0;ck=(1-(EE+zplus)/(2*t0));ka=acos(ck);
    sL=-t0*exp(1i*ka)*eye(2);sR=sL;
    s1=-t0*exp(1i*ka)*(eye(2)+P1(1)*sx+P1(2)*sy+P1(3)*sz);
    s2=-t0*exp(1i*ka)*(eye(2)+P2(1)*sx+P2(2)*sy+P2(3)*sz);

    sigL=kron(L,sL);sigR=kron(R,sR);
    sig1=kron(L1,s1);sig2=kron(L2,s2);
      gamL=1i*(sigL-sigL');gamR=1i*(sigR-sigR');
      gam1=1i*(sig1-sig1');gam2=1i*(sig2-sig2');
```

```
    G=inv(((EE+zplus)*eye(2*Np))-H-sigL-sigR-sig1-sig2);

    % {1 L} {2 R} = {a} {b}

    TM1L=real(trace(gam1*G*gamL*G'));
TML1=real(trace(gamL*G*gam1*G'));
      Taa=[0 TM1L;TML1 0];

    TM12=real(trace(gam1*G*gam2*G'));
TM1R=real(trace(gam1*G*gamR*G'));
    TML2=real(trace(gamL*G*gam2*G'));
TMLR=real(trace(gamL*G*gamR*G'));
      Tab=[TM12 TM1R;TML2 TMLR];

    TM21=real(trace(gam2*G*gam1*G'));
TM2L=real(trace(gam2*G*gamL*G'));
    TMR1=real(trace(gamR*G*gam1*G'));
TMRL=real(trace(gamR*G*gamL*G'));
      Tba=[TM21 TM2L;TMR1 TMRL];

    TM2R=real(trace(gam2*G*gamR*G'));
TMR2=real(trace(gamR*G*gam2*G'));
      Tbb=[0 TM2R;TMR2 0];

    Taa=diag(sum(Taa)+sum(Tba))-Taa;Tba=-Tba;
    Tbb=diag(sum(Tab)+sum(Tbb))-Tbb;Tab=-Tab;
      if abs(sum(sum([Taa Tab;Tba Tbb]))) > 1e-10
        junk=100,end

    V=-inv(Tbb)*Tba*[1;0];
    VV2(ii)=V(1);VVR(ii)=V(2);
    angle(ii)=theta/pi;
    I2(ii)=TM21;IL(ii)=TML1;IR(ii)=TMR1;

    Gn=G*(gam1+V(1)*gam2+V(2)*gamR)*G';
Gn=Gn([2*N2-1:2*N2],[2*N2-1:2*N2]);
    A=i*(G-G');A=A([2*N2-1:2*N2],[2*N2-1:2*N2]);
    g2=i*(s2-s2');XX2(ii)=real(trace(g2*Gn)/trace(g2*A));
end
```

```
X2=VV2;XR=VVR;max(X2)-min(X2)

hold on
h=plot(angle,X2,'k');
set(h,'linewidth',2.0)
h=plot(angle,XX2,'ro');
set(h,'linewidth',2.0)
set(gca,'Fontsize',36)
xlabel(' \theta / \pi ---> ')
ylabel(' V_{2} ---> ')
grid on
```

∎∎∎

```
%Fig.22.7, Spin Precession
clear all
hbar=1.06e-34;q=1.6e-19;m=0.1*9.1e-31;a=2.5e-
9;t0=(hbar^2)/(2*m*(a^2)*q);
sx=[0 1;1 0];sy=[0 -1i;1i 0];sz=[1 0;0 -1];zplus=1i*1e-12;

Np=50;N1=5;N2=45;X=1*[0:1:Np-1];
  L=diag([1 zeros(1,Np-1)]);
    R=diag([zeros(1,Np-1) 1]);

  L1=0.1*diag([zeros(1,N1-1) 1 zeros(1,Np-N1)]);
  L2=0.1*diag([zeros(1,N2-1) 1 zeros(1,Np-N2)]);

ii=0;for al=[0:0.005:0.3]*t0
  P1=[0 0 0.7];P2=[0 0 0.7];ii=ii+1;
  alph=al*1;% Rashba
  BB=al*0;% Hanle

  H0=diag(ones(1,Np));
    HR=diag(ones(1,Np-1),1);HL=diag(ones(1,Np-1),-1);
  beta=t0*eye(2)+1*i*alph*sx;
    alpha=2*t0*eye(2)+1*BB*sx;

H=kron(H0,alpha)-kron(HL,beta')-kron(HR,beta);

EE=t0;ck=(1-(EE+zplus)/(2*t0)));ka=acos(ck);
```

```
sL=-t0*exp(1i*ka)*eye(2);sR=sL;
s1=-t0*exp(1i*ka)*(eye(2)+P1(1)*sx+P1(2)*sy+P1(3)*sz);
s2=-t0*exp(1i*ka)*(eye(2)+P2(1)*sx+P2(2)*sy+P2(3)*sz);

sigL=kron(L,sL);sigR=kron(R,sR);
sig1=kron(L1,s1);sig2=kron(L2,s2);
  gamL=1i*(sigL-sigL');gamR=1i*(sigR-sigR');
   gam1=1i*(sig1-sig1');gam2=1i*(sig2-sig2');

G=inv(((EE+zplus)*eye(2*Np))-H-sigL-sigR-sig1-sig2);

% {1 L} {2 R} = {a} {b}
TM1L=real(trace(gam1*G*gamL*G'));
TML1=real(trace(gamL*G*gam1*G'));
   Taa=[0 TM1L;TML1 0];

  TM12=real(trace(gam1*G*gam2*G'));
TM1R=real(trace(gam1*G*gamR*G'));
   TML2=real(trace(gamL*G*gam2*G'));
TMLR=real(trace(gamL*G*gamR*G'));
    Tab=[TM12 TM1R;TML2 TMLR];

   TM21=real(trace(gam2*G*gam1*G'));
TM2L=real(trace(gam2*G*gamL*G'));
  TMR1=real(trace(gamR*G*gam1*G'));
TMRL=real(trace(gamR*G*gamL*G'));
   Tba=[TM21 TM2L;TMR1 TMRL];

  TM2R=real(trace(gam2*G*gamR*G'));
TMR2=real(trace(gamR*G*gam2*G'));
   Tbb=[0 TM2R;TMR2 0];

  Taa=diag(sum(Taa)+sum(Tba))-Taa;Tba=-Tba;
  Tbb=diag(sum(Tab)+sum(Tbb))-Tbb;Tab=-Tab;
    if abs(sum(sum([Taa Tab;Tba Tbb]))) > 1e-10
       junk=100,end

V=-inv(Tbb)*Tba*[1;0];
VV2(ii)=V(1);VVR(ii)=V(2);
```

```
alp(ii)=al*2/t0;% eta/t0/a

Gn=G*(gam1+V(1)*gam2+V(2)*gamR)*G';A=i*(G-G');
VV(ii)=real(trace(gam2*Gn)/trace(gam2*A));
end

hold on
h=plot(alp,VV2,'k');
set(h,'linewidth',2.0)
h=plot(alp,VV,'ro');
set(h,'linewidth',2.0)
set(gca,'Fontsize',36)
ylabel(' Voltage ---> ')
grid on
```

▪▪▪

% Fig.22.9, Spin Hall Effect

```
clear all

%Constants (all MKS, except energy which is in eV)
hbar=1.06e-34;q=1.6e-19;m=0.1*9.1e-31;qh=q/hbar;zplus=i*1e-12;
sx=[0 1;1 0];sy=[0 -1i;1i 0];sz=[1 0;0 -1];
%inputs
a=2.5e-9;t0=(hbar^2)/(2*m*(a^2)*q);eta=1e-11;
NW=25;n=0;Np=2*n+1;NWp=NW*Np;
L=zeros(Np);R=L;L(1,1)=1;R(Np,Np)=1;

%Hamiltonian
al=4*t0*eye(2);by=-t0*eye(2)-(1i*eta/2/a)*sx;
bx=-t0*eye(2)+(1i*eta/2/a)*sy;
alpha=kron(eye(NW),al)+kron(diag(ones(1,NW-1),+1),by)+...
kron(diag(ones(1,NW-1),-1),by');
beta=kron(diag(ones(1,NW)),bx);

H=kron(eye(Np),alpha);
if Np>1
H=H+kron(diag(ones(1,Np-1),+1),beta)+...
kron(diag(ones(1,Np-1),-1),beta');end
```

```
ii=0;A0=zeros(NWp,1);N0=A0;Nx=A0;Ny=A0;Nz=A0;
for EE=[0.05:0.005:0.05]*t0
ii=ii+1,ig0=(EE+zplus)*eye(2*NW)-alpha;
if ii==1
gs1=inv(ig0);gs2=inv(ig0);end

   change=1;
   while change >1e-4
     Gs=inv(ig0-beta'*gs1*beta);
     change=sum(sum(abs(Gs-gs1)))/(sum(sum(abs(gs1)+abs(Gs))));
     gs1=0.5*Gs+0.5*gs1;
   end
   sig1=beta'*gs1*beta;sig1=kron(L,sig1);gam1=i*(sig1-sig1');

   change=1;
   while change >1e-4
     Gs=inv(ig0-beta*gs2*beta');
     change=sum(sum(abs(Gs-gs2)))/(sum(sum(abs(gs2)+abs(Gs))));
     gs2=0.5*Gs+0.5*gs2;
   end
   sig2=beta*gs2*beta';sig2=kron(R,sig2);gam2=i*(sig2-sig2');

   G=inv((EE*eye(2*NWp))-H-sig1-sig2);
   A=i*(G-G');Gn=G*gam1*G';Tcoh(ii)=real(trace(gam1*G*gam2*G'));

   S0=kron(eye(NWp),eye(2));Sx=kron(eye(NWp),sx);
   Sy=kron(eye(NWp),sy);Sz=kron(eye(NWp),sz);

   A0=A0+kron(eye(NWp),[1 1])*diag(A*S0);
     N0=N0+kron(eye(NWp),[1 1])*diag(Gn*S0);
       Nx=Nx+kron(eye(NWp),[1 1])*diag(Gn*Sx);
        Ny=Ny+kron(eye(NWp),[1 1])*diag(Gn*Sy);
         Nz=Nz+kron(eye(NWp),[1 1])*diag(Gn*Sz+Sz*Gn);
   E(ii)=EE/t0;
end
   F0=real(N0);Fx=real(Nx); Fy=real(Ny);Fz=real(Nz);

Y=[0:1:NW-1];
hold on
```

```
%h=plot(Y,F0([n*NW+1:(n+1)*NW]),'k-o');
%h=plot(Y,Fx([n*NW+1:(n+1)*NW]),'b');
%h=plot(Y,Fy([n*NW+1:(n+1)*NW]),'k');
h=plot(Fz([n*NW+1:(n+1)*NW]),Y,'k');
set(h,'linewidth',[3.0])
grid on
ylabel(' Width (nm) --> ')
xlabel(' S_{z} (arbitrary units) --> ')
```

F.5.   Lecture 23, n(μ) from Fock space calculations

% Fig.23.7, n versus mu, single dot
```
clear all
%define constants
eps = 10; U = 20;

%define N and H matrices
N = diag([0 1 1 2]);
H = diag([0 eps eps 2*eps+U]);

ii=1;for mu = 0:0.1:50
   p = expm(-(H-mu*N));
   rho = p/trace(p);
   n(ii) = trace(rho*N);X(ii)=mu;ii=ii+1;
end

G=diff(n);G=[0 G];G=G./max(G);
hold on;
grid on;
h=plot(X,n,'k');
set(h,'linewidth',2.0)
set(gca,'Fontsize',36)
xlabel('\mu / kT --->');
ylabel(' n ---> ');
```

■■■

% Fig. 23.8, I versus V, single quantum dot

```
clear all
%define constants
eps = 10; U = 20; g1=1; g2=1;N = [0 1 1 2];

ii=1;mu1=0;for mu2=0:1:50
    f1a=g1/(1+exp(eps-mu1));
    f2a=g2/(1+exp(eps-mu2));
    f1b=g1/(1+exp(eps+U-mu1));
    f2b=g2/(1+exp(eps+U-mu2));

W1=[0 g1-f1a g1-f1a 0;
    f1a 0 0 g1-f1b;
    f1a 0 0 g1-f1b;
    0 f1b f1b 0];W1=W1-diag(sum(W1));

W2=[0 g2-f2a g2-f2a 0;
    f2a 0 0 g2-f2b;
    f2a 0 0 g2-f2b;
    0 f2b f2b 0];W2=W2-diag(sum(W2));

W=W1+W2;
[V,D]=eig(W);diag(D);
P=V(:,1);P=P./sum(P);
I1(ii)=N*W1*P;
I2(ii)=N*W2*P;
X(ii)=mu2;ii=ii+1;
end

grid on;
%h=plot(X,I1,'k');
%set(h,'linewidth',2.0)
h=plot(X,I2,'k');
set(h,'linewidth',2.0)
set(gca,'Fontsize',36)
xlabel('qV/kT --->');
ylabel(' Normalized current ---> ');
```

■■

% Fig. 23.9, n versus mu, double quantum dot

```
clear all
%define constants
eps1 = 20; eps2 = 20; t = 10;
U = 20;

%define N and H matrices
N = diag([ones(1,1)*0 ones(1,4)*1 ones(1,6)*2 ones(1,4)*3
ones(1,1)*4]);
H0 = 0;
h11 = [eps1 t;t eps2];H1 = blkdiag(h11,h11);
h21 = [2*eps1+U 0;0 2*eps2+U];
h22 = [eps1+eps2 0;0 eps1+eps2];
H2 = blkdiag(h21,h22,h22);H2(1:2,3:4)=t;H2(3:4,1:2)=t;
h31=[eps1+2*eps2+U t;t 2*eps1+eps2+U];H3 = blkdiag(h31,h31);
H4 = 2*eps1+2*eps2+2*U;
H = blkdiag(H0,H1,H2,H3,H4);

ii=1;for mu = 0:60
   p = expm(-(H-mu*N));
   rho = p/trace(p);
   n(ii) = trace(rho*N);X(ii)=mu;ii=ii+1;
end

hold on;
grid on;
box on;
h=plot(X,n,'k');
set(h,'linewidth',2.0)
set(gca,'Fontsize',36)
xlabel('\mu / kT --->');
ylabel(' n ---> ');
```